# PRINCIPLES OF FOOD TOXICOLOGY

# PRINCIPLES OF FOOD TOXICOLOGY

Tõnu Püssa

CRC Press is an imprint of the
Taylor & Francis Group, an **informa** business

CRC Press
Taylor & Francis Group
6000 Broken Sound Parkway NW, Suite 300
Boca Raton, FL 33487-2742


Printed in the United States of America on acid-free paper
10 9 8 7 6 5 4 3 2 1

International Standard Book Number-13: 978-0-8493-8090-7 (Hardcover)



**Library of Congress Cataloging-in-Publication Data**

Pussa, Tonu.
Principles of food toxicology / by Tonu Pussa.
p. ; cm.
Includes bibliographical references and index.
ISBN-13: 978-0-8493-8090-7 (alk. paper)
ISBN-10: 0-8493-8090-1 (alk. paper)
1. Food--Toxicology. I. Title.
[DNLM: 1. Food--toxicity. 2. Food Contamination. WA 701 P987p 2008]

RA1258.P87 2008
615.9'54--dc22 2007013979

**Visit the Taylor & Francis Web site at**
**http://www.taylorandfrancis.com**

**and the CRC Press Web site at**
**http://www.crcpress.com**

# *Dedication*

*To my wife, Tiia*

# Contents

# Preface

Food, as an extremely complex and complicated system, consists of practically an endless number of high- and low-molecular substances, mostly of natural origin. A majority of these compounds are indispensable for the normal functioning of the human organism, either as a source of energy or building material or normal source of pleasure, and their function is to turn eating into a pleasure and to improve digestion. Some of the food components also make food healthier and safer, as well as prolong its storage life.

On the other hand, food also contains substances that are capable of evoking smaller or bigger health disorders, that is, food can sometimes be toxic. Poisonous compounds may not only originate from the raw material of food but may also get into food during its processing, transportation, or storage. Toxic substances may also be the compounds, often synthetic, that are intentionally added to food. Although nowadays these substances called food additives are subjected to exhaustive toxicological examination, one can never be absolutely sure that a long-known food constituent can be regarded as safe in a new environment, where it can turn toxic by itself or synergistically enhance the toxicity of another so far nontoxic food component. Food is never ready; various physical and (bio)chemical processes are continuously going on, which may result in the formation of new and not always harmless substances. The so-called health-promoting functional additives may also elicit toxicological problems.

This textbook is an attempt to put into one pot the principles of general and food toxicology and to spice them up with the most important and vivid examples of food-related poisons and poisonings from all over the world. Owing to the rapid development of food toxicology, it is not usually possible to present the ultimate truth about toxic effects and their mechanisms. And this is good, because it makes the reader think with us. Special attention is paid to the (bio)chemical mechanisms of the toxic effects as much as they are known. Knowledge of the mechanisms helps toxicologists perform risk assessment scientifically.

The first part of the book is dedicated to introduction of the principles of toxicology at the molecular, cellular, as well as organism level, related as closely as possible to food. At times, examples from the second part are drawn to illustrate the principles. The second part is a systematic characterization of the most important foodborne toxicants, closely interconnected with the first part of the book.

This textbook is a thoroughly revised and updated translation of the respective book written in Estonian, which is being used in the author's course of food toxicology at the Estonian University of Life Sciences. It may be of interest for students of food science and technology, for professional food scientists, manufacturers, and regulatory agency personnel.

# *Acknowledgments*

The author would like to express his gratitude to his colleagues who were helping him at the time of writing the Estonian version as well as in translation. Special thanks to associate professor Ain Raal for his reading of the Estonian manuscript and his valuable advice concerning the medicinal side of toxicology and the textbook, to master student Piret Raudsepp for preparation of the figures, and my daughter, Triina, for the linguistic proofreading of the translation.

# *Author*

Tõnu Püssa is an associate professor of food toxicology at the Department of Food Science and Hygiene of the Estonian University of Life Sciences in Tartu, Estonia. After graduating from Tartu University as an organic chemist in 1969, he received his PhD in chemistry at the same university in 1973. During his work at the Department of Organic Chemistry and Laboratory of Chemical Kinetics and Catalysis at Tartu University as a research fellow and associate professor, his research interests were connected with chemical and enzymatic catalysis, algal carbohydrates, and proteins of the endocellular matrix. For a year, he was a guest researcher at the Finnish Red Cross Blood Transfusion Service in Helsinki. During 1996 to 2001, Tõnu Püssa worked at the Estonian University of Life Sciences (former Agricultural University) as the Head of Laboratory of Environmental Chemistry. He has taught student courses of organic chemistry, analytical biochemistry, hydrochemistry, and food toxicology. His present scientific interests are connected with functional foods, particularly with the mechanism of interaction between herbal antioxidant polyphenols and the oxidation system of polyunsaturated fatty acids producing mutagenic epoxyacids. He is married, and has two daughters and two grandsons.

*section one*

# *Basics of toxicology connected to food*

*chapter one*

# Introduction

## 1.1 What is toxicology?

The term *toxicology* is composed of two Greek words: *τοξικον* = poison, toxicant and *λογος* = word, reason, science. The word *τοξικον* meant the toxic substance into which arrowheads were dipped, whereas *τοξικος* meant bow.

A *toxicant* or *poison* is a chemical substance that, after entering an organism, is capable of causing smaller or larger adverse changes in the functioning of cells, tissues, or even the whole organism, perhaps resulting in the death of this organism. *Toxicant* denotes, above all, a synthetic substance causing adverse health effects.

A similar term "toxin" usually refers to any proteinaceous poison produced by living organisms, especially microorganisms such as bacteria in the body of a host. By a broader approach, a toxin is any poisonous substance produced by an organism, irrespective of the molecular size and structure, and when used nontechnically, the term "toxin" is often applied to any toxic substance.

The term *venom* (from Latin *venemus*) or zootoxin denotes a poisonous matter (toxin) normally secreted by snakes, scorpions, bees, and the like.

*Homo sapiens*, similar to other living organisms in the contemporary world, is exposed to *diverse foreign substances or compounds—xenobiotics*, which may be simple metals or inorganic compounds as well as organic molecules of a different complexity and size. These molecules may be either *natural* or *anthropogenic*, that is, produced during or due to diverse human activities.

*Toxicology is a science* that studies the

- *Formation, composition,* and *properties* of harmful or toxic substances, the mechanisms of action of these substances on biological systems (organisms) and the ways of assessment and reduction of this adverse effect, methods of prophylactics, and in case of necessity, the medication of the toxic results
- *Toxicokinetics* and *toxicodynamics* of the substances, elaborates sensitive and exact methods for the determination of threshold doses as well as concentrations in case of both acute and chronic effects of xenobiotics

- Cellular toxicity, mutagenicity, teratogenicity, carcinogenicity, and other properties of xenobiotics

*Toxicology is related to pharmacology*. Toxicology makes great use of the methods of pharmacology for estimating substance toxicity, investigating its action mechanism, and obtaining data characterizing the disorders caused by the interaction of a xenobiotic with an organism. *The main difference is that while pharmacology studies desirable changes in an organism, toxicology studies undesirable changes*. Whereas toxicology primarily studies the manifestations of an extreme action of substances often with symptoms preceding the death of a test animal, pharmacology mostly investigates much smaller alterations in the functioning of an organism. In the latter case, the research focuses on substances designed for therapy. At the same time, a modern toxicologist must be interested in the effect of small doses of a toxicant just as a pharmacologist is supposed to determine the toxicity of the drugs. If, during the past centuries, toxicology mainly studied highly toxic substances, most of which had a lethal effect, then in recent times, the main subjects are the substances causing relatively weak, yet adverse, physiological changes that lower the quality-of-life, but usually do not pose serious threats to one's life. More and more, the term "morbidity" is used instead of "mortality" in the toxicological literature. It has become possible essentially because of the fast development of methods of chemical analysis, which provide an opportunity to measure smaller concentrations of toxic substances in various matrices. And conversely, the growing needs of pharmacology as well as toxicology have contributed to the fast development of chemical analysis.

*Toxicology has always been closely connected* with disciplines such as physiology, pathology, epidemiology, biochemistry, molecular biology, genetics, immunology, microbiology, ecology, statistics, and so forth.

Toxicology as an extensive and highly practice-oriented science has a number of subdisciplines or divisions such as analytical, biochemical, clinical, environmental, industrial, forensic, and juridical toxicology, and so forth. In this long list, *food toxicology occupies an important position*.

Food toxicology investigates

- Ways and mechanisms of entrance of toxic substances into food or their generation during food processing and storage, and the ways of avoiding or reducing food contamination
- Methods of assessment of toxicity and risk of food components
- Adverse effects on an organism produced by harmful components of foods and beverages that can lead to functional disturbances of varying degrees in the organism, even death.

*Food is a material*, consumed in natural or processed form to satisfy the substantial energy needs of the consumer. Food consumption can cause a variety of health problems, in which either the whole organism or at least a definite part of it can be shifted from its normal physiological state.

This shift may, in a simple case, cause only some discomfort, but in more serious cases, be life threatening. An initial light discomfort can strengthen over time and develop into a situation where the question "to be or not to be" is to be answered. In these cases, we are dealing, in a broader sense, with the *nonconformity of the food with the organism eating it*. The nonconformities can be divided into *toxic* and *oversensitivity phenomena*. The latter may either be immune-system dependent or independent, such as lactose or milk intolerance (hypolactasia) caused by enzyme lactase deficiency (see Section 12.2).

Our prehistoric ancestors obviously tried to eat as much as possible of the various plants they picked and the animals that they managed to catch to optimize their food intake. So, by the method of trial and error, which was sometimes fatal, they learned which food more efficiently satiated hunger without causing any illness or death. The criteria of choice also certainly included the pleasantness of food, its smell, and taste. In this way, our ancestors developed their nutritional habits, enabling them to survive and grow. Owing to their sedentary way of life, their habits and customs certainly depended on local possibilities. Later on, the food components, especially additives such as spices, turned into articles of long-distance trade. With time and increased trade came a homogenization of sorts, as foodstuffs that were once unique to specific locales are now made widely available.

At the beginning of the twenty-first century, the incidence of food-related illnesses is increasing again after a long period of decline. *Food intoxications have turned into global issues*; some of them, although considered to be defeated, are growing into problems again. Agricultural production methods and consumption habits have been changed; modern technologies of crop harvesting, processing, and packaging may often cause an emergence of new food-borne pathogens. Intensified traveling and trade facilitate the spread of pathogens and outbreak of diseases. All communities in a nation are served by components of the same food system, even in geographically large countries such as Canada and the United States. Often, consumer preference for fresh food means a large-scale transport of such commodities from tropical and subtropical areas that are prone to both infectious agents and fungal toxins. Biochemical terrorism can turn into a serious food toxicological issue (Möller, 2000).

## 1.2 Short history of toxicology

Early historical writings tell us that many centuries ago, humans were already acquainted with the toxic effect of a number of natural materials.

The oldest recipes, more than 800 (!), of poisons such as belladonna (deadly nightshade) and opium alkaloids, lead, copper, and antimony can be found in the Ebers Papyrus written in 1552 BC. The ancient Egyptians were able to distill hydrocyanic acid (HCN) from peach stones, even in those days.

The medical science of ancient India was aware of poisons such as arsenic, opium, and extract of wolfsbane or monkshood (*Aconitum napellus* L.).

The latter was used as an arrow poison in ancient China. The American Indians still use plant seed extracts containing poisonous glucosides as a weapon.

In ancient Greece, where intentional poisoning was quite an everyday issue, both poisons and their antidotes were known. King Mithridates VI (132–63 BC) used criminals to search for poisons and antidotes (mithridatics). He protected himself against poisoning by a mixture of 50 different antidotes. According to legend, he developed serious problems owing to self-poisoning that it became necessary to commit suicide. *It was in Greece that the embryos of contemporary toxicology sprouted.* The writings of *Hippocrates* (460–377 BC), an outstanding physician of those times, demonstrate a true professional knowledge of poisons and toxicology among the Greeks. The most famous victim of intoxication is Socrates (470–399 BC), who was killed by the poison of hemlock (*Conium maculatum* L.) containing *conium* and other alkaloids as toxic components.

In the Middle Ages, political poisoning turned into a real cult in Italy. For example, the Town Council of Venice or the Council of Ten (*Il Consiglio dei Dieci*) concluded contracts of poisoning of their political opponents. The proceedings of this council are available from detailed notes that include the victim's name, the contractual partner, type and quantity of the poison used, and the outcomes. There was a historically well-known family of Borgia (Cesare, Lucretia, etc.), which poisoned spouses, lovers, political opponents, clergymen, and others.

Toxicology must also be thankful to "scientists" such as *Catherine de Medici* (1519–1589), the spouse of French King Henry II, who prepared poisons and tested them on the poor and sick of France, noting all the clinical signs and symptoms.

All of this helped the Swiss doctor Paracelsus (1493–1541) to formulate his famous postulates, such as: *Was ist das nit gifft ist? alle ding sind gifft/und nichts ohn gifft/Allein die dosis macht ein ding kein gifft ist*—(Martinetz, 1982) or:

> *What is not poisonous? Everything is a poison/and there is nothing without poison/ Only the dose permits something not to be poisonous.*

Paracelsus developed the meaning of the term "dose"; modern methods of risk assessment of chemicals, including terms such as threshold dose, safe and nontoxic levels, are based on his postulate. Considering it very broadly, there is not a single substance that does not possess any toxic action above its threshold dose. Even water is a poison if one drinks too much of it. The oxygen we breathe and that which is absolutely necessary for our life is poisonous for anaerobic organisms and even for us via the oxidative stress. Aerobic organisms have, during their evolution, created special mechanisms to control the toxicity of the free radicals produced by oxygen. Paracelsus also believed that illnesses are localized in definite organs and that the poisons act in target organs as well. Unrecognized during his lifetime, the role of Paracelsus in the development of medicine as well as toxicology is enormous (Pappas et al., 1999).

*Toxicology as an individual discipline was first recognized* by Spanish scientist *Matthieu Joseph Bonaventura Orfila* (1787–1853), the physician-in-ordinary of Louis XVIII, king of France, who elaborated the first methods of chemical detection of toxic substances. He was the founder of analytical toxicology. Orfila used dogs as test animals. His main achievement in toxicology was the discovery that ingested toxins do not stay and accumulate in the stomach, but, that after absorption move by blood into other organs. Since then, toxicology started to develop as a science, dealing also with investigation of the mechanisms of the toxicity of substances. So, the French scientist *Claude Bernard* (1813–1878), one of the founders of physiology and experimental pathology, believed that by studying the toxic effects of substances on biological systems, we obtain much useful data about the structure and functioning of these systems. From a later period, a very good example of the success of this approach is the study of the action of cyanide ions on cells, providing researchers with data about the functioning of the electron transport chain in mitochondria (see also Section 8.4). It is known from biochemistry that it is specifically the study of inhibitor reactions (quite often a poison is an inhibitor of a definite enzyme) that provides us with much more information about the structure of an enzyme-active centre and the mechanism of enzymatic catalysis than the respective substrate reaction.

An important milestone in the history of toxicology was the *discovery and characterization of ricin*, the toxic protein of the castor oil plant (*Ricinus communis*) at Tartu University in Estonia in 1888 by Peter Herman Stillmark and Eduard Rudolf Kobert. Some years later, under the supervision of Professor Kobert, the second toxic plant *lectin abrin* from *Abrus precatorius* was isolated by Heinrich Hellin. For more about lectins, see Section 8.1.

## 1.3 *Toxicity dose and response*

*Toxicity* is

1. The capacity of a chemical substance to cause adverse or deleterious effects on a living organism or on a part of it
2. The degree to which a substance is toxic

Toxicity of a substance depends on many different factors, such as

- Chemical structure of the compound
- Route of administration (i.e., the substance applied to the skin, ingested, inhaled, or injected)
- Time of exposure (a brief encounter or a long-term one)

- Number of exposures (a single dose or multiple doses over time)
- Physical form of the toxicant (solid, liquid, or gas)
- Genetic constitution of an individual, an individual's overall health, and so forth

To reveal toxicity, the organism must be exposed to the substance. The exposure to (or contact with) the toxicant can be *acute, subacute, subchronic,* or *chronic*. Among the other parameters, toxicity depends on the *dose* of the toxic substance.

*Dose*—total amount of a biologically active (toxic) compound administered to the organism, expressed usually in micrograms ($\mu$g) or milligrams (mg) per kilogram of body weight (bw); in the case of a toxicant, it is one of the most important determinants of its toxicity. The dose of the compound can enter the organism (to be administered) perorally, intrapulmonarily, intravenously, intramuscularly, percutaneously, intraperitoneally, and so forth.

*Response* (effect)—alteration of a biochemical or physiological parameter of the organism, exposed to the biologically active (toxic) substance, which has reached the specific (super) molecular *action points* in the organism. Intensity of the response depends on the concentration of the toxic substance, or its active metabolite in this action point. Response can either be all or nothing as the death of the organism or graded, for example, alteration of activity of an enzyme or hormone.

*Dose can be either external or internal*. In routine toxicological studies, the net amount of a compound administered per kilogram of the test animal weight is regarded as the dose. It can be called *external dose*. Most substances, excluding local irritants or oxidants, will cause the toxic effect only after they have

1. Absorbed into the general bloodstream. The absorbed part of the external dose is called the *absorbed* or *internal dose*
2. Reached by the bloodstream to the action point(s) or targets in the organism

A number of reasons exist as to why a foreign substance administered to an animal by any of the aforementioned routes does not reach this action point at all or reaches it only partly.

*Bioavailability* is a part of the external dose that is absorbed into the bloodstream as an original compound. Understandably, a better correlation exists between the internal dose and toxic response of the substance. Nevertheless, as it can be more precisely determined, in toxicity determinations of a chemical compound, the external dose is usually regarded as an organism-independent toxicological parameter.

Every substance has low and safe doses, and high and for-certain toxic doses. *The threshold dose* is the minimal amount of a substance, the administration of which will cause adverse alterations beyond the borders of physiological adaptation in the animal organism, that is, the start of a concealed, temporarily compensated disease.

It should be remembered that in case of any substance, no single dose exists that will cause toxic symptoms in all of the test animals. Actually, an interval of doses exists in which different individuals of a test group respond similarly, that is, develop similar symptoms of intoxication.

*What can be the effect we could measure*? The measurable effect can be either discontinuous like the death of an organism or continuous like an alteration of blood pressure or respiration rate or concentration or activity of a physiologically relevant molecule such as a hormone or enzyme. Very often a question arises—if the adverse changes have occurred simultaneously in several organs, which of them is the most important?

*Exposure to the toxicant (contact with) can be*

1. *Acute*: contact time below 24 h, mostly by a single exposure, but there can also be repeated exposures during 24 h
2. *Subacute*: usually repeated contacts during 1 month
3. *Subchronic*: continuous contact during 1–3 months
4. *Chronic*: continuous contact during more than 3 months, most often a daily contact via food, air, and so forth. In the case of animal experiments, it means the lifespan of the organism.

In some textbooks on toxicology, a slightly different list of the types of exposure can be found; sometimes subacute and subchronic exposures— are considered together as one type, sometimes only two—acute and chronic exposures—are presented.

*In the case of absorption of sufficiently high doses of a toxicant, an adequate adverse response* of the organism emerges. The following responses may occur:

1. *Acute:* develops quickly, usually with severe symptoms. For example, an exposure to sufficiently high doses of potassium cyanide is followed by death within a few minutes.
2. *Subacute:* the effects are generally the same as in the case of an acute response, but with weaker symptoms that establish over a longer period (during some weeks).
3. *Chronic:* symptoms develop slowly in the case of a systematic long-term absorption of relatively small amounts of the toxicant. At the time of diagnosis, confusion with other pathologies may occur. For example, a tumor initiated by inhalation of asbestos may develop into cancer over decades.

*Acute contact* or *exposure* may cause either an acute or chronic response (intoxication). Acute intoxication is an illness, which forms after a single or repeated exposure of an organism to the toxicant during a short period (some days). Well-known examples are severe anoxemia, caused by carbon monoxide, and tetrodotoxin intoxication emerging after intake of toxic puffer fish (see Section 13.1) or botulism (see Section 16.4). Many acute poisonings can result in permanent health disorders even if there is general recovery. Therefore, after acute carbon disulfide intoxication, reflex, excitability, visual, and psychical disorders may persist over a long period of time.

*Subacute intoxication* may occur in case of frequent short-term contacts with the toxicant during a period of 28–90 days. It may happen quite often as in the case of agricultural workers dealing with pesticides.

*Chronic exposure* or *contact* may cause either a chronic or acute response (intoxication). Chronic intoxication is a disease, formed by a long-term (months or years) effect of small, separately taken harmless daily doses of a toxic substance. Since no signs of poisoning would be visible after a single or even multiple administration of the substance, it is often very difficult to estimate the genuine reason for chronic intoxication. A fast-developing chronic intoxication is also referred to as subacute.

*Chronic intoxication* develops in the case of either an accumulation of a toxic substance in the organism (*material cumulation*) or due to an accumulation of initially unimportant functional alterations in the organism (*functional cumulation*). In most cases, the second situation prevails. This is the mechanism of the effect of fat-dissolving substances such as chlorinated hydrocarbons, benzene, or trinitrotoluene. Long-term exposure to low doses of organophosphorus inhibitors depresses cholinesterases to such a low level that symptoms of intoxication start to appear. In the case of toxic heavy metals (Pb, Hg, Cd, etc.), which have a long half-life in the organism, mainly material cumulation occurs, although functional cumulation also plays a definite role in the development of an intoxication. Chronic intoxication occurs chiefly as an occupational disease in the case of workers at factories producing these toxic metals; in the case of environmental toxicants such as DDT, PCB-s, cadmium, or mercury, the possible array of the sufferers is much broader.

Some substances such as HCN, due to their fast modification or excretion, never cause chronic intoxication. On the contrary, compounds such as silicon dioxide, which dissolve only slightly in the body liquids, are capable of causing only chronic intoxication.

*Response* (intoxication) can be *local* or *systemic, immediate* or *delayed,* and *reversible* or *irreversible.*

*Local response* emerges in the same organ that was in direct contact with the poison, and is *systemic* in the other distant organs. Many compounds cause damage in the region of the skin or mucous membrane directly exposed to the toxicant. The general symptoms of the intoxication either reflect the development of pathological processes (like leucocytosis or inflammation in the case of corrosion) or are a result of a disruption in the functioning of the injured organ (such as anoxia in the case of pulmonary edema). *Many poisons have only a systemic effect.* These compounds do not cause any tissue damage before they reach blood or tissue liquid.

*An immediate response* follows very shortly after contact with the toxicant, whereas *a delayed response* occurs considerably later. Thus, for example, the symptoms of radiation sickness such as loss of hair or malignant tumor may develop only over a long period after contact with acute irradiation.

Toxicants can predominantly produce either:

1. A selective effect on one or another organ or system or
2. A simultaneous injury of several organs or systems (polythropic effect)

In the case of many poisons, a molecular or supermolecular site, *receptor*, exists in the cell, and this should either be *reversibly* or *irreversibly* contacted by the toxicant to cause a toxic response (see Section 1.6.3). The intensity of this response depends on the toxicant concentration in the vicinity of the receptor, which in turn depends on the dose of the toxicant. A toxicant can have the same point of attack in case of acute and chronic intoxications; these points can also be different. For example, if in the case of acute benzene intoxication, the central nervous system (CNS) becomes damaged, then a chronic intoxication with benzene injures the hematopoietic system.

*A repeated attack* of a toxicant may cause development of *tolerance*, a phenomenon of a reduced response to the action of the toxicant in case there has been a preceding contact with the same (or similar) compound. This tolerance can be caused either by:

1. Induction of enzymes metabolizing the toxic compound (see Section 2.3.4) or
2. Change of the number or binding capacity of specific receptor groups

The second mechanism is connected with (partly) irreversible changes in the spatial structure of the binding site of the receptor during the first contact. As a result, the receptor molecule becomes "damaged" and does not enable as close a contact with the toxic molecule as it did the first time. Both changes are induced by the first contact with the toxicant molecule.

On the other hand, repeated contact with a toxicant can sometimes promote physiological accumulation and, hence, amplification of the toxic effect.

*The simplest acute toxic effect to estimate is the stopping of the organic life of an organism—its death.* Death is the severest parameter, which actually should be used as little as possible. Too frequent use of animal tests in general conflicts with the Animal Welfare Act enforced by the United States Department of Agriculture (USDA) and by the European Union's Directive 86/609/EEC on the protection of animals used for experimental and other scientific purposes as well as with ethical norms. *In addition, use of a lethal dose (LD) provides usually less information* about the mechanisms of a toxic effect (see also Section 5.2). Most often it is not necessary to know the LD, but it is necessary to study what happens to the test animals at those doses of a toxicant that can be contacted by humans or animals in their real life. For assessment of a chemical's toxicity, very often it is more beneficial to use a nonfatal biochemical or pharmacological change in the organism such as a decrease of enzyme activity or appearance in blood of a molecule, characteristic of a definite pathological disorder.

*Table 1.1* Approximate $LD_{50}$ Values for a Selection of Chemical Compounds

| Chemical compound | $LD_{50}$ (mg/kg, rat, orally) |
|---|---|
| Ethanol | 10,000 |
| Sodium chloride | 4,000 |
| Fe-sulfate | 1,500 |
| Malathion | 1,200 |
| Lindane | 1,000 |
| Morphine sulfate | 900 |
| Sodium phenobarbital | 150 |
| DDT | 100 |
| Arsenic | 48 |
| Dieldrine | 40 |
| Picrotoxin, aflatoxin $B_1$ | 5 |
| Strychnine sulfate | 2 |
| Nicotin | 1 |
| *d*-Tubocurarine | 0.5 |
| Tetrodotoxin | 0.1 |
| Tetrachlordibensodioxin (TCDD) | 0.001 |
| Botulinum toxin | 0.00001 |

A *lethal dose* is the amount of a substance (usually in logarithmic scale) causing the death of an animal in the absence of antidotes or treatment. Absolute ($LD_{100}$), minimal ($LD_{min}$), and median ($LD_{50}$) lethal doses (Table 1.1) can be distinguished.

In experimental toxicology, $LD_{50}$ causing the death of 50% of animals of a test group after an acute contact with a toxicant has been the most popular, $LD_{100}$ belongs to the most durable, and $LD_{min}$ to the weakest animal. Instead of doses, concentrations can be used—the respective parameters will be $LC_{50}$, $LC_{100}$, and $LC_{min}$ (mmol/L, $\mu g/m^3$).

*The classical $LD_{50}$ test,* which requires torture and sacrifice of a large number of test animals and which had caused long scientific and social debates, *was finally cancelled by the end of 2002.* Alternative animal acute-oral toxicity tests, such as the Fixed-Dose Procedure (FDP), Acute Toxic Class (ATC) method, and Up-and-Down Procedure (UDP), have been developed (see Section 5.5).

For estimation and comparison of acute toxicities of substances still very often dose–response curves are used, where the percentage of the response against logarithm of the dose is plotted (see Figure 1.1).

What do the location and shape of the curves tell us?

- *The slope of this curve speaks about predictability.* For example, if two toxicants have the same $LD_{50}$, but the slope of curve B is deeper than in the case of toxicant A (Figure 1.1), a change in dose necessary to produce the same response is smaller than in the case of toxicant A.

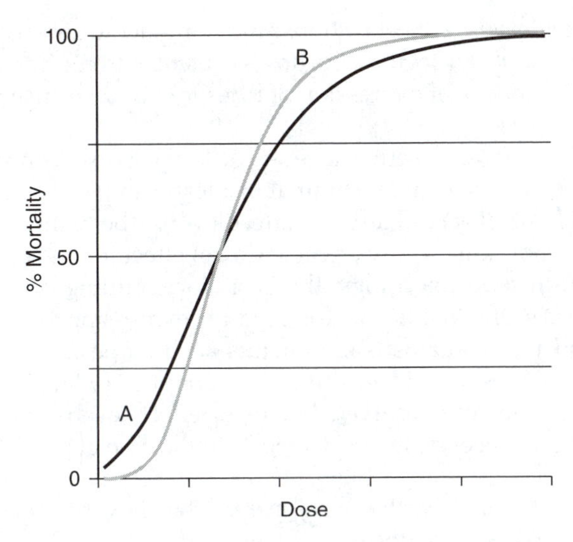


*Figure 1.1* Relationship between percentage of mortality and log dose.

In case of equality of all other parameters, the degree of predictability is higher in the case of toxicant B.

- *The location of the curve with respect to the x-axis* shows the potency of a toxicant to induce toxic responses, that is, its *toxicity*. The curve, representing the substance with a higher toxicity is located more left.
- Sometimes, instead of the usual S-shaped curve as shown in Figure 1.1, *a curve resembling the shape of a saxophone* appears. In this case, the adverse effects of the toxicant do not appear at its very low doses. On the contrary, low doses cause favorable effects, turning into adverse ones only at higher doses of the chemical. Such a phenomenon is called *hormesis*.

Botulinum toxin, which can emerge in food as a result of contamination with bacterium *Clostridium botulinum* (Section 16.4.1), is the most toxic substance known hitherto.

Instead of death, any other exactly fixable physiological (biochemical) parameter can be used for toxicity assessment. *The corresponding statistical parameter will be* $ED_{50}$—dose that evokes 50% of the possible alteration of the parameter or $IC_{50}$ in case of inhibition of some physiological (biochemical) processes. $IC_{50}$ corresponds to the inhibitor dose causing 50% of the maximum inhibition.

In case of oral intoxications as it occurs with food-borne poisons, the toxicant doses are expressed either in milligram or microgram per kilogram body weight or per whole conventional person with a body weight of 70 kg (in the case of pesticides—50 kg). For estimation of the severity of an intoxication and efficacy of the treatment, it is important to measure the toxicant concentration in the blood plasma as well as in other different tissues.

*The threshold doses* (concentrations) inducing acute severe intoxications are always *remarkably higher* than the doses causing chronic intoxications. The most often used modes of expression of threshold doses/concentrations are

- *NOEL* (no observed effect level)—the highest dose/concentration, not causing any effect in any animal in a test-group
- *NOAEL* (no observed adverse effect level)—the highest dose/concentration, not causing any adverse (toxic) effect in any animal in a test-group, in food toxicology, the basis for counting of *acceptable daily* intakes (ADI). ADI is a factor used for expressing safe consumption of food that contains contaminants such as pesticide or veterinary drug residues or food additives (see Chapters 6, 14, 15, and 17)
- *LO(A)EL* (lowest observed [adverse] effect level)—the lowest dose causing an observable effect, at least in the case of part of the animals

*In the case of medicines,* the *safety margin* that is very important is connected with three essential doses:

1. ED—effective or therapeutic dose
2. TD—toxic dose, causing an adverse side effect
3. LD—lethal dose

Ratio $TD_{50} = LD_{50}/ED_{50}$ is called the *therapeutic index* in medicine. The safety margin (conservative estimate) is the ratio $TD_1/ED_{99}$.

## 1.4 Interaction of toxic substances

Very often an organism can be attacked simultaneously by several potentially toxic compounds. Such a situation may occur, if, for example, a patient takes several drugs or in case of contact of a human being or an animal with environmental or industrial toxicants. The resultant toxic effect of such a mixture can be essentially different from the effect if every toxicant acted separately and very often unpredictably.

*The following situations can occur in this case*:

1. *Additivity of responses*. Often the toxic effects can be simply summarized, as $1 + 1 = 2$. This situation occurs when the mechanisms of action of the compounds are similar.
2. *Synergism of responses*. In this case, the combined effect of two different substances is bigger than the simple sum of the separate effects ($1 + 1 > 2$). The toxic effect of a chemical substance can be even *potentiated* by another compound, which otherwise is harmless in the concentration used.

   Examples:

   - Piperonyl butoxide (PB) is added to some insecticides (pyrethrins, pyrethroids, rotenone, carbamates) to enhance their insecticidal effect. PB inhibits *metabolic detoxification of the pesticide by the insect*, thus reducing the amount of the active ingredient required.

- Carbon tetrachloride and ethanol together have much higher hepatotoxicity than either of the substances taken separately.
- Sulfirame, used for treatment of alcoholism, potentiates the toxicity of ethanol. By inhibiting the aldehyde dehydrogenase catalyzed oxidation of acetaldehyde, sulfirame promotes its accumulation in the organism (see Section 16.3). The process is accompanied by an extremely poor feeling.

3. *Antagonism of responses*. The toxic effect of chemical A can be weakened (inhibited) by the addition of another toxicant B (antagonist)—$1 + 1 < 1$. Compound-antagonists are often used as antidotes (see Section 2.6). There may be various biochemical mechanisms of antagonism.
   - *Functional antagonism*—simple mutual balancing of toxic effects, such as in the case of high doses of caffeine and phenobarbital.
   - *Chemical antagonism*—chemical reaction between the antagonist and the toxicant results in the formation of a new, less toxic substance. For example, dimercaprol chelates toxic heavy metals such as arsenic.
   - *Receptor antagonism*—antagonist binds to the specific cellular receptor, thus preventing binding of the toxicant molecule as in the case of atropine and organophosphorus insecticides.
   - *Dispositional antagonism*—another substance changes the fate of the toxicant in the organism. For example, cholestyramine is able to prevent absorption of organic compounds by binding to the compound molecules.

## 1.5 Classification of toxicants

Toxicants can be classified according to

1. *Their chemical structure*: aromatic amines (aniline), halogenated hydrocarbons (PCBs, methylene chloride), organophosphorus compounds (malathion), and so forth
2. *Their usage*: pesticides (atrazine), solvents (benzene), food additives (saccharin), and so forth
3. *The physical state*: gases (carbon monoxide—CO), liquids (methanol), solids (asbestos), and so forth
4. *The toxic effect*: mutagens (diethylnitrosamine), carcinogens (benzo[$\alpha$]pyrene), hepatotoxins (chloroform), and so forth
5. *The biochemical mechanism* of a toxic effect: cholinesterase inhibitors (malathion), methemoglobin-forming substances (nitrite), and so forth
6. Others.

# 1.6 Some toxicology-related principles of cellular biology and biochemistry

## 1.6.1 Structure of cellular membranes

Every biological cell and its organelles are surrounded by biomembranes. Therefore, a toxic substance originating from the environment (food, air, soil, etc.) must permeate several membranes in the organism to cause a response inside a cell. Although the cells of different organs differ from each other, their structures have much in common.

*Plasma membrane* or plasmalemma surrounds every eukaryotic cell determining the cell bounders and guarantees preservation of the differences between the internal and external milieus of the cell. The membrane also has the responsibility of being a selective filter for substances that enter the cell or leave it. Plasma membrane also contains definite systems that enable an active transport, governing the transport of nutrients into the cell and that of wastes out of the cell. Plasmatic membrane generates concentration gradients between the inner and outer environment (interior and surroundings) of the cell, which are of vital importance for any cell. These gradients are used, for example, to start the synthesis of adenosine triphosphate (ATP), a high chemical energy carrying molecule, to produce and relay electrical signals in nerve and muscular cells, and so forth. The plasmatic membrane receives the signals coming from outside the cell enabling an adequate reaction of the cell. To fulfill this task, the membrane has specialized macromolecular sites—*receptors* (Section 1.6.3). One more highly important function of the plasma membrane is communication with the neighboring cells in the organ, tissue, or organism.

A *biomembrane* (including plasmalemmas) is a thin (about 10 nm thick) structured bilayer of *phospholipid* and *protein* molecules. This bilayer is polar and hydrophilic on both sides, and its interior is mainly nonpolar and hence hydrophobic. The nonpolar part of a phospholipid molecule consists of two fatty acid chains (one saturated and the other unsaturated) with a length of 12–20 carbon atoms. The hydrophilic "head" of the phospholipid molecule is formed by glycerophosphate bound to a choline, serine, ethanolamine, or inositol group. In many mammal cell membranes, one can also find *cholesterol* that reduces the permeability of the lipid bilayer for small hydrophilic molecules and enhances the mechanical stability of the membrane. *Lipid molecules* are actually "fluid," that is, they are in permanent lateral diffusion in the frames of one and the same layer. *Lipid bilayer* functions as a solvent for membrane proteins that simply adhere to the membrane (peripheral proteins) or reside within it or even penetrate it (integral proteins), having domains on both sides of the bilayer. Glycosidic groups may have been attached to the extracellular domain of the membrane proteins. A relative amount of protein depends on the type of membrane. For example, if in the myelin membrane of a nerve cell axon the protein content is below 25%, then in mitochondrial and chloroplast membranes, where energy transformation takes place, the protein content is about 75%. Membrane proteins are also capable of lateral diffusion, although they are 10–100 times slower than lipids. These proteins, partly glycoproteins, have the following functions:

- Transport of molecules through membranes
- Generation of ion gradients
- Acceptance and transmission of signals coming from outside the cell
- Mediation of fastening of the cytoskeleton to the cellular membrane
- Mediation of the contact with the extracellular matrix and the neighboring cells

There are also *carbohydrates in the composition of glycoproteins or glycolipids* on the cellular membrane surface. The carbohydrate-rich part of the outer surface of the cell is called *glycocalix*. The carbohydrate groups of the glycocalix are important for intercellular contacts and interactions as well as for cell adhesion. Since a majority of membrane glycoproteins and glycolipids contain sialic acid groups, the outer surface of most cells bears a negative resulting charge at the physiologically relevant pHs.

### 1.6.2 *Transport of substances across biomembranes*

The cell membrane is semipermeable. It means that not all molecules can go unhindered in or out of the cell. Molecules that can penetrate the membrane do it in one of the two main ways:

- *Passively* through diffusion in the direction of the concentration gradient
- *Actively* against the concentration gradient

The rate, extent, and mechanism of movement through biomembranes depend on parameters such as the molecular weight, electrical charge, solubility, chemical, and metabolic reactivities of the substance. Lipid-soluble, relatively nonpolar molecules dissolve well also in membranes and, therefore, they diffuse relatively easily through membranes. In contrast, ionizable comparatively polar molecules do not easily enter the lipid bilayer of the membrane; it is only the equilibrially nonionized part of the compound molecules that is able to diffuse through membranes. As an exception, there are very small hydrophilic molecules that use the so-called water-channels, *anticipated* for transport of water molecules through the membranes.

*Passive transport* of molecules through the cell membrane that needs no chemical energy can be divided into three types.

1. *Filtration* with the help of osmotic or hydrostatic force *along water-pores* or *channels* formed by the membrane proteins.
2. *Passive diffusion through a hydrophobic nonpolar phospholipid bilayer*. This is the most important absorption mechanism of xenobiotics,

including toxicants. Molecules, successfully and relatively nonselectively diffusing along the concentration gradient through lipid–protein membranes, must be lipid-soluble (nonpolar) and nonionized. Passive diffusion is a first-order process, its rate can be calculated by Fick's equation:

$$\text{rate} = KA(C_2 - C_1) \tag{1.1}$$

in which $K$ is the constant of proportionality, $A$ is the membrane surface area, $C_1$ and $C_2$ are the concentrations of the diffusing substance near both surfaces of the membrane, whereby $C_2 > C_1$. Fick's relationship applies to a system at a constant temperature and for diffusion over the unit distance. In the case of hydrophobic substances, factor $K$ depends mainly on three factors:

- *On the relative lipid solubility of this compound,* often characterized by its water–octanol distribution constant $K_{ow} = C_o/C_w$, where $C_o$ and $C_w$ are the compound's equilibrium concentrations, respectively, in *n*-octanol and water. The higher the $K_{ow}$, the easier and faster the molecules of this substance diffuse through the lipid membrane. Usually Log $P$ (decimal logarithm of $K_{ow}$) is used. Log $P$ can be calculated directly from the molecular structure of the compound using various mathematical models that utilize molecular fragmentation schemes and other structure-based indexes.
- *On the dimensions of the diffusing molecule.* Small molecules penetrate the membrane faster.
- *On the degree of ionic dissociation of the diffusing molecule.* Many substances exist in two forms—ionized and nonionized. Diffusion of the ionized form through the membrane is relatively hindered, since ions adsorb better to the polar membrane surfaces and move slower in the lipid phase.

In a normally functioning dynamic biological system, there is always a difference (gradient) in the substance concentrations between the sides of the membrane. The formation of this gradient is essentially caused by differences in $H^+(H_3O^+)$ ion concentrations (pH values) of the liquid in these spaces, governing the dissociation of the ionizable molecules.

The degree of ionization of acidically dissociable molecules is described by Henderson–Hasselbach equation:

$$\log \frac{[A^-]}{[HA]} = \text{pH} - \text{p}K_a, \tag{1.2}$$

where p$K_a$ is the negative decimal logarithm of the acidic dissociation constant of an acid HA. [HA] and $[A^-]$ are the equilibrium concentrations of the nonionized and ionized forms of the substance, respectively. If pH = p$K_a$, then half of the substance molecules are in a dissociated ionic form; increase or decrease of pH causes an increase or decrease of the number of molecules in ionized form, respectively.

In the case of basically ionizing molecules,

$$\log \frac{[AH^+]}{[A]} = pK_a - pH\,, \tag{1.3}$$

where [A] and $[AH^+]$ are the equilibrium concentrations of the nonionized and ionized forms of a base A, respectively. The increase of pH now causes a reduction of the number of ionized molecules and vice versa.

Since in both cases the membranes are preferentially penetrated by more lipophilic nonionic particles, the weak acids are concentrated in a space with a high pH and the weak bases in a space of low pH. Particles, having strongly (pH-independently) ionizing groups such as sulfate or tetra-ammonium ions, are not able to pass the membranes by passive diffusion at any physiological pH.

3. *Facilitated (favored) diffusion*. Since here, in addition to the concentration gradient, a transporting molecule is needed, the process can be saturated at high concentrations of the diffusing substance. As in the case of other diffusions, chemical energy is not required. This is how usually endogenous (organism-borne) substances and structurally similar exogenous substances (xenobiotics) are transported through the biomembranes. So for example, the transport of glucose from enteric cells to the bloodstream is organized.

*Active transport* of molecules through the membrane, carried out against the concentration gradient is a selective process requiring metabolic energy. Two main mechanisms are known:

1. *Simple active transport*. A molecule passes a membrane in complex with a specific transport molecule residing on the surface of the membrane. Similar molecules must compete for their common transporters. Since there can be a shortage of the carrier molecules in the case of higher concentrations of molecules to be transported, the process is saturable and preferably of zero order and is inhibitable by the metabolic poisons.
   The transport molecules can be *uniports, symports,* or *antiports*. The uniports transfer one molecule in one direction, symports and antiports two molecules, respectively, in the same or opposite directions. This type of membrane transport is designed by the organism for specific endogenous and nutritious compounds, but in the same way, similar exogenous molecules and ions, such as lead from the stomach can be transported. This is also an essential mechanism in the elimination process of toxic substances from the organism.

2. The subtypes of *endocytosis* are phagocytosis and pinocytosis. *Phagocytosis* is *anticipated* for transport of large particles (microorganisms, cell debris, etc.), *pinocytosis* for transport of dissolved macromolecules. Endocytosis is a complex type of transportation, starting with envagination of the membrane part named coated pit to enclose a particle or droplet, followed by detachment in the form of a coated vesicule from the remaining membrane and entering the cell. Many molecules and particles, endocytozed by a cell, are transported into lysosomes. In the cell, the membrane bag deliberates its content and returns to the outer membrane forming a new coated pit there. In most animal cells, endocytosis represents a very selective mechanism of concentration of compounds, used by the cell to collect specific macromolecules from its environment. Actually, the forming vesicule also includes extracellular liquid and nonspecific, highly concentrated molecules. Obviously, such transport cannot be very extensive, but is still considerable. For example, a macrophage is able to internalize in 30 min a membrane with an area equal to its own surface area. This is the way the particles of asbestos and uranium dioxide enter the pulmonary cell, and are transported through membrane peptides and antigen–antibody (AG: AB) complexes. Most eukaryotic cells are pinocytoting continuously; phagocytosis is characteristic only of specialized cells such as mammal macrophages and neutrophiles, both of which have been developed from the same ancestor. Phagocytosis, differently from pinocytosis, requires induction, for example, by antibodies.

### 1.6.3 Receptors

*On the outer surface of a cell,* there are numerous biochemically sensitive and active sites called *receptors*—consisting of proteins (protein complex, glucoproteid, lipoproteid)—that usually extend throughout the membrane to cytosol. *Toxicologically important receptors may also be located in the cytoplasma.* Binding of a specific complementary (matching in dimensions and in charge distribution) molecule called a *ligand to the binding site of this receptor complex* will change the conformation of this site. This transformation initiates consecutive alterations (signal transductions) along the whole integral receptor complex up to its intracellular end. This triggers a chain of biochemical reactions leading to alterations in the physiological (biochemical) state of the whole cell. If this alteration is necessary and favorable for this cell or respective organ or even for the whole organism, we have a normal regulation of cell and organism life. It is also beneficial if the ligand is a drug molecule, the right dose of which enables to correct the physiological state of a sick cell. But the same receptor complex can be attacked by substances, capable of giving a signal for taking the cell significantly out of its normal physiological state. In this case, we have a toxicant–receptor interaction, which, depending on the toxicant's dose and other circumstances, can lead to the death of the cell, organ, and even the organism.

*The ligand–receptor interaction* and the respective response can be described by the following simplified scheme:

$$L + R \overset{K_{LR}}{\leftrightarrows} LR \rightarrow \text{RESPONSE}\,, \tag{1.4}$$

where $K_{LR}$ is the equilibrium constant of the ligand–receptor complex LR.

The target molecules of the ligand–receptor interaction, followed by conveying of the conformational alterations in the receptor, can be

- Cellular enzymes (reduction or magnification of their activity)
- Other cellular proteins (tubulin, carrier molecules)
- Other cellular macromolecules (DNA, RNA)
- Other receptors located at the same cellular membrane

Ligand–receptor complex LR can be formed by a combination of the following bonds (in brackets are the respective approximate bond energies in kcal/mol):

1. *Covalent bond* (100)—a common electron cloud between two atoms, seldom in LR, irreversible
2. *Ionic bond* (5)—between oppositely charged atoms, usual in LR, reversible
3. *Hydrogen bond* (2–5)—binding through a common H-atom, usual in LR, reversible
4. *Hydrophobic interaction* (1)—a weak binding between nonpolar groups, entropic, usual, reversible
5. *van der Waals bond* (0.5)—weak electrostatic attractive power, usual in LR, reversible

The higher the bond energy, the stronger and more stable, both chemically and thermally, is the bond. The covalent bond is by far the strongest of them and it forms the basis of the stable structure of a molecule. Formation of a covalent bond between a ligand and receptor most often means that this receptor molecule gets irreversibly blocked and it is very difficult, if not impossible, to regenerate the initial molecule, capable of building up native LR complexes. In the case of normal physiological processes, there is no place for covalent bonds in LR complexes. Weaker noncovalent bonds of types 2–5 can be complementarily combined to form a physiological receptor–ligand complex. The mutual binding sites of a receptor and its genuine ligand molecules are related to each other similar to a key and the respective lock. The differences from a usual door lock and key are as follows:

1. The cellular lock is not too precise, enabling formation of complexes with slightly different molecular keys.
2. The molecular key is able to modify slightly the structure of the cellular lock.

Just as these properties form the basis of the ability of a ligand molecule to direct intracellular processes from outside, a not too high specificity of a receptor enables the existence of antagonists of the ligand–receptor interaction. *The degree of fitness of the ligand binding site* with the binding site of a receptor is characterized by the term *affinity*, a measure of which is the constant $K_{RL}$ (see Equation 1.4). The bigger the $K_{RL}$, the stronger is the link between the receptor and ligand. Actually, the link must be of physiologically substantiated right strength.

In addition to specific ligand(s), a receptor also has specific *agonists* and *antagonists*.

*Agonists* are exogenous substances that also selectively bind to the receptor and activate it, thus triggering a response in the cell. They mimic the action of endogenous biochemical molecules (such as hormones or neurotransmitters) that bind to the same receptor. *Antagonists* are in turn substances that, binding too tightly to the receptor, do not activate it, and since they hinder ligand binding, they inactivate the receptor.

The role of receptors in the development of toxic responses will be discussed in Section 3.3.2.

## References

Martinetz, D. (1982). *Arsenik, Curare, Coffein. Gifte in unserer Welt*, Urania-Verlag, Leipzig.

Möller, L., Ed. (2000). *Environmental Medicine*, Karolinska University Press, Stockholm.

Pappas, A.A., Massoll, N.A. and Cannon, D.J. (1999). Toxicology: past, present and future, *Ann. Clin. Lab. Sci.*, **29**, 253–62.

## chapter two

# Routes of xenobiotics in an organism

After its entry into an organism, a xenobiotic finds itself in the physiological block-scheme *absorption-distribution-metabolism-(new distribution)-excretion*. And then its toxicity depends on many factors including the speed and exact manner of moving along this scheme. Since the free toxicant or its active metabolite is in the action point (tissue) in a dynamic equilibrium with the free toxicant in plasma, in most cases the toxic effect is proportional to the toxicant's concentration in blood. The strength of the toxic effect depends on the pharmacokinetics of the xenobiotic, that is, how the compound enters the organism and the bloodstream (*absorption*), how and in what shape it moves along the organism (*distribution* and *metabolism*), and how it and its metabolites leave the organism (*excretion*). The pharmacokinetics of a toxicant is substantially influenced by the age, gender, manners, and nutritional condition of the individual, tissue (liver, kidneys) functions, and so forth.

Three of the blocks in the aforementioned scheme—absorption, distribution, and excretion are connected with the disposition of the substance in the organism, which needs crossing of *biomembranes*.

## 2.1 Entry and absorption of foreign compounds

*Xenobiotics can enter a mammal via three main gates* (portals)—skin, lungs, and gastrointestinal (digestive) tract, besides intravenously through injections. Since most foreign substances, including toxicants, enter perorally, it is just the digestive or gastrointestinal tract that is the most important gateway, where the *primary toxic effect* manifests. *Respiratory ways* including lungs are the entry portal and site of absorption of gaseous toxicants such as carbon monoxide, nitrogen oxides, and so forth. *Transdermal entry* is essential for organic solvents, detergents, and other fat-soluble lipophilic substances, which can, due to dissolution of the skin fats, cause local irritation and dermatitis.

Toxic compounds may exert their primary adverse effects at the entry itself, but to cause a systemic effect, they must first pass through membranes into the bloodstream by absorption. The role of tissue liquids in the transport of substances is negligible. For crossing of membranes, a substance selects the optimal way (see Section 1.6.2). Mammals have no special system of absorption of xenobiotics; they cross membranes by the same way as the endogenous compounds. Most toxicants are absorbed by diffusion. Absorption can be a complicated multistep process, during which the substances may be subjected to biotransformation that may cause a lowering of their *bioavailability*.

### 2.1.1 Digestive tract

Various food components and drugs enter the organism via the oral cavity and the digestive tract (esophagus, stomach, small and large intestines, rectum) connected with it, which is specifically adapted for digestion and absorption of food. The digestive tract forms the first line of defense in the body against the main load of xenobiotics; the gastrointestinal mucosa has several mechanisms for modification of foreign compounds. *The digestive tract is also the site of absorption of food toxicants.* Lipophilic substances, including toxicants such as phenols and cyanide ions that easily cross the biomembranes, are usually absorbed already in the oral cavity. There is a positive correlation between the parameter $K_{ow}$ (see Section 1.6.2) and sublingual absorptivity of alkaloids. To achieve the "desired" effect, the sublingual dose of lipophilic cocaine must be twice the hypodermic dose, and for a considerably more hydrophilic morphine, the respective ratio is ten. Oral absorption eliminates the destructive effect of gastric and intestinal juices as well as the hepatic metabolism of a toxicant that sometimes may increase the toxicity of the xenobiotics.

Here again we encounter (see Section 1.3) the important toxicological term *bioavailability* that denotes the part of the oral dose (possible interval 0%–100%) that reaches the bloodstream. The fact that bioavailability is generally smaller than 1% (or 100%) is caused by both partial enzymatic destruction in the gastrointestinal tract and incomplete absorption of the compound.

The gastrointestinal epithelium is only one cell thick and has a large surface area in different parts. Since, by moving along the tract, pH substantially changes, it is possible to find a suitable absorption site even for the ionizable substances.

At the very beginning of their journey down the alimentary canal, the ingested substances contact the oral cavity, where pH is near 7 in the case of humans and more basic in the case of rats and some other species. *Ethanol, hydrocarbons, and other neutral nonionizable substances* already begin to absorb in the mouth. *Weak acids* diffuse across the membrane, best of all in the stomach, where pH is approximately 2 in the case of humans and many other mammals; *weak bases* diffuse across membranes in the intestine, where pH ≈ 6. But if the molecule of a weak acid for some reason was not absorbed in the stomach, it can be absorbed in the intestine, where there is a good supply of blood and, due to the folding of the intestine wall and villi, it has a large surface area. *A lipid-soluble nonionizable substance* absorbs equally well in any part of the digestive tract, and its absorptivity increases by an increase in the distribution coefficient $K_{ow}$, that is, the lipophilicity of the substance. This rule does not apply for the lipophilic ($K_{ow} > 3000$) molecules that form in the stomach and intestinal supramolecular colloidal solution, the dispersed particles of which, micelles, are too large to cross the membrane. Molecules with a mass of over 3000 Da are not able to cross the digestive tract wall by simple diffusion.

*For nutrients and endogenous substances* (saccharides, amino acids, amines, inorganic salts, etc.) specialized transport systems exist in the membranes of the digestive tract. For example, glycoprotein P takes part in the absorption of many compounds. Dependence on such membranous systems may reduce the degree of absorption, and, vice versa, factors such as saturation and competition can increase the absorption. *Metals are absorbed* mainly in the first part of the small intestine—chromium, manganese, and zinc in the ileum; iron, copper, mercury, thallium, and antimony in the jejunum. *Alkali metals* are absorbed quickly and completely, and *alkaline earth metals* to the extent of 20%–30% as hydroxides or poorly soluble phosphate complexes. Metals may, during the absorption process, change their valence form; for example, bivalent iron can be oxidized into trivalent, and insoluble inorganic salts of lead may be changed into lead-organic substances, which will be much better absorbed.

*Absorption is favored* by ulcers and irritations of the gastrointestinal tract as well as starvation, but hindered by ample food and vomiting. On the other hand, food can accelerate the absorption of a substance that dissolves in fat, especially in milk fat. The rate of absorption is also influenced by the intensity of the blood supply in the stomach mucosa, by formation of mucus, and by peristalsis. In the case of drugs, the vehicle added to facilitate suspending or dissolving of the active substance may also have a significant effect on the toxicity of the medicine, influencing its absorption and distribution in the body.

*The exact absorption site has a significant role in the fate of a xenobiotic.* Therefore, a substance can be hydrolyzed in the acidic content of the stomach that may result, for example, in the inactivation of a snake toxin or other proteinaceous toxin. A foreign compound can be hydrolyzed by the enzymes produced by the intestinal wall cells, pancreas, or the intestinal bacteria.

For the majority of compounds, oral administration will be followed by absorption into the portal vein supplying blood from the region of the digestive tract to the liver.

### 2.1.2 *Lungs*

*Lung tissue has a very large effective area* (50–100 $m^2$ in the case of humans) and *an excellent blood supply*. The pulmonary alveolus is separated from the blood vessels by a layer with a thickness of one to two cells. Therefore, *the absorption in the lungs is very fast and effective*. The air we breathe contains various toxic substances of different potencies and particles such as carbon monoxide (CO), nitrogen oxides ($NO_x$), sulfur dioxide ($SO_2$), vapors of organic solvents, aerosols, and solid particles (asbestos, pollen). Two essential factors influencing the absorption of substances from the pulmonary alveoli are the speed of breathing and the speed of blood flow. In the case of substances that are poorly soluble in blood, the speed of the blood circulation is crucial; in the case of well-soluble substances, the speed of breathing is more important. The capability of the substance to form complexes with blood plasma (especially transport) proteins is also a very important factor. Small fat-soluble molecules are absorbed very fast. For these molecules, the pulmonary route is the most important way of entering the organism. So the water-insoluble particles of uranium dioxide can, being absorbed via lungs, initiate liver damage. Lead is also absorbed from the air through the lungs. Pulmonary absorption of compounds and particles can occur by pinocytosis or phagocytosis, respectively (see Section 1.6.2). Phagocytosis can harmlessly leave toxic asbestos particles in the wall of pulmonary tissue for a long period. Carcinogenesis may start much later without any extra alarm.

### 2.1.3 *Skin*

*The skin is a complex membrane structure* consisting of thick layers of various cells and intercellular cavities. Although the skin is potentially in maximum contact with gases, solvents, and dissolved substances and its area is quite considerable (approximately 2 $m^2$ in the case of a medium-sized human), due to its multiple cell layers (epidermis, derma, and subcutaneous adipose tissue) with a poor blood supply, skin serves as an *efficient barrier to penetration of substances.*

Human skin permits two distinct routes of absorption:

1. Through the epidermis (in most cases)
2. Through the sweat glands and follicula (e.g., Pb)

The skin can be permeated in considerable amounts only by lipophilic solvents like dimethyl sulfoxide (DMSO). The skin itself can be damaged by corrosive substances such as strong acids or alkalis. Ionic substances are either not able to penetrate the skin or they do it extremely slowly. Nevertheless, cases of percutaneous intoxication with pesticides (e.g., with the insecticide parathion) have been reported. Percutaneous absorption is favored by warm skin, sweating, rash, scratches, and injuries of the skin.

In the case of lipid-soluble substances, the route of entry is of high importance for the starting moment of the toxic effect, since the time after the exposure, during which it is possible to weaken the toxic effect depends on the absorption site. Table 2.1 provides some very approximate information about this issue.

***Table 2.1*** Dependence of the Interval between the Contact with a Toxic Compound and Death of the Organism on the Absorption Site

| Absorption site | Interval (min) |
|---|---|
| Lungs | 15–30 |
| Eye | 15–30 |
| Digestive tract | 30–120 |
| Skin | 60–240 |

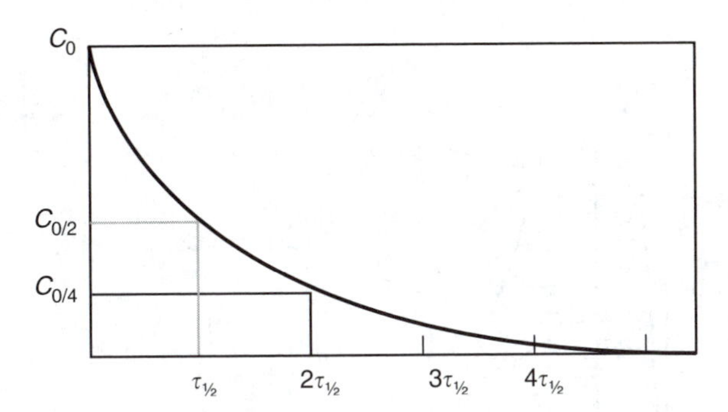


***Figure 2.1*** Explanation of the term "half-life." $C_0$ is the concentration of a substance at the moment 0.

## 2.2 *Distribution of xenobiotics in an organism*

### 2.2.1 *Blood supply and membrane barriers*

The further movement of substance particles (molecules, ions) following absorption into the bloodstream is governed by the properties of various cellular membranes along the circulatory system and binding in the body liquids. Differences in the blood supply of different tissues are also of high importance in the distribution of foreign compounds. The largest volume of blood crosses the liver, kidneys, brain, and skin, while much less blood circulates in fat (adipose) tissue and bones. In the absence of selectivity, the highest amounts of a foreign substance should reach the first four tissues. Relatively, these organs and also other small organs like the thyroid gland and pancreas get much blood owing to their small weight.

*Distribution of xenobiotics is a dynamic process.* It depends on the ratio of the rates of absorption, metabolism, and elimination (excretion), on the physical and chemical properties of the substance, and the particular environment in various body compartments.

Knowing its concentration in plasma, it is possible to calculate such important pharmaco- or toxico-kinetic parameters of a substance as its *half-life, area under the curve, volume of distribution,* and *the body burden.*

*Half-life* ($\tau_{1/2}$) is the time, during which one-half of the initial dose of a substance becomes degraded, metabolized, or eliminated from the organism (Figure 2.1). Half-life is determined by metabolism and excretion rates.

*Area under the curve* (AUC) is the area in a plot of concentration of a xenobiotic in plasma (*y*-axis) against time (*x*-axis). AUC has the unit (mass/volume) × time and expresses the total amount of a xenobiotic absorbed by the body, irrespective of the rate of absorption (Figure 2.2).

*Volume of distribution* $V_D$ (liters) is the volume of body liquid in which a foreign compound is distributed.

$$V_D = \frac{\text{Dose (mg)}}{\text{Plasma concentration (mg/l)}} \tag{2.1}$$

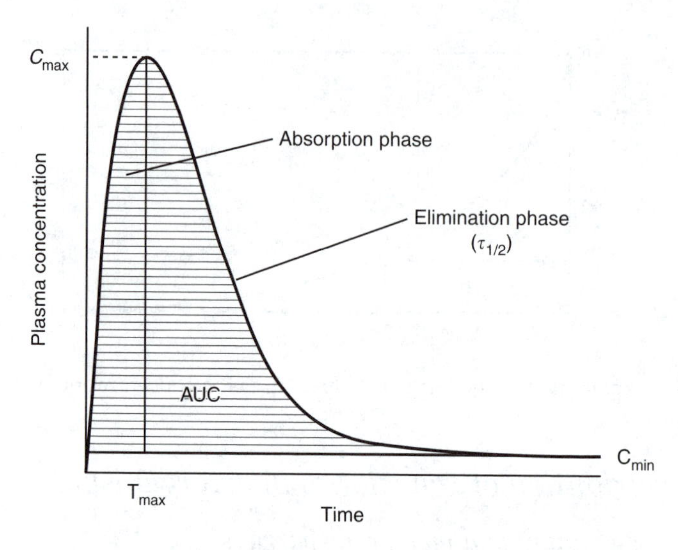


***Figure* 2.2** Explanation of the term "area under curve".

*Body burden* is the total amount of the chemicals that are present in a body at a given point in time. Sometimes, it is also useful to consider the body burden of a specific single chemical, such as lead, mercury, or dioxin.

*Organisms have metabolizing membrane barriers* such as placental (feto-maternal) or blood–brain (hematoencephalous) barriers that hinder penetration of various foreign substances into the fetus and brain, respectively. Unfortunately, these barriers are not perfect. So, the blood–brain barrier is easily crossed by toxicants such as organic methyl mercury, lead, aluminum, and ethanol. The last substance even enhances the permeability of this barrier, thus increasing the toxicity of other poisons to the brain cells. Since, in *newborns*, the brain has not fully established yet, they are especially sensitive to the toxic effects of chemical substances. The placental barrier is crossed by simple diffusion by small ($M < 200$) lipophilic nonelectrolytes. Metals like mercury or selenium permeate the placental barrier and deposit in the fetus. Some organophosphorous insecticides like methamidophos (*O,S*-dimethyl phosphoramidothioate) and viruses like *Rubella* can also cross this membrane barrier and elicit teratogenic effects (see Section 3.2.6).

## 2.2.2 *Binding of xenobiotics to charged particles*

Xenobiotics are transported from one tissue to another mainly by blood. Blood plasma contains various proteins (albumin and other transport proteins) capable of forming complexes with, and transport, both positively and negatively charged low-molecular compounds. A substance becoming inactive through this complexation is no longer capable of crossing the membranes, including the renal membrane, necessary for its elimination. This, generally noncovalent dynamic complex, held together by ionic, hydrogen, hydrophobic, and van der Waals bonds (see Section 1.6.3), can be saturated at high concentrations of the xenobiotic and participate in the exchange with other compounds. This complex causes the existence of the *threshold concentrations* (*levels*) of toxic substances. An appearance of a very small extra amount of the substance in the bloodstream can already cause a toxic effect. A toxicant bound to a plasma protein circulates in the bloodstream until it will be bound to either another macromolecule or a tissue component, for example, a receptor. Analogically with the metabolic enzymes (see Section 2.3) *de novo* biosynthesis of transport proteins can be induced by xenobiotics capable of acting as blockers of the binding sites of these proteins.

### 2.2.3 Bioaccumulation of xenobiotics

Every foreign substance has its specific half-life in the organism. These half-lives can be very different. For example, calcium is deposited in bones and even if a bone cell dies, the remaining calcium will be used to build new bone cells. So, the organism economizes the substances it needs in big quantities, but which are not easily assimilable.

*Most xenobiotics have a short half-life in the living organisms*; they are rapidly metabolized and/or excreted. Nevertheless, a continuous exposure to such chemicals can create a "persistent" body burden. Arsenic, for example, is mostly excreted within 72 h after the exposure. But some compounds can stay in the organism for a longer period and accumulate. *Bioaccumulation* means a stay and deposition of foreign, mostly lipophilic, substances in different tissues of the organism. These substances are accumulated not only in the adipose tissue but also in bones, semen, muscle, brain, liver, or kidneys. So has *ligandin*, a hepatic cytoplasmic protein, high affinity to organic acids and carcinogenic azo-dyes, another hepatic and renal protein, binds cadmium. Bones are the site of accumulation of fluorides, lead, strontium, radium, and tetracyclines. Chlorinated pesticides, such as dichlorodiphenyltrichloroethane (DDT), can remain in the body for 50 years.

*Bioaccumulated substances cannot be acutely toxic* but they may cause chronic intoxications. Very often, a toxicant deposited in fat or bone tissue does not exert any toxic effect over a long period, but results in "lethargy." Under certain circumstances, such as long-term starvation, especially of overweight persons, the volume of adipose tissue is substantially reduced and part of the "lethargic" toxicant deposited escapes into the bloodstream and further to different tissues, where it may become revived. Then the substance becomes toxic. It should be taken into account that fat makes up 50% of the body weight of a stout person or 20–30% of a lean person.

Many environmental poisons like DDT, polychlorinated biphenyls (PCBs), and dioxins are concentrated in the lipid fraction of mother's milk.

The fatter the milk, the more lipophilic toxicants it can hold. Lipophilic environmental toxicants concentrate also in fat-rich egg yolk. Several toxicants and drugs deposit in the organs selectively—for example, adrenaline mainly in the heart, iodine in the thyroid gland, trichloroethylene in the brain, and chloroform in the adrenal gland.

The accumulation site of a substance can change with time—the *substance can be translocated* in the organism. A good example is lead, a substantial part of which is at first bound in the liver and kidneys by erythrosine. About a month later, lead is transported to the bones, replacing the $Ca^{2+}$ ions there. By this route, about 90% of the absorbed lead ends up in the bones.

*Consequently, the longer the half-life of a toxic substance in the organism, the higher is the danger of the intoxication*, even if the concentration of this substance in the environment is low at the very moment. Such problems can emerge in the case of occupational exposition to toxic compounds. Since a person spends a large part of his or her life at the working place, even very low doses of a potent toxicant accumulated chronically may turn out to be toxic. Fat-soluble methyl mercury that tends to accumulate in the brain, and tetraethyl lead, preferentially accumulated in the adipose tissue can serve as good examples of such compounds (see Sections 10.1.1 and 10.1.2). Strontium $Sr^{90}$, which appears in the environment as a result of testing of atomic weapons, deposits in the bones, replacing calcium, where its radiation can cause tissue damage over a long period (see also Section 10.2).

*Natural toxins can also be accumulated*—therefore, extremely toxic are the algal toxins accumulated by different crustaceans and sea fish living in the coral reefs (see Chapter 13).

Organisms can accumulate up to the toxic concentration levels also substances that are harmless and sometimes even useful, in lower, physiologically relevant concentrations. A classical example is vitamin A that becomes highly concentrated in the liver of carnivores—an especially powerful concentrator is the polar bear as an obligate carnivore (see Chapter 18).

*Biomagnification* is a process of the increase of a contaminant concentration in organisms along the food chain. As a result, the concentration of a contaminant is the highest in the organism of a predator dwelling at the very top of the food chain.

## 2.3 Metabolism of xenobiotics

### 2.3.1 General principles

*Metabolism or biotransformation is an extremely complicated network of biochemical reactions* taking place in any organism to guarantee its normal functioning. The term *metabolism* originates from the Greek μεταβολισμός "metabolismos," meaning a change or overthrow. *All metabolic reactions are catalyzed by the enzymes,* whose activity is kept under permanent control (i.e., is regulated) by the neurohormonal system of an organism.

An ingested or inhaled foreign compound also enters this network, where it will be converted into new compounds called *metabolites.* As a rule, the metabolites are less lipophilic and hence, better soluble in water than the parent compound. Simultaneously, its molecular weight and the speed of elimination from the organism increase, causing a reduction of the half-life. In the case of toxic compounds, it means a decrease of the undesired contact time with the organism and probability of bioaccumulation.

*In some cases, the new derivative is less water-soluble* and eliminable at a rate slower than the parent compound. For example, conjugation with an acetyl-group (acetylation) can reduce the water-solubility of drugs like sulfonamides. The formed acetates tend to crystallize in the renal tubules, causing the development of renal necrosis.

The toxicity of a compound in a whole animal can be caused either by:

- The parent compound
- Metabolites, formed at the site of entry into the organism or in the liver that are sufficiently stable to reach other target organs (systemic toxicity)
- Metabolites, formed in the target organ

*Biotransformation* of foreign substances is carried out *mainly in the liver* and also in other tissues such as intestinal epithelium, kidneys, lungs, brain, and skin in *two interrelated phases* (*phase I and phase II*) by the help of enzyme complexes (see Figure 2.3). The substances that are initially metabolized, for example, in the liver may be subjected to further metabolism once they have reached the target organs. The liver and kidneys together contain more than 90% of the metabolic activity of a mammal organism, in this way, determining the speed of elimination of the xenobiotic and its bioavailability.

The enzymes participating in the biotransformation are strictly localized in a specific part of the cell—many of them can be found in the endoplasmic reticulum, part in cytoplasm, or in other organelles like mitochondria.

The main chemical reactions of phase I are oxidation, reduction, or hydrolysis; that of phase II is bioconjugation. This scheme of two phases

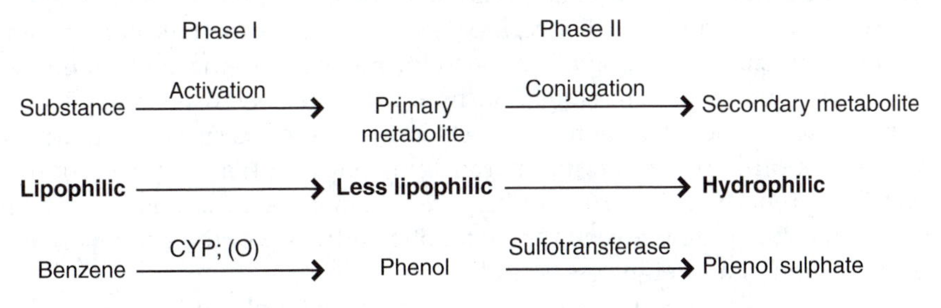


***Figure 2.3*** Two-phase scheme of a xenobiotic biotransformation.

is actually a simplification. It can be illustrated by the example of benzene. Inhaled lipophilic benzene, which is poorly soluble in water, is oxidized in the liver in the course of phase I by enzyme cytochrom P450 monooxygenase complex (CYP—see Section 2.3.2) over benzene oxide mainly into phenol, which is much more polar and hydrophilic than the initial compound. Since the water solubilities at 25°C are 7 and 0.8 g/L, respectively, for phenol and benzene, and at body temperature the difference in solubilities is still bigger, phenol is much faster eliminated from the organism than benzene. Besides phenol, other minor metabolites such as phase I reaction products such as hydroquinone, catechol, trihydroxybenzene, and benzoquinone that are more hydrophilic than phenol are formed.

Phenol, catechol, and hydroquinone can be further metabolized into even more hydrophilic and better-soluble-in-water *sulfates and glucuronides* using the respective *phase II* enzymes as catalysts (see Section 2.3.2).

*But this is not all*—various ring-opened products like *muconates* and *mercapturic acid derivatives* are formed from benzene oxide. These products together with the hydroxylated derivatives of benzene are transported to the *bone marrow* where a subsequent secondary metabolism occurs. The benzene toxicity, which is actually a combined toxicity of its metabolites, involves both bone marrow depression and leukemogenesis caused by damage to multiple classes of hematopoietic cells and a variety of hematopoietic cell functions. Two potential mechanisms by which benzene metabolites may damage cellular macromolecules to induce toxicity include the covalent binding of reactive metabolites of benzene to cellular macromolecules and the capacity of benzene metabolites to induce oxidative stress (Ross, 1996; Snyder and Hedli, 1996).

All chemicals are potentially susceptible to biotransformation and all cells and tissues are potentially capable of carrying out metabolic processes. Species, strain, gender, individual, organ, and age-specific differences may have sufficient and important effects on the biotransformation of xenobiotics.

*Metabolism usually reduces the toxicity of an absorbed compound.* In the case of most substances, phase I generally results in a stable and nontoxic (or at least substantially less toxic) hydroxy-group containing metabolites of the hydrophobic RH. But numerous cases can be observed, when a substantially more chemically active and, hence, more toxic substance than the parent compound (or primary metabolite) has been formed during phase I (sometimes even during phase II) of the metabolism. It can cause necrosis, mutagenesis, carcinogenesis, and immunotoxic effects in the tissue. As suitable examples, the desulfation of parathion into paraoxon (Chapter 14) and epoxidation of benzo[$\alpha$]pyrene (Chapter 16) can be given. Sometimes, there is a competition for the substrate between the metabolic pathways of the activation and detoxification of a foreign compound.

Some chemical compounds can be further metabolized by reactions of phase III.

**Table 2.2** Reaction Types of Phase I and the Respective Enzymes

| | |
|---|---|
| Oxidation | Dehydrogenation |
| Cytochrome P450 monooxygenase (CYP) | Alcohol dehydrogenases |
| Flavin-containing monooxygenase (FMO) | Aldehyde dehydrogenases |
| Peroxidases | Radical recombination |
| Amine oxidase | Superoxide dismutase (SOD) |
| Monoamine oxidase | Hydrolysis |
| Dioxygenases | Carboxylic esterases |
| Reduction | Amidases |
| Cytochrome P450 monooxygenase (CYP) | Hydration |
| Ketoreductase | Epoxide hydrolase |
| Glutathione peroxidases | Dehalogenation |
| | Dehalogenases |

## 2.3.2 *Phase I reactions*

Phase I or activation (functionalization) consists of the addition of either an electrophilic (unsaturated carbonyls, epoxides) or a nucleophilic group (alcoholic or phenolic hydroxyls, amino, sulfhydryl, or carboxylic groups) to the parent molecule. A number of phase I reaction types exist, the most important of which are presented in Table 2.2.

The enzymes of phase I can be divided into two main groups—*oxidoreductases* and *hydrolases*. Oxidoreductases introduce an oxygen atom into or remove electrons from the substrates; hydrolases hydrolyze esters, amides, epoxides, glucuronides, halogenides, and so forth.

### 2.3.2.1 *Enzyme superfamily CYP*

A majority (over 90%) of the phase I reactions are catalyzed by the enzyme *cytochrome P450 monooxygenase complex* (abbreviated *CYP* for mammalian/plant and *P450* for bacterial species), (EC 1.14) that

- *Is a superfamily of mostly membraneous isoenzymes*, a perfect example of the evolution of enzymes defending organisms against toxic chemicals. Humans have, for example, over 50 CYP isoenzymes belonging to 18 families and 42 subfamilies.
- *Is a hemoprotein*, that contains, in addition to about 500 amino acids, a heme group, with an iron atom in the center. The most famous hemoprotein is hemoglobin, the oxygen transporting protein in the blood.
- *Can be found in most of the tissues*, if not in all, but in very high concentrations in the liver.
- Is located not only on the membrane of the smooth endoplasmic reticulum, but also in other organelles.
- *Switches an oxygen atom into the substrate molecule*, that is, oxidizes the substance. In certain cases, it is capable of catalyzing also the opposite reaction—reduction of a substrate.

- *Requires a helper enzyme*—NADPH-dependent cytochrom P450 reductase (EC 1.6.2.4), donating electrons to the CYP to reduce $Fe^{3+} \rightarrow Fe^{2+}$ in its heme.

*CYP is a real multigene superfamily of enzymes*, the so-called mixed-function oxidases (MFO), and it consists of at least 27 gene families. Mammals have over 200 functioning genes of CYP. The amino acid sequence of over 100 CYPs has been determined.

*CYP isoenzymes are divided into families*. In the case of mammals, three of these families (CYP1, CYP2, and CYP3) are involved in the metabolism of xenobiotics. CYP4 is, in addition, responsible for the fatty acid metabolism. Several CYP isoenzymes have a genetic polymorphism, a phenomenon that has its own effect on the metabolism of xenobiotics. Relative amounts of different isoenzymes not only depend on the tissue type, but also on species, gender, and so forth. Human beings may substantially differ from each other in the metabolizing ability of foreign substances.

*CYP is an extremely versatile enzyme*. It is capable of catalyzing over 60 reaction types. The common parameter of the numerous substrates is their lipophilicity. CYP substrate can be either aromatic, alicyclic or aliphatic hydrocarbon, and also a heterocyclic (*N*-, *O*-, or *S*-containing) substance. The reaction also needs NADPH, molecular oxygen, and magnesium ions.

The actual reaction catalyzed by CYP isoenzymes can be hydroxylation (including *N*-hydroxylation), epoxidation, heteroatom (*N*-, *S*-) oxygenation, heteroatom (*N*-, *S*-, *O*-) dealkylation, dehydrogenation, isomerization, ester cleavage, replacement by oxygen, or even reduction under anaerobic conditions. Products of these reactions will either be secreted or metabolized further by phase II reactions.

The mechanism and the stoichiometry of a CYP-catalyzed reaction can be described by the scheme:

$$\text{RH (substrate)} + O_2 + \text{NADPH} + H^+ \rightarrow \text{ROH (product)} + H_2O + \text{NADP}^+$$

*From a toxicological point of view, the epoxidation* of compounds containing carbon–carbon double bonds catalyzed by CYP is very important. The epoxides formed are usually unstable and chemically reactive. They can interact with nucleophilic groups in cellular macromolecules and have a deleterious physiological effect. Good examples are epoxidation of benzo[$\alpha$]pyrene (Figure 16.1 and Section 16.2), aflatoxin $B_1$ (Section 11.2), and epoxyderivatives of polyunsaturated fatty acids (PUFAs) (Section 16.1).

#### 2.3.2.2 *Other enzymes, catalyzing oxidation*

1. *The flavin-containing monooxygenase (FMO) system* is used for oxidation reactions in drug metabolism. The major differences between the FMO and CYP systems are that the FMO system does not oxidize

carbon atoms and is not induced by exposure to drugs or environmental pollutants. On the other hand, both systems require NADPH and oxygen. The FMO system also contains multiple isoenzymes that are also localized in the endoplasmatic reticulum.

2. *Peroxidases.* PUFAs are converted into bioactive metabolites (hydroperoxides, epoxyacids, etc.) by three different enzymes—cyclooxygenases (COX), lipoxygenases (LOX), and CYP (Kuroda, Maeba, and Takashio, 2003).
3. Human *amine oxidases* (AO) are flavin-containing enzymes that are present on the outer mitochondrial membrane. In contrast to CYP or FMO, they have only two isoenzymes. AOs catalyze the oxidative deamination of biogenic neurotransmitters such as serotonin, noradrenalin, and polyamines like spermines as well as xenobiotic dietary primary, secondary, and tertiary amines such as tyramine or phenylethylamine, into aldehydes. AOs need the FAD cofactor. The reaction can be summarized as:

$$RCH_2NH_2 + H_2O + O_2 \rightarrow RCHO + NH_3 + H_2O_2$$

The resultant hydrogen peroxide is the source of hydroxyl radical (OH), the most toxic free radical known.

### 2.3.2.3 *Examples of phase I reactions*

1. *Reactions catalyzed by CYP*
   a. *Hydroxylation* of aliphatic or aromatic carbon:

Propoxur (carbamate pesticide) → *para*-Hydroxyropoxur

Toluene → *para*- and *ortho*-Hydroxytoluenes

   b. *Oxidation* of a heteroatom (N or S)

Aniline → Phenylhydroxylamine (H-hydroxylation)

   c. *Dealkylation or deamination* of heteroatom (N, S, or O). N and O are dealkylated easier than S. Ethers R–O–R can be also dealkylated, the reaction is similar to the *N*-dealkylation. The reaction products are an alcohol and aldehyde as in the *N*-dealkylation

$$R - NH - CH_3 \rightarrow R - NH_2 + HCHO$$

$$R - N - (CH_3)_2 \rightarrow R - NH - CH_3 + HCHO$$

$$R - S - CH_3 \rightarrow R - SH + HCHO$$

d. *Desulfation.* CYP catalyzes desulfation and dehalogenations, whereby a heteroatom is replaced by O atom

Parathion → Paraoxon

This reaction is catalyzed also by paraoxonase (see Figure 14.3). The formed paraoxon is much more toxic than the initial parathion.

e. *Reductions* are an interesting type of reaction consisting of electron transfer from $Fe^{2+}$ to the substrate. Examples are the reduction of a nitro group and azo group and of arene oxides and reductive halogenation. These reactions are usually studied *in vitro* in anaerobic conditions in the presence of isolated microsomes and NADPH. The role and relevance of these reactions in *in vivo* processes is not completely clarified yet. They can proceed, for example, in the microsomes and in the intestine in participation of microflora. An opinion has been expressed that these reactions can also take place *in vivo* in a cell in some conditions of a low partial pressure of oxygen.

i. Reductive halogenation

DDT → DDE (see Chapter 14)

ii. Azo-reduction

Azobenzene → Aniline

2. *Hydrolysis of esters and amides*

These reactions are catalyzed by *esterases* (EC 3.1.1 or 3.5.1), a group of enzymes, characterized again by a broad substrate specificity. Carboxylic esters are, in general, hydrolyzed more rapidly than the corresponding amides. The conversion of both types of substrates is actually catalyzed by the same enzymes, despite the assignment of different enzyme class numbers due to the two different activities. The esterases fulfill numerous endogenous functions in the organism but they are also able to metabolize foreign compounds. According to their reactivity with toxic organophosphorous compounds, the esterases are historically divided into *A, B,* and *C-esterases.*

*A-esterases* prefer esters of aromatic- or aryl-carboxylic acids as substrates, but are also capable of hydrolyzing organophosphorous esters. Many insects have the level of A-esterases, which hydrolyze these insecticides, much lower than that of the mammals. This phenomenon enables to synthesize organophosphorous insecticides, which are relatively harmless to humans and other warm-blooded animals.

*B-esterases* that represent a large heterogeneous group of enzymes prefer esters of aliphatic carboxylic acids; organic esters of phosphoric acid are their inhibitors. Toxicity of the well-known organophosphorous and carbamate insecticides is based on the inhibition of acetylcholinesterase, belonging to the B-esterases.

*C-esterases* prefer esters of acetic acid (acetates) and do not interact with phosphoric acid esters.

Some examples of hydrolysis reactions catalyzed by the esterases are:

- Phosphoric acid esters. Enzyme A-esterase

Paraoxon → respective phosphoric acid + *para*-nitrophenol

- Acetylcholine. Enzyme acetylcholinesterase (AchE, a B-esterase)

Acetylcholine → Choline + Acetate

- Carboxylic esters. Enzyme C-esterase

Vinyl acetate → Acetate + Acetaldehyde

3. *Reactions catalyzed by epoxide hydrolase* (*hydration*)

Organic chemistry teaches that the most stable intramolecular cycles are the 5- and 6-linked ones, the geometry of which fits best with the angles between the valence orbitals of carbon and other bioelements like O, N, or S. Epoxides with their 3-linked strongly strained oxirane cycles are, on the contrary, very reactive electrophilic particles; the cycle is rather easily opened either spontaneously ($S_N1$) or through the attack of a nucleophilic particle according to the nucleophilic substitution ($S_N$) mechanism of $S_N2$. When the nucleophil is an important cellular macromolecule like DNA or some protein, a covalent bond is formed between the DNA and the nucleophil. The consequences of this chemical reaction can be deleterious for the cell. This primary biochemical injury may initiate a long chain of damages, leading, for example, to the initiation of cancer. Serious acute and/or subchronic intoxications can be also initiated by the reaction between an epoxy-group and nucleophiles such as protein molecules.

Organisms are fortunate to have several ways for detoxification of the epoxides, such as:

- Hydration by the help of an epoxide hydrolase. This is the main way of detoxification of an epoxide
- Spontaneous degradation of epoxides ($S_N1$)
- Nonenzymatic binding to glutathione (see Section 2.4.3)

Epoxide hydrolases (EC 4.2.1.63) are located primarily in the endoplasmatic reticulum, and also in the cytosol of the hepatic cells. They catalyze *trans*-binding of a water molecule to the epoxy group, resulting in the formation of significantly less reactive and more water-soluble *trans*-diols. Owing to the localization of a part of this enzyme in the membranes in the vicinity of CYP, the lipophilic epoxides formed by CYP can be quickly hydrolyzed and hence rendered relatively harmless. This enzyme helps deactivate many labile "strong carcinogens," at the same time helping to activate "hidden" carcinogens such as benzo[α]pyrene (see Section 16.2). Epoxide hydrolases are induced mostly by the well-known CYP inducers (see Section 2.3.4). In the case of humans, high dependence of hydrolase activity on the individual, caused both by the high inducability of this enzyme and differences in life-styles, is observed.

### 2.3.3 *Phase II reactions*

The main reaction types of biotransformation phase II and the respective catalyzing enzymes are presented in Table 2.3.

In this phase of biotransformation, further modification of xenobiotics called *conjugation* is performed. Here the water solubility, molecular weight, and molecule dimensions are generally increasing. The reaction product, conjugate, is still more easily excreted from the organism. These conjugates are usually already nontoxic, but sometimes, metabolically active compounds can be formed that can be even more mutagenic and carcinogenic than the initial compound.

The molecule of a xenobiotic or its phase I metabolite can be conjugated with the following.

1. *Sulfate group*. The respective enzymes, found in procaryotes, plants, as well as animals are *sulfotransferases* (EC 2.8.2). They need *coenzyme 3-phosphoadenosine-5′-phosphosulfate* (PAPS). Since the tissue concentration of PAPS is relatively low (4–80 ng/g), a shortage in the coenzyme may easily develop in the case of high activity of a sulfotransferase. A limiting factor of the synthesis of PAPS is the sulfate that is obtained either directly from food or by decomposition of sulfur-containing amino acids methionine or cysteine.

Sulfotransferases help sulfatate a large number of various endogenous and foreign substrates. Part of them are membrane-bound, and part are

**Table 2.3** Phase II Reaction Types and Respective Catalyzing Enzymes

| | |
|---|---|
| Sulfation | Amino acid conjugation |
| Sulfotransferases | Acylase |
| Glucuronidation | Methylation |
| Glucuronyl transferases | *O*-methyltransferases |
| Glutathione conjugation | *N*-methyltransferases |
| Glutathion *S*-transferase | *S*-methyltransferases |
| Glucosylation | Acetylation |
| Glucosyl transferase | *N*-acetyl transferase |
| Thiolation | Acyl transferases |
| Thiol transferase | Thiosulfate *S*-transferase or rhodanase |
| Amide synthesis` | |
| Transacylase | |

located in cytosol. The first group of sulfotransferases is located mainly in the membranes of the Golgi apparatus, where their main role is the sulfation of endogenous compounds such as steroids, proteins, carbohydrates, and neurotransmitters. These sulfotransferases are not switched into the metabolism of xenobiotics, which are sulfated by cytosolic enzymes. Sulfation directs the movement of lipophilic compounds into more polar environments such as body liquids and active centers of the enzymes. Substances containing either an aliphatic or aromatic hydroxy group are most easily sulfated but compounds that need a preliminary activation (e.g., oxidation) during phase I are also sulfatable:

$$\text{Phenol} \xrightarrow{\text{Sulfotransferase, PAPS}} \text{Sulfobenzene} + \text{PAP}$$

$$\text{Toluene} \xrightarrow{\text{CYP; [O}_2\text{]}} \text{Benzyl alcohol} \xrightarrow{\text{Sulfotransferase, PAPS}} \text{Sulfobenzene} + \text{PAP}$$

Conjugation with a sulfate group via O-sulfonation, catalyzed by sulfotransferases can be the basis for metabolic activation of a number of promutagens and procarcinogens such as acetylaminofluorene or secondary nitroalkanes like 2-nitropropane. *The latter case is a good example of a rodent promutagen.* Since sulfate acts in chemical structures as an electron-acceptor (withdrawing) group, the sulfates formed can be chemically unstable and decay nonenzymatically with the formation of an electrophilic product, easily reacting with DNA with the formation of a respective adduct. Unlike rodents, humans have a parallel and safer metabolic way—denitrification, catalyzed by CYP. This is the probable reason why no direct link between human exposure to 2-nitropropane and cancer has been established (Plant, 2003, p. 34).

Up to now, over 40 cytosolic sulfotranferases have been discovered, 11 of them are from the human organism. Cytosolic sulfotransferases are divided into two main families: phenol-(SULT1) and steroid-(SULT2) transferases. This division is actually not very strict. As in the case of CYP, supergene families of SULTs exist. *No data are available concerning the possible induction of sulfotransferases*. Nevertheless, a number of *synthetic inhibitors* of sulfotransferases have been found.

2. *Glucuronic acid*. This is the largest group of the phase II reactions. The respective enzyme is *UDP-glucuronyltransferase* (UGT, EC 2.4.1.17), that represents an enzyme superfamily (humans have 16 different molecules), coded by similar genes, and requiring *uridyl-diphospho-α-D-glucuronic acid* (UDP-GA) as a coenzyme. The concentration of UDP-GA in the hepatic tissue is about 200 ng/g. The UGT concentration is the highest in the liver, but the enzyme can also be found in the kidneys, spleen, intestine, lungs, and skin. Unlike CYP, the enzymes of glucuronidation are not complexed with each other, but are only mutually dependent. For example, glucose—the initial compound of the coenzyme synthesis is converted into UDP-GA in the course of an enzymatically catalyzed two-step process in cytosol; the coenzyme obtained is subsequently conjugated with the substrate in the lumen of the endoplasmatic reticulum.

The substrates of UGT can be substances, containing a nucleophilic heteroatom (mostly O or N, and also S) such as alcohols, phenols, enols, hydroxylamines, aliphatic and aromatic carboxylic acids, amines, carbamates, thiols, and sulfonamides. An oxidation product of phase I can also take on the role of the substrate.

Glucuronidation is very largely used by organisms—in addition to xenobiotics, a number of endogenous substances exist that have the functional groups necessary for such a conjugation. Glucuronides, synthesized in the liver, are secreted either into the intestines in the content of the bile or into urine via the kidneys. The actual manner of secretion depends largely on the molecular weight of the xenobiotic or its metabolite—smaller molecules go to the urine, larger ones into the bile, the boundary being at 250–350 Da, depending on the animal species. Thus, glucuronides of morphine, chloramphenicol, and endogenous steroids are transported to the intestines with the bile. But further *enterohepatic circulation* can occur, that may cause, after hydrolysis of the glucuronide by $\beta$-glucuronidase, unwanted in the case of toxic compounds, a lengthening of their half-life in plasma (see Section 2.4). This circular repetition essentially enhances the exposition time of the target organs such as the liver to the foreign compound.

$$\text{UDP-GA} + \text{Phenol} \rightarrow \text{Phenyl glucuronate}$$

3. *Glutathione.* Catalyzing enzymes are *glutathione S-transferases* (GST), which are found in many evolutionarily diverse species—bacteria, insects, plants, fishes, birds, and mammals. According to their location in the cell, the *GST-enzymes are divided into two superfamilies—cytosolic* and *microsomal.*

Conjugation with glutathione is an important way of detoxification of versatile hydrophobic compounds containing an electrophilic carbon atom, such as aromatic, heterocyclic, alicyclic, and aliphatic epoxides, isothiocyanates, $\alpha$, $\beta$-unsaturated carbonylic compounds, aromatic halogenides, and nitrocompounds such as 2,4-dinitrochlorobenzene or bromobenzene, unsaturated aliphatic compounds and highly reactive intermediates. Also, electrophilic N-, S-, or O-containing substances can be glutathionylated. Conjugation with glutathione increases the molecular weight and water-solubility of substances; simultaneously, the excretion rate with the bile or urine is increased and the probability of a toxic interaction with molecular targets is reduced. *The reaction is catalyzed by CYP* (phase I) *and cytosolic GST* (phase II), the latter is accompanied by *reduced glutathione* (tripeptide Glu-Cys-Gly; GSH with $\gamma$-peptidic bonds) as *a coenzyme.* This conjugate is decomposed in the kidneys with the formation of a respective cysteine derivative. The highest concentrations of GST-enzymes are localized in the liver, but they can also be found in the kidneys, intestine, lungs, and other tissues. Transferases catalyze nycleophilic attack of the glutathione thiolate ion ($GS^-$) on an electrophilic atom of the substrate.

*The substrates of GST can react,* although much less slowly, *with glutathione also nonenzymatically.*

In addition to the aforementioned enzymatic reaction, GST is able to bind a substrate molecule on its surface, including the formation of a covalent bond with the active center of the enzyme. This binding may or may not inhibit the catalytic reaction. But, in any case, it prevents the foreign compound from interaction with such toxicologically critical intracellular molecules like nucleic acid and proteins. GSTs possessing such a property are called *ligandins*, and the respective process is called *suicide inactivation* or detoxication. The same mechanism exists also in the case of CYP.

So, making use of three different mechanisms, glutathione and GST are able to render harmless the large amounts of various toxic compounds, which are either produced by the organism in the course of normal metabolism or have entered the organism from outside. The cellular concentrations of this relatively slow enzyme as well as the coenzyme are rather high—the concentration of glutathione in a hepatic cell is 10 mM and GST can constitute up to 10% of the hepatic cell protein. Nevertheless, a deficit of both compounds can occur at high concentrations of xenobiotics, thus needing conjugation.

Glutathione conjugates can sometimes, as in the case of the flame retardant *tris*-(2,3-tribromopropyl)phospate, be even more toxic than the initial compound or the respective phase I product.

Glutathione conjugates can be further metabolized to mercapturic acid during phase III.

4. *Amino acids*. Organic acids, foreign to the organism, can be conjugated with amino acids or glucuronic acid with the help of mitochondrial enzyme *acylase*. The actual amino acid used, which most often is glycin, depends on animal species. A carboxylic group of the foreign compound first reacts with coenzyme A (CoA) and then with the particular amino acid.

5. *Water molecule*. A good example of a hydration process is the enzymatic hydrolysis of benzo[$\alpha$]pyrene-7,8-epoxide formed from benzo[$\alpha$]pyrene with the help of CYP (phase I). The process is catalyzed by *epoxide-hydrolase* and results in benzo[$\alpha$]pyrene-7,8-diol, which is further converted into a potent carcinogen benzo[$\alpha$]pyrene-7,8-diol-9,10-epoxide (see Section 16.2). Hence, epoxide-hydrolase, which mostly has the aureole of a carcinogen neutralizer, can also act as a moderator of the formation of a new toxicant (Plant, 2003, p. 41).

6. *Methyl group*. It is possible to methylate a hydroxy, amino, or thiol group in a molecule; the catalyzing enzyme is, respectively, *O-, N- or S-methyltransferase* and the *coenzyme S-adenosylmethionine* (SAM). First of all, endogenous as well as exogenous substances are methylated.

A very important toxicological example is the methylation of heavy metals like mercury, which can be done by the environmental microorganisms. As a result, mercury turns from a water-insoluble inorganic ion into a lipid-soluble organic substance. Simultaneously, the target organ of the toxicant is altered—if the inorganic mercury is toxic to the kidneys, then organic mercury is toxic to the nervous system.

7. *Acetyl group*. The catalyzing enzyme is either *acetyl-* or *N-acetyltransferase*, which can be found in hepatic and gastric mucosa cells and also in the

cytosol of leukocytes. These enzymes act together with *cofactor acetyl-coenzyme A* (CoA). The substrates of acetyltransferases can be aromatic amines, sulfonamides, hydrazines, or hydrazides. Rabbits and also probably humans have two acetyltransferase isoenzymes that are substantially different in their catalytic activity. Having one or another isoenzyme, that is, being a fast or slow acetylator, is genetically determined. This phenomenon, demonstrating the importance of genetical factors in toxicology, has a substantial role in the toxicity of several drugs like hydralazine, isoniazide, and procainamide. Both acetylation and methylation are unusual reactions for phase II, since their products are generally less soluble in water than the respective initial compounds.

*In addition, conversion of cyanide* ($CN^-$) into thiocyanate ion ($SCN^-$) by the mitochondrial enzyme *thiosulfate S-transferase* or *rhodanase* is worth mentioning. This two-step reaction ensures the detoxification of cyanide.

### 2.3.4 *Induction and inhibition of metabolic enzymes*

Induction and inhibition are the two most important regulation mechanisms of an enzyme activity. CYPs, for example, are often the rate-limiting enzymes in the biotransformation process, and therefore, they have an important role in the determination of *in vivo* kinetics and interactions. CYP is a unique enzyme also for the high versatility of its regulation mechanisms.

#### 2.3.4.1 *Induction of enzymes*

It has been observed that if an animal has had a direct contact at least once with a foreign compound, the metabolic enzyme systems (first of all, CYP) of this animal become more capable of metabolizing versatile xenobiotics. The level of many components of the metabolic system can be substantially increased by the exposure of the organism to a number of drugs or other foreign compounds. Such a metabolic induction causes an increase of the concentration of CYP as well as other microsomal proteins. An additional synthesis of RNA and respective proteins are prerequisites of the metabolic induction. *Such an induction* occurs either due to increased transcription, or translation, or as a result of stabilization of enzymes, or RNA, for example, by mediation of the ArH receptor and *is a slow regulatory process* (Lin, 2006).

As a result of this potentiating phenomenon, an increase in the speed of elimination of xenobiotics and reduction of half-life in the organism will take place. Enzyme induction with one substance generates temporary resistance against the toxic effects of many substances. This is not correct for substances that need metabolic activation to become toxic. The toxicity of these substances can, on the contrary, increase due to enzyme induction. The final physiological result of the enzyme induction depends on the particular xenobiotic—on its metabolic pathways and on the balance of the reactions that increase and decrease toxicity.

*The list of metabolic inducers,* containing representatives of different compound classes—PAH, halogenated hydrocarbons, nicotine and other alkaloids, antioxidant food additives, food dyes, and natural components of various foods, is growing day by day. All these substances are

1. Lipid-soluble, localizable in the smooth endoplasmatic reticulum
2. Either substrates of CYP or other microsomal enzymes or otherwise capable of binding to these enzymes

Several types of CYP inducers have been found:

1. *Glucocorticoid type.* Steroid biosynthetic CYPs are induced by adrenocorticotropic hormone (ACTH) that stimulates the production of cAMP that activates a protein kinase that phosphorylates some unidentified protein, leading to an increase in gene transcription.
2. *Peroxisome proliferators* such as clofibrate that act through a binding protein called the peroxisome proliferator activated receptor (PPAR).
3. *Polyaromatic hydrocarbons* induce CYP1 family via activation of a specialized arylhydrocarbon receptor AhR (see Section 3.3.2).

Other chemicals may also induce CYPs. Ethanol induces the CYP2E enzymes and phenobarbital induces the rat CYP2B 40–50-fold through a phenobarbital receptor called CAR (Waxman, 1999).

The contact of test animals with various inducers differently alters the velocity of hydroxylation of various compounds. For example, 3-methylcholanthrene and similar compounds, which induce the synthesis of the isoenzyme $CYP1A_1$ with a slightly shifted absorption maximum (448 nm), accelerate the hydroxylation of polycyclic aromatic hydrocarbons (PAH). Phenobarbital, in turn, induces the synthesis of $CYP2B_1$, thereby accelerating the metabolism of hexobarbital and aminopyrine.

Normal components of food can also alter the activity of other metabolic enzymes. One important group of food, remarkably enhancing microsomal enzymatic activity, is the *cruciferous plants* like cabbage, cauliflower, broccoli, Brussels sprout, and so forth. These vegetables cause approximately a 100-fold increase in the activity of CYP1A in the cells of small intestine wall of rats. *The factual inducers* are several indole groups containing substances such as indole-3-carbinol. This compound is not active in its initial form, but becomes activated by gastric acid, converting carbinol into indolo[3,2-$\beta$]carbazole. The compounds originating from cruciferous plants also induce glutathione S-transferases.

Although CYP induction has been known for a long time, our understanding of the molecular mechanisms of the induction is still in a fledgling stage but is growing fast because of the recent advances in molecular biology and biotechnology (Lin, 2006).

*Epoxide hydrolase* is activated in rat liver by phenobarbital, 3-methylcholanthrene, and butylhydroxyanisol (BHA). Many such compounds also activate *UGT*.

Nowadays, it is still difficult to prognosticate the final effect of such inducers on the toxicity of different xenobiotics. These agents often influence simultaneously several enzymes of the metabolic system and the final result is a combination of different effects. For example, induction of CYP should, due to the formation of reactive epoxides, cause an increase of the toxicity of PAHs. But because of a parallel activation of at least one epoxide detoxification and one conjugation system, the excretion of the conjugate is sped up and the toxicity of PAHs conversely decreases. Revision of numerous experimental data shows that most of the inducers help reduce the toxicity of the foreign compounds. For example, the aforementioned indoles inhibit the neoplasy of murine forestomach, promoted by benzo[$\alpha$]pyrenes and development of rat mammal tumor, induced by 7,12-dimethylbenzo[$\alpha$]-anthracene.

#### *2.3.4.2* Inhibition of enzymes

A possible hindering (inhibition) of the CYP system, very essential for xenobiotic metabolism, may have dramatic consequences from the toxicological point of view. The results of the inhibition lower the elimination rate as well as increase the half-life of all compounds in the organism, potentially metabolizable in the active cooperation of CYP. It can lead to an increased concentration of these compounds in plasma and, consequently, to an increase of their toxic effect. For example, simultaneous administration of several drugs can cause unexpected adverse effects. The reason lies in the inhibition of the metabolism of one drug by the other. The inhibition of enzymes can be one of the following types:

1. Competitive inhibition
   a. Two xenobiotics, both substrates of the same, for example, CYP isoenzyme, compete for the enzyme active site. The metabolism of both compounds is retarded.
   b. The second xenobiotic is an irreversible competitive inhibitor of the same isoenzyme, blocking its active center and switching this enzyme molecule out of metabolism.
2. Noncompetitive inhibition
   a. *Product inhibition.* The product of a reaction catalyzed by an enzyme has a higher affinity to the enzyme active center than the substrate. The next substrate molecule is not capable of excluding the product molecule from the active center, which remains occupied and the whole enzyme molecule is inactivated.
   b. The reaction product or metabolite is highly reactive and binds often covalently and in essence, irreversibly to the heme group of CYP apoprotein, damaging its structure and activity. These types of substrates are called *suicide substrates.*

In addition to the ones discussed, less usual inhibition mechanisms based on, for example, the modification of synthesis of enzyme protein or heme, or on the disruption of the cofactor accessibility, or on the inhibition of the enzymatic activity of the helper enzyme NADPH P450 reductase can occur. Foreign compounds also induce and inhibit other enzymes involved in their metabolism.

### 2.3.5 *Participation of enteric microflora in the metabolism of xenobiotics*

Microorganisms, living in the animal intestine, are also important metabolizers of xenobiotics. These metabolic reactions depend on the growth substrate and on the environment where the microbes live. Microorganisms habitating in *aerobic conditions* are capable of destroying otherwise stable aromatic cycles, which they can use as the single source of carbon in the oxidative biosynthetic reactions, necessary for their life. Microbes, growing in *anaerobic conditions such as in the intestine* are orientated to the *reductive metabolism*. The enteric microflora, containing over 500 species, is capable of changing the bioavailability and, hence, the toxicity of xenobiotics, either increasing or decreasing their absorptivity. This microflora can continue the biotransformation of the products of xenobiotic metabolism that have been secreted into the intestines directly from blood or that have reached the enteric cavity in the content of bile or saliva or ingested the respiratory tract mucus (for the enterohepatic circulation see Section 2.4).

*The digestive tract of mammals contains versatile microorganisms*; their species, location, and abundance depend on the particular animal. The intestinal microbes have an especially important role in the metabolism and energy supply of an organism in the case of ruminants; humans and other monogastric animals have only the large intestine heavily colonized with bacteria. Most of the mammals have an ascending gradient of microflora, both in species richness and abundance along the intestines. Most of the scientific research concerning microbial metabolism is focused on the large intestine of humans just to satisfy the needs of toxicology.

Microflora, inhabiting human intestine, has an essential role in the decay of plant cell wall material, resistant to the enzymes of the host organism; that is, the dietary fiber provides the large intestinal bacterial populations with the metabolic energy that influences the bacterial metabolism of xenobiotics. Components of the dietary fiber-like pectin may, by creating favorable conditions for bacterial growth, affect the toxicity of these foreign compounds that need metabolic activation by the anaerobic intestinal microflora to exert their toxicity.

Examples of the conversions carried out by the intestinal bacteria are

- *Hydrolysis*, including that of sulfamates such as cyclamate
- *Dehydroxylation*, including that of primary bile acids
- *Decarboxylation* of amino acids
- *Dehalogenation*, including that of the pesticide DDT (see also Section 14.1)
- *Reduction*, including that of unsaturated fatty acids and azo groups in the azo-dyes such as tartrazine

- *Formation of nitrosamines* from nitrite and secondary amines (see also Section 16.6)

A recent review of xenobiotic metabolism by intestinal bacteria has been published by R.R. Scheleni (Scheleni, 2006).

### 2.3.6 Influence of diet on metabolism

Substance toxicity is influenced by various parameters of the organism—species, gender, stem, age, circadian rhythms, and so forth. All these factors are genetically predetermined and may cause, for example, an accelerated metabolism, imperfection of the metabolic system of the organism, errors in membranes structure, and so forth. The real toxicity of a compound can also be influenced by an *extraneous factor* such as *diet*. Study of the diet effects may reveal the peculiarities in metabolism of an individual. Theoretically, a shortage in any of the nutrients, necessary for operation of phases I or II, can lead to reduction of the activity of the whole metabolic system. Quite often, the changes in administered nutrients can cause unexpected changes in the toxic effects of a xenobiotic.

Examples

- A short-term deficiency in *vitamin* $B_2$ *(riboflavin)* increases the CYP level in mice, causing an acceleration of the oxidation of the respective substrates. A long-term (up to 7 weeks) riboflavin deficiency results in a substantial decrease in the level of CYP. A 10–15 day supplementation of food with riboflavin is necessary to normalize the situation.
- Since *vitamin E (tocoferols)* is a regulator of heme synthesis and heme is an important building block of CYP, the deficiency in dietary vitamin E reduces the speed of some oxidative methylation reactions in rats.
- The levels of both CYP and NADPH-cytochrom P450 are influenced by the deficiency in *vitamin C (ascorbic acid)*. The mechanism behind this effect is unknown. It is interesting that CYP isolated from the liver of ascorbic acid–deficient animals is more sensitive to the effect of ultrasound, dialysis, and iron-chelating substances than the CYP of animals fed with a vitamin C supplement.
- Metabolism of many compounds is influenced by *dietary protein* deficiency. As a result, the toxicity of various compounds can either increase or decrease. Protein deficiency causes, for example, a reduction of the level of some CYP isoenzymes. Addition of protein to these deficient diets can accelerate the biotransformation of some, but not all of the compounds. The reasons for these differences are not clear.
- The effect of dietary fats on the metabolic system of an organism depends on the type of the fat, its amount, and on the compound to be metabolized. Thus, enhanced concentrations of *saturated fats* may

cause an acceleration of aniline oxidation, having no effect on the oxidation of hexobarbital. Addition of *unsaturated fats* to the diet causes an acceleration of the oxidation of both the mentioned substances, simultaneously increasing the concentration of CYP and decreasing the activity of glucose-6-phosphate dehydrogenase.
- Modification of *mineral substances* supply to the test animals can also influence the rate of metabolism of different compounds. For example, *deficiency in iron* accelerates the hepatic metabolism of many compounds. At the same time, it has no effect on the levels of CYP as well as cytochrome P450 reductase in the liver. The activity of CYP in the small intestine is, on the contrary, highly dependent on dietary iron. Deficiency in this metal abruptly reduces the activity of several CYP-linked enzymes.
- *Magnesium* deficiency causes a remarkable retardation of metabolism of foreign compounds in the liver, whereby, both CYP and cytochrome P450 reductase levels become reduced. The concentration of both enzymes is also lowered in case of malnutrition.

A recent review of the effect of dietary factors on the expression and functioning of CYP has been published by Murray (2006).

## 2.4 Elimination of xenobiotics and their metabolites from organism

The two main methods applicable for reduction of xenobiotic level in the bloodstream are

1. Metabolism, leading to an accelerated elimination
2. Artificial elimination from the organism

The speed of elimination of a toxic compound from the organism is one of the most important determinants of its toxicity. The higher the speed of elimination, the shorter is the contact time with the xenobiotic and, in the case of a poison, the lower is its total toxicity. The parameter that best of all characterizes the speed of elimination of a foreign compound is the so-called *whole body half-life*, that is, the period required for half of a xenobiotic to leave the body (see Section 2.1).

Toxicants and their metabolites are eliminated from the body through organs and tissues such as

- Kidneys — urine
- Liver — bile and possible enterohepatic recirculation
- Intestines — feces
- Lungs — exhaled air
- Mouth (vomiting)

- Mammary gland — milk
- Skin — sweat, fat-soluble compounds in the case of amphibians, casting off the cuticula in the case of insects
- Eyes — tears

### 2.4.1 Kidneys

Water-soluble, relatively low-molecular substances are eliminated through the kidneys. Large molecules like proteins are not able to pass through intact glomeruli and fat-soluble molecules like bilirubin are reabsorbed from renal tubules. This is the most important route of elimination, as 25% of the arterial blood passes through the kidneys, 20% of which is filtrated through glomeruli. For this reason, the kidneys are also an *important site of exposure to foreign compounds and their metabolites.*

Substances are moved from the blood to the kidneys by one of the following mechanisms:

1. Filtration through glomerular pores—small molecules
2. Passive diffusion into renal tubules—fat-soluble molecules
3. Active transport into tubular liquid—separate systems for anions and cations

The renal structure is adapted to and hence is favorable for removal of substances from the blood. *The main structural unit, the nephron,* enables most of the small molecules ($M < 20$ kDa) to move from glomeruli to the tubular ultrafiltrate through relatively large (diameter 70–80 Å) capillary pores. Proteins and protein-bound small molecules are forced to stay in plasma. Lipid-soluble molecules diffuse passively along the concentration gradient from blood through the glomerular membrane. If those molecules are not ionized at pH of the tubular liquid, they will be reabsorbed by passive diffusion into the blood. Reabsorption is also important for saving a number of nonionized compounds like glucose from excretion, which are essential for the body. Water-soluble molecules, which are ionized in tubular liquid, are not reabsorbed and leave the body with urine.

Some molecules such as *para-aminohippuric acid* (a metabolite of *para*-aminobenzoic acid) are transported from the blood to the renal tubules by a *special anion-transporting system.* Analogical systems also exist for the transportation of cations. The existence of separate transport systems for different types of ions may have an essential role in the determination of the relationship between the speed of elimination of substances from the bloodstream and concentration in the blood. In case of passive transport, the amount of substance passed in time unit through the membrane by diffusion is proportional to the concentration of the substance in the blood. In case of active transport, which needs special carrier molecules, the speed of elimination at high concentrations of the substance is constant, due to the saturation of the carrier system. If the dose of this substance rises above the saturation boundary, its concentration in the blood begins to increase. This situation is typical, for example, of ethanol, continuous consumption of which leads to the well-known adverse effects on the central nervous system (CNS).

*Elimination through the kidneys can be hindered* also by binding of the substance to a plasma protein that is not able to pass through the renal tubular membrane. Binding to a protein has no effect on the active transport of a small molecule into the renal tubules. So it happens, for example, in the case of *p*-aminohippuric acid, 90% of which is bound to the plasma proteins. In spite of that, this compound is already removed from the blood during the first passage through the kidneys.

*A very important elimination parameter is the urine pH*, which normally is lower than the blood pH. For example, an acidic drug such as phenobarbital ionizes in tubules at alkaline pH and a basic drug such as amphetamine at acidic pH. These phenomena are used in case of intoxication with barbiturates and aspirin (*ortho*-acetylsalicylic acid) when the pH of plasma and urine is turned alkaline by the infusion of glucose-containing solution of sodium hydrogen carbonate. If necessary, the urine pH can be turned acidic, for example, by the infusion of ammonium chloride. The pH of urine can also be influenced by diet; for example, the pH of the urine is reduced by a protein-rich diet. Elimination of a compound via the kidneys can also be accelerated by facilitating diuresis, for example, by drinking a lot of water (see Section 2.6.1).

### 2.4.2 Liver

Elimination in the content of bile formed in the liver is especially important for large polar molecules. Bile is secreted by hepatocytes into the gall bladder, from where it flows along the biliary tract to the intestine. Substances excreted into bile are actually eliminated from the organism with feces. Polar compounds with a molecular weight around 300 Da and higher, like glutathione conjugates, are eliminated from the organism by this route. Transition from urine to the biliary route generally proceeds in the molecular weight interval of 325–500 Da, depending on the animal species. A remarkable biliary elimination starts at $325 \pm 50$, $440 \pm 50$, $475 \pm 50$, and 500 Da in the case of rats, guinea pigs, rabbits, and humans, respectively. Excretion into the bile is an active process—the liver has three specific transport systems, separately for the neutral molecules, anions, and cations. Saturation of these systems leads to the rise of the concentration of these substances in the liver; in the case of a toxic compound, it can cause liver damage.

Animal tests have proved that, for example, chlororganic pesticides such as DDT, eldrin, and dieldrine form an *enterohepatic cycle* in the mammal organism. *The scheme of this cycle is as follows.* During the biliary excretion, the xenobiotic or its metabolite contacts intestinal microflora that can modify it to a less polar (more fat-soluble) compound. The formed compound can be reabsorbed from the intestines and, via the portal vein that transports 65%–75% of the hepatic blood, reach the liver again. In the course of this recirculation, the toxicity of the initial compound can increase. If a substance has been administered by the oral route, it is transported from the intestines by the same route to the liver, where it can be switched into the bile. It can happen that the initial compound will never reach the systemic blood circulation. Further, this substance meets intestinal microflora, becomes metabolized into a more toxic product that can be reabsorbed, and so forth. So are hydrolyzed, for example, perorally administered *cycasin* (methyl-*O,N,N*-azoxy-*β*-D-glucopyranoside)—the glucoside of king sago palm (*Cycas revoluta*) and queen sago (*C. circinalis*) seeds, by intestinal bacteria into a *potent carcinogen methylazoxymethanol*. The latter substance, being carcinogenic to the liver, kidneys, intestine, and lungs, causes fatal gastrointestinal and liver damage in cattle.

*Summary*: The biliary route of elimination can substantially increase the half-life of a compound in the organism, lead to the formation of toxic compounds in the intestines, increase the exposure of the liver to a toxicant, and cause serious hepatic damage due to saturation of the excretion system.

### 2.4.3 *Intestines*

In the previous paragraph, the biliary excretion that takes a foreign compound or its metabolite to the intestines was discussed. Xenobiotics can also reach the intestines by diffusion from blood, saliva, and pancreatic juice. These routes are actually not very important, although the situation can turn essential if a xenobiotic is metabolized into a more toxic compound by the intestinal bacteria. Pertinent examples are nitroaromatic compounds such as 2, 4-dinitrotoluene or the synthetic sweetener cyclamate (sodium and calcium salts of cyclohexylsulfamic acid c-Hex-$NHSO_3H$). For example, the hydrolysis of cyclamate into a presumable bladder carcinogen cyclohexylamine is carried out by the intestinal microflora. Thereby, owing to the metabolic induction, the extent of such a metabolic process increases at repeated consumption of cyclamate. Cyclohexylamine, which is more lipophilic than cyclamate, is readily absorbed into the blood and excreted via the kidneys, therefore unavoidably contacting the urinary bladder (see also Section 17.1.3).

### 2.4.4 *Lungs*

The respiratory tract that ends with the lungs is a very important route of elimination of volatile compounds such as benzene, toluene, alcohols, anesthetics, carbon tetrachloride, and gaseous metabolites of foreign compounds. For example, more than half of benzene is eliminated via the respiratory tract. Passive diffusion of the substance molecules from the blood to the pulmonary alveoli takes the volatile compounds to the lungs. Since the blood capillaries and membranes of alveoli are very thin and very close to each other, this route is a very effective way of elimination of lipid-soluble compounds. In addition, due to the fast removal of gases, there is always a favorable concentration gradient in lungs. These circumstances are extremely important factors in case of medical treatment of intoxication with carbon monoxide (CO). Fish have gills as the respective excretory organ, through which both hydrophilic and lipophilic substances are eliminated from their organisms.

## 2.5 *Biomagnification*

Biomagnification is a process of raising the substance concentration in an organism higher than it was in the food and so up to the top of the food chain. The higher the trophic level or level in the food chain of an organism, the higher the concentration of this substance in its body. Biomagnification involves, first of all, nonpolar hydrophobic substances that tend to bioaccumulate in the adipose (fat) tissue.

$$\text{Biomagnification factor (BMF)} = \frac{\text{Concentration in organism}}{\text{Concentration in the food}}$$

For terrestrial animals, including humans, food is the most important contact route with environment-borne contaminants. Compounds that are stable in one environment can be nonstable in another. Water-, soil-, as well as airborne toxicants can move along the food chain. There are two well-known types of food chain—*herbivorous* and *detritivorous*. In the first one, herbivores eats grass; carnivores eat herbivores; and so on. The second one involves organisms after their death. Here the organisms are small and their dimensions do not increase in the food chain. The entire food chain is rather a big and complicated network that enfolds both main and other food chains.

A vivid example of biomagnification is the case of pesticide DDT (see also Section 14.1). DDT is a chemically stable substance, both in the organisms and in the environment, that only slightly dissolves in water but dissolves well and readily accumulates in the body fat. DDT is slowly absorbed by skin and is metabolized in animal organisms by several routes. Its main metabolite *dichlorodiphenyl-dichloroethene* (DDE) is even more persistent and toxic than DDT. Although, due to its low solubility in water, the concentration of DDT in rivers and lakes is low, it is still high enough to enable small water organisms such as plankton or water flea (*Daphnia*) to absorb it from water, either passively or by filter-feeding. The concentration of DDT in the adipose tissue of these organisms can exceed its concentration in the surrounding water by hundreds or thousands of times. Since daphnia is a food for insects and small fish, DDT reaches their bodies, which are in turn eaten by larger organisms, and so on. In this way, DDT, carried once somewhere out from the fields or forests to the water bodies, can reach humans. A case was reported in California, in which the concentration of DDT was 4 ppm in plankton, 138 ppm in bass, and finally reached 1500 ppm in the meat of the perch-eating grebe (*Podiceps crestatus*) (Timbrell, 2002, p. 112).

DDT has an adverse effect on the equilibrium of the hormonal system of birds, reducing the thickness and, hence, the mechanical strength of the eggshell. Breast milk may also contain DDT due to the biomagnification effect. It has been shown that when nursing mothers received a daily dose of DDT 0.0005 mg/kg, their milk contained 0.08 ppm of DDT. In turn, the babies received a daily dose of 0.0112 mg/kg, which exceeds the mother's dose by about 22 times (Timbrell, 2002, p. 113).

The higher the organism's weight, the smaller the food consumption per weight unit and the lesser should have been absorbed of the toxicant per weight unit. As the average body weight of an individual increases moving upward along the food chain, there should be no biomagnification at all. *The real reason for biomagnification is the time factor*. Organisms dwelling at higher levels of the food chain live longer and, hence, have more time for bioaccumulation of the xenobiotics.

## 2.6 Antidotes

Antidotes are compounds, of which the fast and correct usage may substantially reduce the adverse effect of a toxicant. Antidotes usually relate to acute intoxications, which most often are caused by incidental or intended peroral overdoses. Acute poisonings can also be caused by a repeated or chronic exposure to some toxicant, whereby the contact can be also by inhalation or percutaneous absorption.

Only some toxicants have specific antidotes; mostly, more general approaches are used to reduce the acute toxic effect of poisonous compounds.

### 2.6.1 General methods

1. If a toxic compound has reached the organism perorally, for example, with food, one should try to remove the food from the digestive tract as soon as possible before intensive absorption starts. The four main methods are known:
    - Use of emetic preparations such as *Ipecacuanha* syrup (molasses), usable also in the case of children
    - Gastric lavage
    - Use of absorbents. Charcoal is a well-known potent absorbent frequently used for treatment of the intoxications
    - Enema, either with or without laxative administration
2. Another possibility is the acceleration of elimination of the toxicant from the organism. It can be performed *either by increasing the urine volume or by changing the urine pH*. The urine volume can be increased by forcing the patient to drink more water or by injecting a physiological solution into the veins. Such a forced diuresis is unfortunately not totally safe.

    Enhancement of the elimination by changing of urine pH is efficient in case of either acidic or alkaline substances or their metabolites.

Acidic compounds are better soluble at alkaline pH and basic compounds at acidic pH. The urine pH can be increased, for example, by feeding solid sodium carbonate or by injection of its solution into the vein. These methods have been successful in case of barbiturate or aspirin intoxications. Urine can be turned more acidic by administration of ammonium chloride. This is the way of acceleration of elimination of amphetamine from the organism. In addition to the alteration of urine pH, the blood plasma pH can also be regulated in a similar way, thus accelerating the exit of a toxicant from the cells to the bloodstream. In this way, it is also possible to reduce the penetration of molecules from plasma to cells, including brain cells.

3. It is possible to remove toxic compounds from the blood also by *hemodialysis* or *hemoperfusion*. In both cases, the blood of the patient is directed through an apparatus where the toxic substance is separated from plasma by diffusion through a semipermeable membrane into another liquid (hemodialysis) or bound to charcoal or some other sorbent (hemoperfusion).

### 2.6.2 *Specific mechanisms of antidote action*

The main mechanisms of the antidote effect are

1. *Antidote as a chelating agent.* These antidotes react with the toxicant forming water-soluble chelate complexes, which are excreted faster than the parent compound. *Examples of chelators are* dicobalt salt of ethylene-diamine-tetraacetic acid (EDTA) in case of cyanide intoxications, penicillamine in case of lead poisonings, and dimercaprol in case of heavy metal intoxications.
2. *Antidote as an accelerator of detoxification of a highly reactive metabolite.* A good example here is *N*-acetylcysteine that is used as an antidote in case of drug paracetamol overdoses. This antidote increases the synthesis of glutathione in hepatocytes; glutathione effectively binds *para*-benzoquinone-imine, a dangerous metabolite of paracetamol, removing it from the complex with a hepatic protein. Cyanide intoxication can be mitigated by administration of thiosulfate, which accelerates the metabolism of cyanide into thiocyanate.
3. *Antidote as a decelerator of toxicant's metabolism.* It is used when the metabolite is more toxic than the foreign substance. Hence, it is reasonable to reduce the biotransformation of the toxicant. For example, conversion (oxidation) of ethylene glycol, used as a component of antifreeze, into toxic oxalic acid is reduced by the administration of ethanol. Ethanol blocks the metabolism of ethylene glycol, enabling elimination of the glycol in the unaltered form. The administered ethanol acts analogically in case of methanol poisonings (see also Section 16.2).

## References

Kuroda, H., Maeba, H. and Takashio, M. (2003). Enzymes that transform linoleic acid into di- and tri-hydroxyoctadecenoic acids in malt, *MBAA TQ*, **40**, 11–16.

Lin, J.H. (2006). CYP induction-mediated drug interactions: *in vitro* assessment and clinical implications, *Pharm. Res.*, **23**, 1089–1116.

Murray, M. (2006). Altered CYP expression and function in response to dietary factors: potential roles in disease pathogenesis, *Curr. Drug Metab.*, **7**, 67–81.

Plant, N. (2003). *Molecular Toxicology*, BIOS Scientific Publishers, Taylor & Francis Group, Oxon and New York.

Ross, D. (1996). Metabolic basis of benzene toxicity, *Eur. J. Haematol.*, **57**, 111–118.

Scheleni, R.R. (2006). Drug metabolism by intestinal microorganisms, *J. Pharm. Sci.*, **57**, 2021–2037.

Snyder, R. and Hedli, C.C. (1996). An overview of benzene metabolism, *Environ. Health Persp.*, 104, Suppl. 6, 1165–1171.

Timbrell, J. (2002). Introduction to Toxicology, 3rd edn., Taylor & Francis, London and New York.

Waxman, D.J. (1999). P450 gene induction by structurally diverse xenochemicals: central role of nuclear receptors CAR, PXR, and PPAR, *Arch. Biochem. Biophys.*, **369**, 11–23.

# chapter three

# Toxic response

## 3.1 Variability of toxic response

The factors influencing the intensity of the response to a toxic substance can be divided into two groups:

1. Individual-related (subjective)
2. Living and working environment-related (objective)

Table 3.1 represents the most important factors influencing the intensity of the response to contact with toxic substances.

Every organism has its own sensitivity to toxic effects. The following groups of people are at a higher than medium toxic risk degree:

- Children and adolescents up to age 18, especially newborns (up to 6 months)
- Elderly people
- Pregnant women
- People suffering from the following diseases:
  - Chronic nonspecific pulmonary diseases
  - Organic central nervous system (CNS) diseases and recurrent diseases of peripheral nervous system (PNS)
  - Blood diseases

***Table 3.1*** Factors Influencing the Intensity of Toxic Response

| Subjective | Objective |
|---|---|
| Age | Physico-chemical properties of the toxicant |
| Gender | Dose of toxicant |
| Ethnic group | Mechanism and ways of action of the toxicant |
| Genetic basis | Previous dose(s) of toxicant |
| Endocrine situation | Synergistic or antagonistic effect of other substances (drugs) |
| Nutritional habits | Nature of job |
| Degree of fatigue | Factors connected with job environment |
| Biorhythms | |
| Hereditary and/or previous diseases and their therapy | |

- Clinically established hepatic diseases
- Nephritis, nefrosis, nefrosclerosis
- Hypertonia and the coronary arterial diseases (CAD)
- Active forms of tuberculosis
- Ulcer of stomach or duodenum
- Diseases of the endocrine system
- Established diseases of the vegetative nervous system
- Mental diseases
- Hereditary allergic diseases

## 3.2 Types of toxic response

### 3.2.1 General principles

According to the physiological extent in the organism, the response to a toxic effect can be

- *Local*, occurring only at the site of exposure of the organism to the potentially toxic substance, for example, on the skin, in the lungs, or digestive tract. Precedes absorption into the blood.
- *Systemic*, revealing itself after distribution of the toxicant via the bloodstream around the affected organism including the target organ or tissue, distinct from the absorption site. This is the way that most substances that are not extremely reactive, act.

Systemic toxicants can predominantly produce either

1. *A selective effect* on one or another organ or system, or
2. *A simultaneous injury* of several organs or systems (polythropic effect)

The main target organs *for the systemic toxicity* of xenobiotics are

- The skin, mucous membranes
- The lungs
- The liver, kidneys
- The bone marrow
- The immune system
- The nervous system, both central and peripheral
- The cardiovascular system
- The reproductive system
- The muscles and bones

The reasons why an organ or tissue is especially sensitive to a particular toxicant are the following:

- The toxicant *accumulates* preferably in this organ or tissue (e.g., cadmium in the kidneys or paraquat in the lungs).

- Inactive pro-toxicant *is activated* in this organ or tissue—a compound enters the organism in an inactive form and is converted into the active form by phase I enzymes occurring in high concentrations just in this tissue.
- One tissue can be, by virtue of having a well-developed repairing system, less sensitive to this compound when compared with another tissue, where the *repairing system is either less developed* (as of nitrosamines in the liver) *or even absent*.
- This tissue has receptors *specific* to this toxicant *receptors* on the cell membranes.
- This tissue has *an elevated physiological sensitivity* to this toxicant—therefore, the nervous system is especially susceptible to substances blocking tissue oxygen supply (e.g., nitrite that oxidizes hemoglobin to methemoglobin) or the cardiac muscle to the cyanide ion that inhibits mitochondrial enzyme cytochrome oxidase and prevents usage of oxygen by the cells.

It should be emphasized that the highest concentration of a toxic substance is not always in the target organ. For example, lead (Section 10.1.2) accumulates in the bones, but its toxic effect reveals itself mainly in the soft tissues like the liver, kidneys, and blood cells. DDT (Section 14.1), which is primarily toxic to the CNS, accumulates in the adipose tissues.

*According to the nature of their adverse effect on the target organ*(s), the chemical substances can be divided as follows (Loit and Jänes, 1984):

- *Irritants* that cause damage to the eyes and other mucous membranes, for example, bromine, chlorine, tear gases (e.g., pepper spray), hydrogen fluoride, ammonia, phosgene, sulfur dioxide, and acroleine
- *Corrosive substances* that corrode the skin and mucous membranes—strong acids and alkalis
- *Substances that cause toxic pulmonary edema*—chlorine, ammonia, and nitrogen oxides
- *Blockers of mitochondrial respiratory enzymes*—cyanides, salicylic acid, and gossypol
- *Inhibitors of thiol enzymes*—heavy metals
- *Blockers of the citrate (Krebs) cycle*—fluoroacetates
- *Emetic substances*—apromorphine, emetine, zinc, and copper sulfates
- *Neurotoxicants*— many substances with various mechanisms of action belong to this group, for classification see Section 3.2.4
- *Cardiotoxicants* that selectively damage the heart—digitoxin, digoxin, gitotoxin, convallotoxin, scillarene, and other cardioglucosides, and aconitine
- *Hepatotoxic substances* that selectively damage the liver—carbon tetrachloride, chloroform, trichloroethane, bromobenzene, and allylic alcohol
- *Nefrotoxic substances* that selectively damage the kidneys—carbon tetrachloride, chloroform, trichloroethylene, mercury, chlorine, and lead

- *Substances that damage the bone marrow and blood cells*—nitrobenzene, benzene, trinitrotoluene, and arsine
- *Asphyxiants*—substances that cause a reduction of the blood's ability to bind and transport oxygen
    - *Simple asphyxiants*—hydrogen, nitrogen, helium, and methane
    - *Complex asphyxiants*—carbon monoxide (formation of carboxyhemoglobin), aniline (formation of methemoglobine), hydrogen sulfide, nitrites, arsine, nitrobenzene, pyridine, and phenylhydrazine
- *Anticoagulants*—substances that disturb blood coagulation, such as dicumarine, neodicumarine, phenyline, syncumare, and heparin
- *Hemolytic substances*—snake venoms, mushroom toxicants, phenylhydrazine, arsine, saponins, nitro- and amidobenzenes
- *Histamine and antihistaminic compounds*—dimedrol, pipolphene, suprastine, ethysine, dinesine, and diazoline

This division is certainly rather relative, many toxicants possess several toxic effects on various target organs.

*Depending upon the character of damage of a cell or an organism*, the effects of the toxic compounds can be divided as follows:

- *Generally toxic*—damage of the organism as a whole (*en bloc*)
- *Dystrophic*—causing the aging of cells or tissues
- *Genotoxic*—alteration of the genetic material (DNA, RNA)
- *Mutagenic*—generation of irreversible changes in the hereditary material (chromosomes, genes) of an organism
- *Carcinogenic*—generation of malignant tumors
- *Gonadotropic*—harming the germ cells, perishing, inhibiting development or reducing their quality
- *Teratogenic*—evoking disorders in the embryonal development of an organism (defects, malformations)
- *Sensibilizating*—making an organism ultrasensitive toward this compound, resulting in allergic reactions and diseases

According to the final result, toxic responses can be divided into the following categories:

- Direct injury of cell or tissue
- Biochemical damage
- Neurotoxicity
- Immunotoxicity
- Teratogenicity
- Genetic toxicity (genotoxicity)
- Carcinogenicity
- Endocrine disruption

There is possible overlapping between these responses, and also some substances are capable of causing more than one response.

### *3.2.2 Direct injury of a cell or tissue*

Here, we often have to deal with the decomposition of cells or *necrosis*. Necrosis is an irreversible process consisting of degeneration of the cell, fragmentation of the nucleus, and denaturation of cellular proteins. During necrosis, the cell disperses, accumulates liquid, and its content flows out. Mechanism of this process may comprise formation of an intermediate that reacts with definite cell components like structural proteins or, for example, have immunological basis. *A highly reactive molecule* can react with the cell membrane and, by breaking it, cause a momentary death of the cell. By this mechanism, the toxicant (e.g., $CN^-$ ion) can interact with the respiratory system of a cell, which usually leads to a fast cell death. Another substance (such as lead) may cause a similar situation that develops much slower. The intermediate stages of this type of necrosis have not yet been clearly understood. Necrosis can be accompanied by injury of the neighboring tissue and inflammatory process that is observable in a light microscope.

As an extreme example of cell and tissue damage, the effect of corrosive chemicals like *strong alkalis* (concentrated water solutions of NaOH and KOH) or concentrated *strong acids* (sulfuric acid, nitric acid, hydrochloric acid, and hydrofluoric acid), hypochlorite (both acid and oxidant) can be exemplified. Acids denaturate and precipitate the cell's membranous proteins within a few seconds. The effect of alkalis is somewhat slower, but over a sufficiently long time, they can damage even the skin. Ammonia and ammonium hydroxide are much weaker alkalis than NaOH or KOH, but because of its volatility, ammonia can cause injury to sensitive mucous membranes of the respiratory organs, which are not caused by a much stronger but nonvolatile alkali. *Organic solvents* such as aliphatic hydrocarbons, benzene, and chloroform are also able to dissolve lipid membranes and dissociate nucleoproteinous membrane complexes.

*Strong oxidizers* like ozone ($O_3$), chlorine gas, bromine, or fluorine are also very harmful to both humans and animals, but particularly to microorganisms. This is the reason why ozone and chlorine gas are used for killing microorganisms in the purification systems of drinking water.

An alternative mechanism of cell death is the programmed death or *apoptosis,* which can be evoked for *medical purposes*. Apoptosis is encoded in every cell development as a normal process for tissue renewal, but it can be evoked by certain substances. During apoptosis, genes of the specific proteins like those of *caspases* are expressed, chromatin and components of cytoplasma are condensed, DNA is fragmented and cell division is blocked. At the end of the apoptotic process, the dying cell loses contact with its neighbors and is thereafter removed from the tissue by phagocytosis without any inflammatory response. It is possible to kill the cells by apoptosis optionally; definite natural compounds act in this manner with abnormal malignant (tumor) cells, which otherwise are immortal. Study of the cell apoptosis mechanisms and use in medicine is a fast developing area of medical biochemistry closely connected with toxicology. One such well-known compound is *trans*-resveratrol and its relatives (resveratrol oligomers and glucosides, other hydroxystilbenes like pterostilbene or piceatannol) that are found in a number of plants;

*trans*-resveratrol became famous after its discovery in red grape wines, but this health-promoting malignant-cell-killing polyphenol can also be found in peanut, garden rhubarb, berries of genus *Vaccinium*, and so forth (Fulda and Debatin, 2006).

### 3.2.3 Biochemical damage

Biochemical injury may lead to small pathological alterations like degeneration of a single cell and they may also touch vitally important functions such as respiration, causing the death of the organism. So, the *cyanide ion* ($CN^-$) can cause the death of the whole organism through an irreversible disruption of the electron transport in mitochondria leading to the incapacity of the cell to use oxygen if the disabled cell is located in one of the vital organs like the heart or brain (see Section 8.4).

Some toxic biochemical effects such as binding of *carbon monoxide* (CO) to hemoglobin are reversible. CO usually does not cause any pathological damage, except at very high concentrations when it becomes impossible to save the victim's life.

*The fatty liver* is a general hepatic damage stemming from the disruption of the hepatic lipid metabolism, caused, for example, by an extensive ingestion of ethanol (see Section 16.3). The same response can be *demonstrated* in other organs like the heart or the kidneys.

A much more specific disruption is *phospholipidosis* (excessive reversible accumulation of intracellular phospholipids), caused by a number of medicines, including antibacterials, antipsychotics, antidepressants, antiarrhythmics, antianginals, antimalarials, anorexic agents, cholesterol-lowering agents, and so forth. All these chemicals that affect cellular phospholipid metabolism belong to the class of cationic amphiphilic drugs (CADs). Phospholipidosis may take place in various tissues but preferably in the lungs and in the adrenal glands (Reasor et al., 2006).

Biochemical injury can cause the death of an organism through a deficiency of several organs. Therefore, an overdose of *acetylsalicylic acid (aspirin)* leads to serious biochemical and physiological disorders (deficiency in ATP, acidosis, hyperthermia) that can kill the organism. The biochemical reason for such a result is the saturability of both the metabolism and elimination of aspirin. Aspirin is converted by esterases into salicylic acid that conjugates with glucuronic acid or amino acid glycine (see Section 2.3.3). *The conjugation reactions are saturable.* Salicylic acid as an inhibitor of the electron transport chain in mitochondria blocks the production of ATP, eliciting an increase in the use of oxygen and production of acidic carbon dioxide ($CO_2$). Metabolic acidosis and a lack of ATP will arise. Acidosis, in turn, changes the distribution of salicylate in the organism. Therefore, due to the increase of aspirin reabsorption in the kidneys, the secretion of acidic salicylate with urine will be reduced. At acidic pH, most of the salicylate is in the unionized form that is preferably absorbed into tissues, especially into the brain, which still enhances the salicylate toxicity through blocking of ATP synthesis.

*All in all, the intoxicated person suffers from a shortage of ATP in the brain and heart*. Since the energy produced in the body is not used for the synthesis of ATP, the body temperature will rise. In addition, *salicylate inhibits some steps of the citrate cycle* and, due to a shortage in ATP, it accelerates glycolysis that once again amplifies acidosis. The doses necessary to evoke the symptoms of aspirin intoxication are still quite high—from 3.6 (tinnitus) to 21 g (respiratory and renal failure and coma) and 30 g (death).

The treatment of aspirin intoxication consists in a reduction of the acidosis by raising the blood pH, compensation of the lacking energy by administration of glucose and acceleration of elimination of salicylate. A detailed description of aspirin toxicity mechanisms and treatment of the intoxication can be found in the book by J. Timbrell (2002).

*Fluoroacetates*, which metabolically block the very important Krebs or citrate cycle, can serve as another example of the role of biochemical damage in the toxicity of chemical compounds (see Section 8.13).

### 3.2.4 Neurotoxicity

Compounds that have a toxic effect on the nervous system can be divided into three categories:

1. Toxicants of the central nervous system (CNS)
2. Toxicants of the peripheral nervous system (PNS)
3. Toxicants of a combined effect

*To guarantee functioning of an organism, it is of primary importance to defend the CNS or the brain*. That is why, in addition to the enzyme system, which is responsible for the detoxification of foreign compounds, the *hematoencephalous membrane barrier* has evolved to defend the brain. This barrier has a very limited network of proteinaceous ion channels; its cells are connected to each other by dense and heavily penetrable junction zones. Such a "firewall" should have prevented all undesirable molecules from free entry into the brain. Unfortunately, this barrier is not solid—many toxic compounds are able to pass through and can cause serious brain impairment. Compounds exerting an undesirable effect on the nervous system can be, according to the mechanism of their effect, divided into the following classes (adapted and translated from Loit and Jänes, 1984):

- *Neurotoxic compounds*—substances that elicit a disturbance in the functioning of the nervous system such as mercury, acrylamide, hexane, methyl-n-butylketone, and carbon dioxide
- *CNS inhibitors*—aliphatic and chlorinated hydrocarbons, benzene, acetone, and diethyl ether
- *Psychomimetics*—substances that elicit a disturbance in psychical activities—mescalin and other phenylethylamine derivatives, harmine and other indole derivatives, and psilocybine

- *Compounds, inhibiting the respiration center*—narcotics, hypnotics, and hydrocarbons
- *Convulsion toxicants*—compounds that cause convulsions of central origin such as corazol, strychnine, tremorine, cholinesterase inhibitors such as organophosphorous pesticides
- *Toxicants, paralyzing transmission of nerve impulses to the muscles*—toxin of *Clostridium botulinum*
- *Toxicants, paralyzing transmission of nerve impulses in the nerves*—tetrodotoxin
- *Neuroparalytic poisons*—anticholinesteratic substances like thiofos, chlorofos, and carbofos
- *Toxicants, acting with mediators or synaptic poisons*, including
  - Substances, stimulating cholinoreactive systems—acetylcholine, cholinomimetics
  - Substances, blocking cholinoreactive systems—atropine, scopolamine, curare
  - Substances, stimulating adrenoreactive systems—adrenaline, noradrenaline, ephedrine
  - Substances, blocking adrenoreactive systems—anaprilin, ornide, fentolamine
  - Inhibitors of monoamine oxidase—ipraside and other hydrazines

### 3.2.5 *Immunotoxicity*

The toxicity connected with the immune system can be expressed in several forms, such as:

1. Supersensitivity or allergic reactions and autoimmunity (indirect immunotoxicity)
2. Immunosuppression and immunostimulation (direct immunotoxicity)
   a. *Supersensitivity or allergic reactions* may appear if the immune system becomes stimulated during individual contacts with a substance (hapten), which is capable of binding to a macromolecule, usually a protein, and changing it. The protein must be large enough to bind sufficient amount of haptenic groups that change the protein molecule to such an extent that it will be recognized as foreign (antigenic) by the immune system of the organism. Allergic reactions can be *either local*, appearing in the contact point (first of all, in the skin or in the lungs) *or systemic*. One of the most well-known elicitors of the allergic reactions is the antibiotic *penicillin*, repeated contact with which may lead to a fatal anaphylactic shock.
   Four types of supersensitivity reactions are known:
      i. Anaphylactic reactions
      ii. Cytolytic reactions

iii. Precipitin reactions
iv. Cell-mediated supersensitivity
The first type of reactions can be caused by chemicals such as *penicillin* or *2,4-toluenediisocyanate*, the component of many polyurethane foams and other elastomers. Sensibilization is evoked at the first contact and the following exposures will cause anaphylactic shock, resulting, for example, in bronchoconstriction or asthma. To produce asthma, it is necessary for the antigen to bind to a blood cell to initiate cytolysis. This is the way of action, for example, of the drug *aminopyrin*. A reaction of the third type can be, for example, the result of the organism's contact with *hydralazine*. A reaction of the fourth type can be the basis of formation of contact dermatitis after skin exposure to *nickel* or *cadmium*.

b. *Immunosuppression* is a result of a direct effect of a toxic compound on the immune system. As the basis, either an injury of thymus producing B-lymphocytes or bone marrow producing blood cells can be suggested. A good example here is dioxin (TCDD—see Section 10.4), which on damaging the thymus, inhibits the production of lymphocytes. On the contrary, an immunostimulating effect is elicited, for example, by novel peptide drugs, produced by the help of recombinant DNA.

*The immune system together with the nervous and endocrine systems are the most critical parts of an organism.* These functionally tightly interconnected systems have a permanent mutual effect, a toxicant of one of them indirectly influences the other two and, in this way, the whole organism. Especially recognizable is such an interaction in the developmental phase of an organism. In addition, one and the same compound may have an effect on several of these sensible systems. It is not easy to elucidate the adverse changes in these systems and their mechanisms, except for the cases when the damage of a system is obvious as in the case of the HI virus.

*Toxicogenomics* is a novel technique consisting in a simultaneous measurement of expression of thousands of genes in response to the exposure of an organism to a toxic substance. The examples of the assessment of the immunotoxicity of substances by gene expression profiling show that the *microarray analysis* is able to detect known and novel effects of a wide range of immunomodulating agents. Besides the elucidation of mechanisms of action, toxicogenomics is also applied to predict the consequences of exposure of biological systems to the toxic agents (Baken et al., 2007). For more about toxicogenomics, see Section 3.4.

### 3.2.6 *Teratogenicity*

*Teratogenesis* (from Greek *teras* = marvel, monster) or formation of embryonal malformations can be caused by different factors like hereditary anomalies, pathologies of the mother, hypovitaminoses, different contagious as well as parasitical diseases, all types of irradiation except visible light, vibration, mechanical traumas, and so forth.

Embryos and fetuses are generally very sensitive to contact with definite substances called *chemical teratogens*. The result of these contacts can be structural or functional abnormalities both of the fetus and the animal developed from it. Although cytotoxic compounds may concurrently be teratogenic, in many cases, the formation of a malformation is not connected with a direct injury of the embryo or fetus, but with the formation of a disturbance in their development.

*Chemical teratogens are usually quite safe for the mother* but affect only the fetus by a specific mechanism. This is the reason why the exact contact period during pregnancy is of extreme importance for the development of the malformations. Owing to the mediation of a mother's organism, the slope of the dose–response curve (see Figure 1.1) can be very deep. The result of a teratogenic effect can be

- Death of the fetus or abortion
- Malformations
- Growth retardation
- Functional disturbances

*Teratogenesis has many different mechanisms,* at times the initial reason can be a deceptively insignificant disturbance in the cell functioning. Sometimes the injury becomes evident only after the birth of the child. *A drastic example is the case of diethylstilbestrol* (DES), a drug taken by pregnant women, that caused the development of vaginal cancer of their female offspring when they reached their puberty (see also Section 15.2).

During the 1960s, a sensation was caused by another drug called *thalidomide,* which when taken during pregnancy caused injuries of 15,000 fetuses, 12,000 of which in 46 countries were born with birth defects. Only 8,000 of them lived longer than 5 years. Most of the survivors are still alive, but almost all of them have defects caused by thalidomide. Later on, it was discovered that the defects of many *thalidomide children* were transmitted by DNA to their offspring. This drug, sold in about 50 countries, but not in the United States, under at least 40 names during the period of 1957–1961 as a sleeping aid and emetic to combat the morning sickness of pregnant women, turned out to be an *angiogenesis inhibitor,* hindering the formation of blood vessels of the fetus, especially if taken on the 25–50th day of the pregnancy. The later court trial established that some toxicity tests with animals were either incorrectly performed or their results were even falsified. Most of the tests were still performed correctly in accordance with the methods of that time, whereby in some of them, in the case of rodents, doses were used that exceeded the normal human doses by more than 600 times.

The animal tests that were improved later, showed a clear teratogenicity of thalidomide. The drug was banned for its initial intended use, but now,

40 years later, thalidomide is again on clinical tests as an *antiangiogenic agent* to be used for treating a number of illnesses like leprosy, AIDS-linked symptoms, progressive multiple myeloma, cancer of prostate, glyoblastoma, and Crohn's disease (Teo et al., 2005).

*Ethanol* is also a teratogen, causing the so-called fetal alcoholic syndrome if taken in large quantities during pregnancy.

*A number of other substances* have demonstrated teratogenicity in animal tests. The teratogenicity of a number of food contaminants like aflatoxins, ochratoxin, dioxin, nitrosamines, and caffeine has been detected in animal experiments. It should be stressed that, in the case of caffeine, the results of different tests are actually controversial (see Section 8.3.3). None of the aforementioned compounds have shown teratogenicity in human tests, although they have in some cases shown in clinical observations. Also, organic solvents such as alkanes, arenes, chloroform, carbon tetrachloride, amides, and so forth have been teratogenic in animal tests. Teratogenic effects can also be caused by a deficiency or excess of some compounds like vitamins A, D, C, and E or nicotinic, folic or panthotenic acid, deficiency of metals like Zn, Mn, Cd, or Co in the food (see Chapter 18), or a number of viral infections.

*Chemical teratogens have an especially strong effect during such critical periods of pregnancy* as implantation (the 6th–8th day of development), period of formation of the gastrula (the 2nd–3rd week), and the period of organogenesis (the 4th–7th week).

In addition to the *embryopathic effects* characteristic of thalidomide, a number of other symptoms of teratogenicity such as deafness, goiter, hemolytic anemia, mental retardation, maldevelopments of the face and heart, and so forth have been observed.

### *3.2.7 Genotoxicity and mutagenicity*

As a result of the interaction between a xenobiotic and genetic material of the cell, alterations in cell genotype or *mutations* can be formed, which are transmitted to the next generation of the cells. Three types of mutations—aneuploidy, clastogenesis, and mutagenesis are known. *Aneuploidy* is a deviation in the multiplicity of single chromosomes, that is, disappearance or addition of chromosomes. *Clastogenesis* is an addition or reorganization of parts of a chromosome and *mutation* is a persistent and transmissible alteration in the structure of DNA or RNA, the carrier molecules of heredity. The formation of mutations is called *mutagenesis* and compounds that elicit this property are *mutagens*. Mutations are of one the following types:

- *Point mutations* mean a replacement of one base pair with another in the molecule of DNA. Most often, the bases of the same type are replaced with each other (a purine with another purine or a pyrimidine with another pyrimidine). This type of replacement or *transition* is caused, for example, by nitrous acid ($HNO_2$). A rarer replacement of a purine with a pyrimidin and the other way round

is called *transversion.* Transversion can render the amino acid code wrong. Point mutation can be reversed by another point mutation (inverse mutation). As a result, the initial structure of the nucleotide will be restored.
- *Insertions* add one or more nucleotides to the DNA molecule.
- *Deletions* remove one or more nucleotide groups from the DNA molecule.

In the case of insertions or deletions, reading of the whole genetic code becomes disturbed (shifted). Here we have the so-called *frame shift mutation.* Large removals or reorganizations can spoil the whole structure of a DNA molecule. Analogically, whole segments of a chromosome can be turned upside-down by *clastogenesis.* As several genes are involved in this process, clastogenesis can result in serious disorders in the functioning of the whole cell.

*Mutagenicity has many specific biochemical mechanisms.* So, for example, an alkylating agent like aflatoxin can react with DNA right in the cellular nucleus, but a compound such as bromacyl can react only during cellular replication. In the case of mammals, mutations in gametes can produce birth defects in the offspring organism; mutations in somatic cells produce tumors.

According to their action mechanism, mutagens are divided into the following groups:

1. *Destructurating*—here belong hydrogen peroxide ($H_2O_2$), nitrates, and nitrites. The last two are mutagenic to bacterial and not to mammal cells
2. *Joining*—here belong substances, containing a highly reactive alkylic group like epoxides, dialkyl sulfates, and lactones
3. *Replacing* such as nucleic acids

*Mutagens can be formed during food processing at high temperatures.* Most important are *Salmonella* mutagens formed in the course of food preparation and storage or in the intestines as presented in Table 3.2. More detailed information about these mutagens is provided in Chapter 16.

*Food mutagens act along the route of movement* (oral cavity, esophagus, gastrointestinal tract) *and also in organs that are distant to this route.* So, the liver is the target organ for aflatoxin $B_1$, the lungs, oropharynx, breast, gastrointestinal and genitourinary tracts for PAHs, gastrointestinal tract, oropharynx, and lungs for *N*-nitrosamines, and colon and breast for heterocyclic amines. In addition to the substances presented in Table 3.3, aflatoxins are also clearly identified food-borne mutagens (see also Section 11.2). Food mutagens are able to course different types of DNA damage, nucleotide alterations, and gross chromosomal aberrations (Goldman and Shields, 2003).

Upon entering the body, food mutagens typically undergo a metabolic activation and detoxification like other xenobiotics. Sometimes the chemically modified mutagens, which are now highly electrophilic, bind together during metabolic phase II to DNA rather than to the hydrophilic carrier molecules such as glutathion or glucuronic acid. Most of these DNA adducts are

***Table 3.2*** Mutagens, Formed during Preparation, Storage, and Digestion of Food

| | |
|---|---|
| Mutagens formed during heating (cooking) | • Polycyclic aromatic hydrocarbons (PAH), formed from organic material during barbequeing and smoking (benzo-[$\alpha$] pyrene)<br>• Heterocyclic amines, such as 2-amino-3,8-dimethylimidazol [4,5,-f]quinoxaline (MIQx), 2-amino-1-methyl-6-phenylimidazo[4,5,-b]pyridine (PhIP), and so forth, formed at a long-term high-temperature cooking of meat<br>• Substances formed by Maillard reaction: furanes, aminocarbonyls, pyrazines, and other promelanoid secondary amines |
| Mutagens formed during storage and digestion | • Hydroperoxides and epoxides formed from unsaturated fatty acids and cholesterol<br>• Products of reaction between secondary amines and nitrosating agents (nitrites), like *N*-nitrosodimethylamine (NDMA), *N*-pyrrolidone (NPYR), *N*-piperidine (NPIP) |

*Sources:* Adapted from Altuğ, T. (2003). *Introduction to Toxicology and Food,* CRC Press, Boca Raton, FL; Wu, K., et al. (2006). *Cancer Epidemiol. Biomarkers Prev.,* **15**, 1120–1125.

***Table 3.3*** Examples from IARC and NTP Lists of Chemicals and Mixtures Carcinogenic to Humans (Group 1)

| |
|---|
| Genotoxic (DNA-reactive) |
| Aflatoxins |
| 4-Aminobiphenyl |
| Analgetics containing phenacetin |
| Asbestos |
| Benzidine |
| Betel quid with or without tobacco |
| *N,N-bis*(2-chloroethyl)-2-naphthylamine (*chlornaphazine*) |
| *Bis*-(Chloromethyl)ether |
| 1,4-Butandiol dimethanesulfonate (*myleran*) |
| Cadmium alloys and salts |
| 1-(2-Chloroethyl)-3-(4-methylcyclohexyl)-1-nitrosourea (*methyl-CCNU*) |
| Chlorambucil |
| Coal tars |
| Chromium (VI) compounds |
| Cyclophosphamide |
| Ethylene oxide |
| Melphalan |
| MOPP and other combined chemotherapy including alkylating agents |
| Nickel and nickel compounds |
| Soots |
| *Sulfur mustard** |
| Tamoxifen |
| Tobacco smoke and products |
| Triethylenethiophosporamide (*thiotepa*) |
| Vinylchloride |
| Viruses (Epstein–Barr, human papillomavirus, hepatitis B and C, HIV type I, etc.) |

*Continued*

*Table 3.3* (Continued)

| |
|---|
| Epigenetic |
| Asathioprine |
| Cyclosporin A |
| Estrogen therapy, postmenopausal |
| Oral combined contraceptives |
| 2,3,7,8-Tetrachloro-dibenzo-*p*-dioxin (TCDD) |
| Unclassified |
| Alcoholic beverages |
| Arsene and arsenic compounds |
| Benzene |
| Diethylstilbestrol |
| Mineral oils |
| Nickel and nickel compounds |
| Oil shale oils |

* *Sulfur mustard* = compounds with general formula $S(CH_2CH_2Cl)_2$.

*Source*: Adapted from IARC (2006). *Monographs on the Evaluation of Carcinogenic Risks to Humans*, Preamble, Lyon; NTP, National Toxicology Program, Reports on Carcinogens, 11th edn., http://ntp.niehs.nih.gov/ntp/roc/toc11.html

repaired by the respective cellular enzyme systems but a few of them can endure and cause point mutations, deletions, insertions, and so forth.

*Some dietary exposures can modify the effects of food mutagens*. For example, alcoholic beverages induce an isoform of CYP, which metabolically activates *N*-nitrosamines. Exposure to food mutagens like PAH can induce another isoform of CYP, causing an increased metabolic activation of PAHs.

Food mutagens can be assessed by different methods such as *in vitro* DNA assays, *in vitro* cell-culture testing, animal bioassays, and epidemiological method. Whether one or another compound is mutagenic for humans can be conclusively proved only in humans.

Biomarker assays are frequently used to determine the exposure to food mutagens and response of the organism to these exposures. Typical phenotypic biomarkers of exposure to aflatoxin and PAH are adducts in DNA, albumin, and hemoglobin as well as urinary metabolites, $O^6$-methylguanine and 7-methylguanine in case of *N*-nitrosamines and DNA, and albumin adducts in the case of heterocyclic amines (Goldman and Shields, 2003).

More about formation and toxicity of these compounds can be found in the sections devoted to the description of the particular compounds (Chapters 11 and 16).

### 3.2.8 Carcinogenicity

*Carcinogenesis* (from Greek *karkinos* = crab, cancer + genesis) *or tumor formation* is the response of an organism to the toxic effect of *carcinogenic substances* consisting in an abnormally fast, uncontrolled growth and multiplication of the somatic cells affected by the carcinogen. A hypothesis exists that most of the human malignant tumors (cancers) are caused by specific chemical compounds—*carcinogens*. In many, but not in all, cases of chemical-induced carcinogenesis, we have to deal with a mutation in a somatic cell. Hence, a chemical substance can be carcinogenic if it influences the intracellular processes through mutations.

*Historically*, soots and coal tar were the first materials that were considered to be carcinogenic as early as the eighteenth century. The reason was that cancer of the scrotum was widely distributed among chimney sweepers who started their unhealthy career at a very young age. The respective studies have proved that the younger a person is at the moment of the first contact with a carcinogen, the more sensitive is the individual to the effect of this compound.

*Formation of tumor or oncogenesis* is a multistep process with the following sequence of main stages:

(1) initiation → (2) transformation → (3) promotion → (4) progression

Every stage is characterized by specific processes.

- *Initiation*. A normal cell meets a carcinogen, for example, an alkylating agent like vinyl chloride or aflatoxin. As a result, the tumor is initiated. Initiation is an irreversible process and usually involves DNA. A genetic alteration (usually DNA adduct) is formed and the cell is changed from normal to *preneoplastic*.
- *Transformation*. In the preneoplastic cell, an oncogen is activated, then the suppressor gene is inactivated and the apoptotic potency of the cell is reduced. Replication of the cell is carried out and the genetic alteration formed during the previous stage is transformed. A *neoplastic* cell is formed from a preneoplastic cell.
- *Promotion*. Usually there follows repeated contact with the carcinogenic substance, called a *promoter*. As a rule, the promoter causes an intensification of the gene expression and formation of a clone of the initiated cell. Repeated contacts with the initiator can result in a tumor formation even without a promoter. From a neoplastic cell a *benign neoplasm* is formed.
- *Progression*. During this stage, the neoplastic cells can change their phenotype and turn into malignant, that is, uncontrollably multiplicating cells. Formation of new blood vessels or neoangiogenesis starts and the apoptoticity of the cells decreases. From a benign neoplasm a *malignant neoplasm* is formed.

*Two institutions* are regarded by the Occupational Safety and Health Administration (OSHA) as authoritative on the classification of carcinogens. These are

1. *IARC*: The International Agency for Research on Cancer—also referred to as CIRC (*Centre International de Recherche sur le Cancer*)

with its headquarters in Lyon, France. The task of this section of WHO (World Health Organization) is to coordinate and conduct research on the causes of human cancer, the mechanisms of carcinogenesis, and to develop scientific strategies for cancer control.
2. *NTP*: The National Toxicology Program headquartered in Research Triangle Park, NC, U.S.A.—a cooperative effort involving the NIH (National Institute of Environmental Health Sciences, the CDC National Institute for Occupational Safety and Health, and the National Centre for Toxicological Research of the FDA). NTP designs, conducts, and interprets the animal assays.

There are two types of classified and an additional class of unclassified carcinogens (see Table 3.3):

1. Genotoxic (interacting with a molecule of DNA)
2. Epigenetic (notoriously not interacting with DNA)
3. Nonclassified carcinogens

*The first group* can, in turn, be divided into *activation independent* (primary) and *activation dependent* (secondary) carcinogens and a heterogeneous group with the common title *inorganic compounds*.

*Activation independent* are, for example, alkylating agents such as dimethyl sulfate or ethylene oxide.

The group of genotoxic *carcinogens needing chemical activation* is much larger—here belong aliphatic halogenides such as vinyl chloride, aromatic and heterocyclic amines, aminoazo dyes, aromatic nitro-compounds, polycyclic hydrocarbons, hydrazines, *N*-nitrosocompounds such as *N,N*-dimethylnitosoamine, mycotoxins such as aflatoxin $B_1$ and so forth. Inorganic DNA-reactive carcinogens are, for example, metals such as cadmium, chromium, zinc, nickel, and mineral asbestos. Inorganic carcinogens have different mechanisms of action. Some of them reveal their genotoxicity through hurting of DNA, the others act epigenetically, for example, by changing the reliability of DNA polymerase. *The hazard of DNA-reactive carcinogens can, in principle, be assessed quantitatively*. The response to their effect may appear already after the first random contact and the effect is cumulative. Genetic tests detect only genotoxic agents and the second and third groups of carcinogens are unfortunately beyond their scope.

*Epigenetic carcinogens* usually act at high contact levels, leading to prolonged physiological abnormalities, a disruption of the hormonal balance, and tissue damage. These include cancer promoters such as various chlororganic pesticides, polychlorinated biphenyls (PCBs) and dioxins, endocrine modificators such as DES or atrazine, immunosuppressors such as purine analogs, and cyclosporin, cytotoxins such as butylated hydroxyanisol (BHA), diallylphthalate, *d*-limonene, peroxisome proliferators such as clofibrate and phthalates.

*Nonclassified carcinogens* are, for example, acrylamide, acrylonitrile, dioxane, furfural, sugar alcohols, and so forth. A more exhaustive list of rodent carcinogens is presented for example in the book edited by A. Wallace Hayes (Wallace Hayes, 2001).

*By far, not all rodents' carcinogens have proved to be carcinogenic to humans.* The list of substances as well as mixtures, for which the direct link between human exposure and tumor formation has been proven, is fortunately considerably shorter and nothing definitive, but is a subject of continuous reassessments (see Table 3.4). The reassessment is conducted by the following agencies:

*NTP*, which has divided carcinogens with respect to their human carcinogenicity into two parts, regardless of the precise mechanism of the carcinogenic effect (NTP, *Reports on Carcinogens*, 11th edn.):

*Part A*. Known as Human Carcinogens (54 substances, groups of substances or processes).

*Part B*. Reasonably Anticipated to be a Human Carcinogen (184, mainly individual substances but also some groups of substances and processes).

*IARC* has elaborated its own classification system of substances and mixtures.

***Table 3.4*** IARC Classification of Substances and Mixtures

| Group (number of listings by January 2007) | Amount of information | Examples |
|---|---|---|
| 1. Carcinogenic to humans (100) | Sufficient for humans | Arsenic, aflatoxin $B_1$, benzene, estrogens, TCDD, cadmium |
| 2a. Probably carcinogenic to humans (68) | Inadequate for humans/ sufficient for animals | Benzo[$\alpha$]pyrene, *N*,*N*-diethyl-nitrosamine, PCBs, acrylamide dimethyl sulfate, epichlorohydrin |
| 2b. Possibly carcinogenic to humans (246) | Limited for humans/ insufficient for animals | Chlordane, DDT, cycasine, carbon tetrachloride, hexachlorobenzene, aflatoxin $M_1$, fumonisin $B_1$, MeIQ, bracken fern, BHA, lead, lindane |
| 3. Unclassifiable as to carcinogenicity (516) | Insufficient both for humans and animals | 5-Azacytidine, diazepam |
| 4. Probably not carcinogenic to humans (1) | None of the carcinogenicity tests shows positive result | Caprolactam |

*Source:* Adapted from IARC (2006). *Monographs on the Evaluation of Carcinogenic Risks to Humans*, Preamble, Lyon.

*US EPA* has elaborated a similar classification of human carcinogens:

1. Group A: Human carcinogen
2. Group B: Probable human carcinogen
3. Group C: Possible human carcinogen
4. Group D: Not classifiable as human carcinogen
5. Group E: Evidence of noncarcinogenicity for humans

As we can see, most of the human carcinogens proved by IARC and NTP are of the DNA-reactive type. This fact demonstrates the essentiality of this mechanism in the case of humans. Except for aflatoxins and tobacco smoke, contact with all the other DNA-reactive carcinogens is of either the occupational or therapeutic type.

*The actual degree of toxicity of various carcinogens is very different.* For example, in tests with rodents, the interval can be up to 10 million times. The largest electronic source of the respective data is the Carcinogenic Potency Data Base (http://www. potency.berkeley.edu/cpdp.html). It contains $TD_{50}$ numbers. $TD_{50}$ is defined as a daily dose of a compound that reduces twice the probability of keeping a test animal without cancer until the end of its standard lifespan. In the case of DNA-reactive carcinogens, there is a positive correlation between the compound boundability to DNA and its carcinogenicity, which can be expressed by the *chemical binding index* (*CBI*) (Lutz, 1986).

*It is considered that about 70% of deaths caused by cancer have food factors.* There are both tumor initiators and promoters among the natural endogenous food toxicants. Table 3.3 contains a selection of naturally occurring carcinogens. A look at this list and at Table 3.2 may cause an impression that it is not possible to eat almost anything without a serious risk of cancer. Fortunately, the same foods (especially that of vegetable origin) most often contain "antidotes," that is, *antimutagens* and *anticarcinogens*, mostly with an unknown or insufficiently known mechanism of their antitoxic effect. An interested reader can find a list of these compounds together with foods where they are found, in the book edited by Wallace Hayes (2001).

Interactions between carcinogens, anticarcinogens, and living cells are actually very complicated and many-sided. A good example is *indole-3-carbinol*, a constituent of cruciferous plants such as cabbage or cauliflower that inhibits tumors of the mammary gland and forestomach in rodents. Indole-3-carbinol reduces the ability of aflatoxin $B_1$ to cause these neoplasms if administered as a predrug, but increases its carcinogenicity if administered after a contact of the organism with aflatoxin $B_1$. It should be emphasized that indole-3-carbinol interacts with the same aryl-hydrocarbon-receptor (AhR) as the extremely hazardous *dioxin*; the difference lies in the fact that the same effect (1 tumor per 1 million individuals) that is caused by a dioxin dose 6 fg/kg/day is caused by 5 mg of indole-3-carbinol that can be found in 100 g of broccoli (Wallace Hayes, 2001, p. 520).

*Not all carcinogens are mutagens* and, vice versa, *not all mutagens are carcinogens*, although a respective correlation exists.

***Table 3.5*** Naturally Occurring Food Carcinogens

| Carcinogen | Main foods where it can be found |
|---|---|
| Ethanol | Alcoholic beverages |
| Allylisothiocyanate | Cabbage, Brussels sprouts, brown mustard |
| Unsaturated and chlolesterol-rich fats—promoters | Vegetable and animal fats |
| Flavonoids (quercetin, rhamnetin, etc.) | Plants, tea, coffee |
| Hormones (estrogen, testosterone, etc.) | Meat (veterinary drug residues) |
| Hydrazines (gyromitrin) | Mushrooms |
| Mycotoxins (initiators) | Cereal grains, peanut, cottonseed |
| Nitrosamines (after conversion of nitrates) | Beet, lettuce, radish, spinach, celery, fennel |
| Pyrrolizidine-alkaloids (initiators) | Herbs, herbal teas, honey |
| Safrole, estragole, piperine, etc. | Black pepper, nutmeg, other spices |
| *d*-Limonene | Citrus fruit juices |
| Furocoumarins (psoralene) (initiators) | Parsley, celery, parsnip, fig |
| Tannins | Tea, wine, plants |
| Gossypol | Unrefined cottonseed oil |

*Sources:* Adapted from Wallace Hayes, A., Ed. (2001). *Principles and Methods of Toxicology*, 4th edn., Taylor & Francis, Philadelphia, London, Table 11.5; Altuğ, T. (2003). *Introduction to Toxicology and Food*, CRC Press, Boca Raton, FL, Table 4.3.

A number of compounds that are added to food in the preparation chain "from field to fork" are carcinogenic to the experimental animals. *These compounds can be divided into contaminants and food additives*. While the first group gets into the food unintentionally, the second group is added intentionally (see Chapter 17). The first group includes, for example, arsenic from soil (Section 9.1), heavy metals from the environment (Chapter 10), asbestos and vinyl chloride, respectively, from cement and PVC water pipes, PCBs (Section 10.3), residues of DDT and other organochlorous pesticides and hormones (Chapters 14 and 15). Carcinogenic or conditionally carcinogenic food additives like amaranth are described in Section 17.1.

*Cooking of protein-rich food at high temperatures*, such as broiling or barbecuing of meat may lead to the formation of versatile potent carcinogens that are comparable to those found in cigarette smoke (Wu, 2006; Zheng, 1998). Precooking of meats in a microwave oven for 2–3 min before broiling can help to minimize the formation of these carcinogens. Microwaves by themselves do not generate carcinogens.

Three groups of potential human carcinogens—*nitrosamines, heterocyclic amines*, and *polycyclic aromatic hydrocarbons* (PAH) exist, which generally are not found in the raw material of the food but are formed during food processing and/or storage. It is strange that nobody has yet proved a direct connection of these compounds with the promotion and development of any cancer; so far, the whole proof has been indirect.

*Nitrosamines* (see Section 16.6) are synthesized during food storage and digestion from precursors—nitrites and nitrates that are present in the raw material. Nitrosamines can be found in salted, marinated, and fermented foods. *High-temperature processing of foods generates PAHs,* that get fixed on the food surface. The carbonized surfaces of meat and fish contain *heterocyclic amines* that have been formed from creatine or creatinine, amino acids and carbohydrates. Most common nitrosamines and PAHs as well as heterocyclic amine 2-amino-3-methylimidazo(4,5,*f*)quinoline have been placed by IARC into carcinogenicity group 2A, the remaining nitrosamines and most of the heterocyclic amines belong to group 2B (see Table 3.3). On the basis of the data published in 1980–2003 in the databases of Medline and EMBASE concerning the content of these three substance groups in 207 different ready-made foods from 23 countries, a new database, EPIC (*European Prospective Investigation into Cancer and Nutrition*) has been compiled (Jakszyn et al., 2004). For further information about these substance groups, see Chapter 16.

*Actually, there is no conclusive evidence* concerning the existence of a causal link between the food cooking or storage methods and cancer risk. The only recognized exception is fish cooked á la China and consumed in early childhood, the role of which as one of the etiological factors of the *endemic nasopharyngeal carcinoma* (NPC) has been proven. In most parts of the world, NPC is a very rare disease. The only endemic areas include the southern parts of China and other parts of Southeast Asia, the Mediterranean basin, and Alaska (among Eskimos). The other accepted etiological factors of NPC are Epstein-Barr virus infection, environmental risk factors including tobacco smoking, and genetic susceptibility (Vokes et al., 1997).

*A small group of human epigenetic carcinogens consists mainly of drugs.* This group also includes *dioxin* and its congeners, more thoroughly discussed in Section 10.4.

An interesting group of nonmutagenic epigenetic carcinogens are the so-called *peroxisome proliferators.* These include many drugs as well as industrial chemicals. For example, drug *clofibrate* and plastifiers like *phthalates* (Section 16.9) are able to cause, in the case of repeated contacts, development of rodent hepatic tumor. Peroxisome proliferators elicit both a numerical exuberance of these intracellular organelles and an increase of peroxisome enzyme activities and hyperplasy of the liver. It is important to know that *these compounds have a carcinogenic effect only for rodents.* This phenomenon is caused by the fact that peroxisome proliferators need a special carcinogen receptor, which only the rodents have. There is considerable evidence that humans are refractory to the carcinogenic effect of PPs. Still, some toxicologists argue that these chemicals should be considered carcinogenic until the opposite has been proven (Bosgra et al., 2005). Another component of carcinogenicity of these compounds is the enhanced oxidative stress, caused by an increase of hydrogen peroxide ($H_2O_2$) in peroxisomes. Since the peroxisome proliferators have a clear threshold dose, it is possible to perform an assessment of their risk.

### 3.2.9 Endocrine disruption

*Endocrine disrupters* (ED) are exogenous substances causing adverse effects to an organism that they contact or its offspring by disturbing the functioning of their *endocrine systems*. EDs can act as estrogen imitators, antiestrogens, or antiandrogens, causing various physiological and pathological effects through a displacement of the hormonal equilibrium of the organism.

*The endocrine system of an organism is a complicated teamwork* of a large number of hormones including the sex hormones (estrogens and androgens), thyroid hormones, and so forth, the different links of which must act in concert to guarantee a normal functioning of the organism.

In recent years, scientists have come to a conclusion that various chemical compounds originating from the environment can influence the endocrine systems of higher organisms, causing disorders in the functioning of the reproductive systems. Many of these disorders, well documented in wildlife, can be experimentally reproduced. It is much more complicated to prove it in the case of humans. Nevertheless, an increase of incidence of breast as well as testicular cancers, decline of sperm count, earlier onset of puberty for girls, and other abnormalities connected with the reproductive system can be caused by the EDs. The incidence of breast cancer has increased by 50% in Denmark in the period of 1945–1980. The respective studies have shown the existence of a direct relationship between the incidence of mammary cancer and exposure to compounds like dioxin, DDT, and PCBs. Also, a lowering of the age of reaching puberty by a month (teratogenic effect) has been shown in the case of exposition of the mother to PCBs or DDE.

*Testicular cancer* is the most common malignancy in 20- to 34-year-old men, and the incidence has doubled over the last 40 years. A major part of this increase has taken place in industrialized countries, that is, in North America, Europe, Oceania, and Japan. And even broader, compounds exhibiting an estrogenic activity may be a factor in the *testicular dysgenesis syndrome* (TDS), a complex of male reproductive tract disorders and the other components being cryptorchidism, low sperm count, and hypospadia (Newby and Howard, 2006). In the United Kingdom, cryptorchidism and hypospadias have become more frequent.

*During the last 20–40 years, the birth rate of boys has decreased in many countries*. The direct reasons of this tendency are not clear but some hints can be obtained from the echo of industrial spraying of toxic chemicals, above all, dioxins in Seveso, Italy, in 1976. Nine married couples, who were most severely exposed to these chemicals during the 8 years following the disaster, had 12 female babies and not a single boy. Or more broader—instead of the typical ratio in humans of 106 males to 100 females, during those 7 years, intoxicated parents produced only 26 boys to 48 girls. The scientists have no explanation for such a shift, but they note the evidence that normal sex ratios are maintained through the hormone concentrations in the parents (Mocarelli et al., 1996).

*Endocrine disrupters can be divided into man-made and natural compounds*. Since most of the EDs have reached the environment due to human activities, the problem has become very important during the last decades. Examples are anthropogenic substances such as chlororganic insecticides (DDT, lindane, and vinclozoline); industrial chemicals such as PCBs and related dioxines, alkylphenols, bisphenol, and phthalates; cosmetic chemicals such as parabene; food additives such as BHA, drugs like DES and synthetic estrogens that reach water-bodies with urine. In addition to synthetic compounds, a number of natural endocrine disrupters such as mycotoxin zearalenone and phytoestrogens like genistein from soya exist. Soya (*Glycine max*) contains many phytoestrogens—some of them are lipophilic and stable, and are prone to bioaccumulation and biomagnification.

*Endocrine disruption is not only a problem for human medicine.* For example, there is data from different parts of the world concerning the hermaphroditism of fish caused by polluted industrial and household wastewaters. Such altered male fishes started to synthesize a specific protein *vitellogenin*, which is normally synthesized only by female individuals in response to the effect of the hormone estradiol. Vitellogenin is a good marker of endocrine disruption in the case of fish; its synthesis by male fishes is caused, for example, by an exposure to alkyl phenols and synthetic estrogen ethynylestradiol (Navas and Segner, 2006).

The biocide *tributyltinoxide* (TBTO) that is used for wood conservation and in the paints for the submarine parts of boats for protection against algae and barnacles causes an endocrine disruption in mollusks at concentrations of even below 1 ng/L in water. In addition to the hormonal equilibrium, the immune systems of mammals are disturbed by TBTO.

*The most notorious example of endocrine disruption of wildlife* is probably the effect of chlororganic compounds, especially DDT and its metabolites on *both male and female alligators* (*Alligator*) in Florida. As a result, the alligator population of Lake Apopka started to decrease. Later it was established that DDT was probably not the only causative agent of this process. The toxicity of DDT and other chlororganic insecticides are also dealt with in Chapter 14.

Serious issues of endocrine disruption are connected with *populations of the ringed seal* (*Phoca hispida baltica*) in the Baltic Sea. The Baltic Sea is a shallow-water sea, low in salinity in the European part of the Atlantic Ocean. In the middle of the last century, the Baltic seal populations started to decrease with catastrophic speed. An investigation revealed that female seals had different injuries in their reproductive organs like closed oviducts, uterine tumors, and diverse injuries of their internal organs. About two-thirds of the Baltic female seals were not able to have offspring. It is considered that, unlike in the case of alligators, the main culprits here are not DDT and congeners but another anthropogenic group of endocrine disrupters, namely, PCBs and related dioxins (see also Chapter 10) (Nyman et al., 2002).

*Of the cases related to human involvement*, mention should be made of those related to the drug DES, which was used for 20 years in the 1950s and 1960s to avoid spontaneous abortions. DES was banned at the beginning of the 1970s when its role in the formation of serious disorders of the offspring's reproductive systems was proved. These disorders were reproduced on experimental animals. Since DES was used also as a growth stimulator of meat producing animals, it could have had a contact with humans through residues in meat.

Metabolites as well as natural estrogens are excreted into urine, through which they can end up in the drinking water.

*The main mechanisms of the endocrine disruption are*

1. *Binding to estrogen receptor with a subsequent activation.* Substances having this mechanism of action imitate the female sex hormone 17-$\beta$-estradiol. Different tissues have different estrogen receptors. Many chemicals act together additively. Their small effects can add up into a larger one. They can act synergistically, for example, some phthalates together with some natural estrogens.
2. *Binding to estrogen receptor without following activation.* Those substances are anti-estrogens.
3. *Binding to other receptors.* The hormonal system has many non-estrogen receptors like the androgen receptor that can also bind xenobiotics with or without activation. As an example of the latter case, dichlorodiphenylethane (*p,p′*-DDE), a stable metabolite of DDT can be mentioned.
4. *Modification of the metabolism of natural hormones.* Compounds like lindane or atrazine can influence the estradiol metabolism so that more various metabolites are formed. Some other compounds can either activate or inhibit the enzymes catalyzing hormone metabolism. So the testes contain specific enzymes that metabolize estrogens. Some exogenous estrogens can in turn inhibit these enzymes, resulting in an increase of the exposure of the testes to an estrogen, which can be especially important in the embryo phase of the organism development.
5. *Alteration of the number of hormone receptors in the cell.* An endocrine receptor can either increase or decrease the number of receptors thus influencing the cellular response to the effect of both natural and synthetic hormones.
6. *Modification of the synthesis of endogenous hormones.* Substances can influence the level of synthesis of endogenous hormones, acting in some other signalization system of the organism, for example, in some other part of the hormonal system or immune system or nervous system.

A number of factors like absorption and distribution, bioavailability, and timing have their role in determining whether a potential endocrine disrupter will have any effect on an organism. First of all, the ED must enter the organism either with food or beverage, or through the skin in the composition of a cosmetic preparation, or be inhaled. Lipophilic EDs like PCBs or DDT accumulate in body fats from which they are mobilized into the blood during stress-ridden periods or starvation or pregnancy. The active level of endogenous hormones like estrogen is being modified by binding to carrier proteins such as the sexual hormone binding globulin (SHBG) or albumin. These proteins bind most of the endogenous hormones in the blood, thus lowering their ability to influence the cell. Estradiol is especially strongly and specifically bound to the SHBG. Since exogenous estrogens like DES, octylphenol, or *o,p′*-DDT bind remarkably weaker to these proteins, most of the ED molecules stay free to bind to the receptor. Lipophilic EDs cross the placental barrier into the embryo.

The endocrine system of an organism is tightly intertwined with two other vital systems, namely, immunosystems and nervous systems. The poisons of one system can indirectly influence the functioning of the other systems and, as a result, that of the whole organism.

## 3.3 Molecular mechanisms of toxicity

Most poisonous compounds elicit their effect through a disturbance of the molecular processes that assure cellular homeostasis. Other effects induced by a toxicant are the alteration of the cell repair mechanism or cellular proliferation as well as general cytotoxicity. The final effect of a toxicant on the organism is actually an intricate combination of different effect pathways consisting of numerous biochemical processes occurring in the whole organism. *The following discussion touches upon the main mechanisms of toxic effects at the molecular level.*

### 3.3.1 Disturbance of cell homeostasis

*Homeostasis is the vitally important property of living systems* to maintain, by the help of coordinated physiological reactions, the dynamic balance of the processes taking place in an organism and to avoid big deviations from this balance that can threaten the existence of the organism. The homeostatic parameters of the cell are divided into two large groups:

1. *Materials that the cell needs* for energy supply, growth, and repair—water, saccharides, proteins, fats, oxygen, $Na^+$, $Ca^{2+}$ ions, and other inorganic components
2. *Environmental factors* that influence the cell activity—osmotic pressure, temperature, pH, and so forth

The starting reaction for a disturbance of cellular homeostasis by a xenobiotic can be an interference with such essential parameters as metabolic rate, cellular growth, or gene transcription. Further disturbances of homeostatic processes can be a chain reaction enfolding alterations in the basic functions of a cell, which in turn are determined by the functions of the target organ to which the cell belongs. Toxicants are usually able to elicit a number of cellular alterations either by changing the intracellular concentration of cell constituents (e.g., electrolytes) or concentration of endogenous substances such as hormones in the vicinity of the cell. So the alteration of the concentration of intracellular calcium first leads to changes in the cytoskeleton and formation of air bubbles, to adherence of cellular chromatin, and condensation of mitochondria. In the later phase, when the calcium concentration raises further, DNA fragmentation and other processes characteristic of apoptosis will follow.

### 3.3.2 Receptor-mediated mechanisms

The toxicity of definite chemicals can be explained by the *receptor-mediated effects at the plasma membrane or cytosolic level.* Although receptors have been attributed with a rather big role in conveying the pharmacological effects, their role in toxic effects is still less pronounced. The following receptors must be mentioned as examples:

1. The neurotoxic effect of the cyclodiene-type insecticides like dieldrine and heptachlor is considered to occur via *antagonistic* interactions with membrane-bound *γ-aminobutyric acid (GABA) receptors*. The GABA-receptors modulate the flow of chloride ions through the membranous chloride channel. The false inhibition of GABA-receptors can lead to neurotoxic effects, such as excitation and convulsions. Well-known biological toxins like *tetrodotoxin* of the puffer fish or *saxitoxin* of dinoflagellata (see Chapter 13) disturb analogically the Na-channels, blocking their action potential. Chlorinated hydrocarbon-insecticide *DDT* alters the repolarization in the excitated membranes by closing the Na-channels.

2. *Halogenated aromatic hydrocarbons* such as the superstrong poison 2,3,7,8-tetrachlorodibenso-p-*dioxin* (see also Section 10.4) bind, after entering the cell, to the *cytosolic AhR*. This receptor, which was identified in 1976, is considered to be one of the main factors of the ligand-activated transcription or RNA-synthesis. The gene of AhR has been identified in all mammals and even in some invertebrates. AhR, like other ligand-activated transcription factors, is located in the cytosol. In the absence of the ligand, AhR is in complex with a molecule of a chaperon. Binding of a ligand molecule to AhR causes changes in the conformation of this complex followed by dissociation of AhR from the chaperon and binding to AhR-nuclear translocator (ARNT) by heterodimerization. The complex thus formed moves from the cytosol to the cellular nucleus, where, by binding to the molecule of DNA and initiation of gene expression, it activates the synthesis of a number of essential proteins. Among those proteins, there are molecules metabolizing the xenobiotics and regulating growth, like the epidermal growth factor (EGF) receptor.

AhR also activates a gene battery connected with phases I and II of foreign compounds metabolism, such as genes encoding the CYP isoenzymes (see Section 2.4).

As a result of receptor activation, the enzyme-catalyzed process should have become more effective and accelerate the elimination of a foreign compound from the organism. Why is the result often still just the opposite—why does the toxicity of the compound still increase? The answer has two factors—the toxicological outcome depends on the chemical structure of a particular ligand and on the fact that different genes of the AhR gene battery are activated by different ligands to a different extent.

*AhR-ligand complexes can be formed* by compounds belonging to one of the following three classes:

1. *Hydrophobic aromatic compounds with a planar molecule* that either completely or partly (the boundable part) matches well with the binding site of AhR. Those are, for example, planar PCBs and PCDD/PCDFs, polychloro-azobenzenes (PCAOB), polychlorinated naphthalenes (PCN), and several PAHs.
2. *Other potential AhR-agonists with a specific stereochemical configuration* like polyhalogenated (Cl, Br, I), mixed halogenated (Cl, Br, I) or alkylated analogs of compounds of the first class as well as polychlorinated xanthenes and xanthones (PCXE/PCXO), polychlorinated diphenyltoluenes (PCDPT), anisoles (PCA), anthracenes (PCAN), fluorenes (PCFL), and so forth.
3. *Weak ligands and short-time inducers of AhR,* the structure of which does not satisfy the traditional criteria of planarity, aromaticity, and hydrophobicity of an AhR ligand. These compounds are quickly degraded by the detoxifying enzyme induced by themselves. Here belong indoles, heterocyclic amines, and some pesticides and drugs with different molecular structures (imidazoles and pyridins).

Since most of the compounds enumerated are metabolized into *highly reactive electrophilic intermediates,* the acceleration of their metabolism causes an increase in the probability of adduct formation between these intermediates and a DNA molecule, that is, increase of mutagenicity of the compounds.

The toxicity of a substance does not depend only on the rate of formation of the reactive particles but also on the elimination rate of both the initial substance and the metabolites from the cell or organism. Unfortunately, phase I enzymes, CYPs, are induced in the case of expression of gene battery AhR at a lower ligand concentration than many other genes coding the other biotransformation enzymes. For example, the expression of the isoenzyme CYP1A1 gene occurs at 1000-fold lower dioxin concentration than the expression of the gene of the phase II enzyme UGT1A6 (isoenzyme of uridine diphosphate glucuronyltransferase). Such a lack of coordination may lead to an accumulation of highly reactive xenobiotic intermediates in the cell.

*A justified question arises—what are the functions of AhR in the cell?* If AhR is a starter button for processes that are potentially harmful for the cell, why does it exist in the cell at all? One theory proposes that nature has invented this receptor for animal cells, namely, to assist in rendering harmless various plant toxins. This hypothesis is in contradiction with the fact that AhR is possessed by evolutionarily distant species like mammals and the nematode *Caenorhabditis elegans*. It has been shown that $AhR^{-/-}$ transgenic mice have a deficiency of liver and blood vessels. On the other hand, the mice with an overexpressed AhR have a shortened lifespan and a higher incidence of gastric cancer. The last circumstance is interesting in the sense that indoles, the degradation products of several food components, are supposed to be endogenous ligands of AhR. No clear endogenous role has been found for AhR yet, but the topic is gaining attention (Bock and Köhle, 2006).

3. Steroid hormones like estradiol and testosterone control various important physiological processes, for example, the development and differentiation of both female and male sexual organs. The first stage of this influence is the binding of the hormone to specific *intracellular hormone receptors*. Many foreign substances such as industrial chemicals, environmental toxicants, and plant estrogens are able to bind to the same receptors, starting a process called *endocrine disruption* (see Section 3.2.9). It results in effects that are qualitatively the same but quantitatively substantially weaker than the effects of real hormones; so do different pesticides like DDT, plant estrogens (lignanes, flavonoids such as daidzein or genistein), and mycotoxins (zearalenone). Some compounds are capable of binding to the hormone receptors to inhibit binding of genuine hormone molecules. For example, two metabolites of fungicide vinclozoline act as antiandrogens, successfully competing with the endogenous androgen for binding to the respective receptor.

### 3.3.3 *Other toxic effects mediated by cellular membranes*

Cellular membranes are a very frequent target site for the toxicants. One of the reasons is that the toxicants necessarily cross the membrane on their way to the cell, where they may interact with *specific components* (receptors) of the membrane (Section 3.3.2). In addition to the receptors, other membrane components can be *nonspecifically attacked*. For example, *organic solvents* (benzene, toluene, acetone, hexane, styrene, xylene, halogenated hydrocarbons, carbon disulfide, etc.) generate depression effects in the CNS by alteration of the membrane fluidity of the nerve cells. The effects of the solvents can be

- *Acute* (after high-level short-term exposures) with variable symptoms, depending on the particular solvent. However, some symptoms such as disorientation, giddiness, dizziness, euphoria, and confusion progressing to unconsciousness, paralysis, convulsions, and death from respiratory or cardiovascular arrest are typical for all solvent exposures. A metabolite may be responsible if the response is delayed. When the exposure ends, the symptoms will abate in most patients.
- *Chronic* (after lower level short- and long-term exposures). The generalized symptoms are headache, fatigue, sleep disturbances, numbness, tingling, mood changes, and others. Symptoms may be of a slow onset and difficult to associate with a concrete chemical exposure.

Here, we usually have to deal with occupational exposures of workers in industries that use these agents, whereas other individuals may have environmental exposures if they live near industrial facilities and have contact with contaminated water, soil, air, or food. Drinking and shower water, outdoor and indoor air, and food, among other sources, are common routes of exposure to environmental toxicants. Inhalation, ingestion, and dermal absorption are also important mechanisms of toxic exposures. The exposures often involve mixtures of solvents. Some of these incidents may occur deliberately when an individual inhales paints, glues, and other products. A review concerning the membrane toxicity of solvents has been published by Sikkema et al. (1995).

The toxic effects of ethanol can also be partially explained by changing membrane fluidity (partial swelling).

### 3.3.4 *Alteration of cell energetics*

Production and usage of metabolic energy is vital for any cell. *Certain brain, heart, and liver cells are supersensitive* to the decrease of their ability to produce molecules of energy carriers, first of all, adenosine triphosphate (ATP), and to use them for metabolic energy transfer. Any compound, directly or indirectly influencing these systems may in principle produce adverse effects. For example, aniline (or triphenylmethane) dye *malachite green* (Figure 3.1), a toxic chemical primarily designed to be a dye and subsequently used to treat parasites, fungal infections, and bacterial infections in fish and fish eggs in aquaculture, is able to produce dispersion (scattering) of the mitochondrial membrane potential, resulting in an increase of the permeability of the membrane and its swelling. Cellular inhalation will become inhibited and its energetic equilibrium shifted. In 1992, it was determined in Canada that a significant health risk may be inflicted upon humans who eat fish contaminated with malachite green. Malachite green was classified by the Food Safety and Inspection Service (FSIS) of the U.S. Department of Agriculture (USDA) as a Class II health hazard as it was found to be toxic to human cells and possibly conducive to liver tumor formation through mutagenic effects (Culp et al., 2002). However, due to its ease of production and low manufacturing costs, it is still used in certain countries with less restrictive laws for nonaquaculture purposes. *Malachite green is a good example of a compound eliciting multiple toxicity mechanisms.*

Here belongs also the toxicity of the cyanide ion (see Sections 3.3.5 and 8.4).

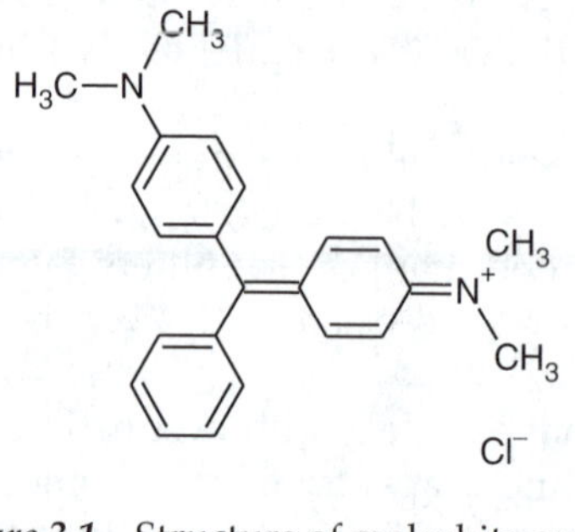

*Figure 3.1* Structure of malachite green.

### 3.3.5 *Covalent binding to essential cellular macromolecules*

A toxicant molecule can be covalently bound to macromolecules like structural proteins of the cell, essential enzymes, lipids, and nucleic acids. In chemical parlance, it usually means a reaction between an electrophilic intermediate of the toxicant and the nucleophilic thiol-, amino-, or hydroxyl-group of a macromolecule. When the frequency of these reactions exceeds the self-repair ability of a cell, such covalent binding can lead to an irreversible formation of a tumor. Binding of a reactive electrophile to the nucleophilic site (nitrogen bases) of DNA leads to the phenomenon of genotoxicity (see Section 3.2.7).

*Example 1.* This is the way of action of mycotoxins called *aflatoxins* produced by molds such as *Aspergillus flavus* and *A. parasiticus*, aflatoxin $B_1$ ($AFB_1$) being the most toxic and carcinogenic. As a result of metabolism of the initially inactive $AFB_1$, its active derivative 8,9-epoxy-$AFB_1$ is formed, which is capable of binding to proteins (phenomenon of cytotoxicity) and DNA (genotoxicity, in a long perspective formation of hepatic tumor). For more about toxicity of aflatoxins, see Section 11.2. A similar genotoxic mechanism is realized in the case of PAH (see Section 16.2).

*Example 2. The cyanide ion* interacts with the mitochondrial membrane protein *cytochrom* $aa_3$, by blocking the electronic respiratory chain. As a result, production of molecules of ATP, which is an important cellular energy carrier, is reduced, although not completely stopped and the cells will die. Especially sensitive to cyanide ions are the brain and cardiac cells; the latter ones have, for example, a reserve of ATP only for 3 min (see also Sections 3.2.3, 3.3.4, and 8.4).

*Example 3. Organophosphorous pesticides and other esters of phosphoric esters and carbamates* bind in the nerve tissue to the active center of the enzyme *acetylcholinesterase* (AChE). The process results in an inhibition of the enzyme. Since the physiological function of AChE is to catalyze the hydrolysis and inactivation of acetylcholine, the forwarder molecule of nerve impulses in the synapsis, this inhibition leads to an accumulation of acetylcholine and persistence of the nerve excitation. This process is accompanied by a number of serious physiological disorders. An irreversible damage of nerve tissue can arise from higher doses of phosphoric acid esters (see also Chapter 14).

### 3.3.6 *Oxidative stress*

*Highly reactive molecular particles—oxidants*—are continuously either entering a cell or being formed in the cell. These particles are able to (per)oxidate cellular lipids into (per)oxides, to decompose proteins and react with molecules of DNA and RNA. Alterations in the structure of these macromolecules of utmost importance for the cell can disturb various cellular systems, such as signal transmission, cellular defense, and repair. In a normally functioning cell, the action of oxidants is balanced by *antioxidants*—molecules hindering formation and/or action of the oxidants. *Oxidative stress* that shifts the *(pro)oxidant–antioxidant equilibrium* to the left in a cell, functioning in aerobic (oxidative) conditions, can cause a serious oxidative damage of the cell.

The final result of oxidative stress will be either adaptation to the stress or extensive injury and death of the cell. Oxidative stress can be caused either by the intensification of the entry or formation of the oxidant or reduction of antioxidant properties of the cell or by a combination of the mentioned effects.

*The particles causing oxidative stress are unstable free radicals* that have one or several nonpaired electrons on their electron orbitals. Such a particle has a tendency to take an electron to its semi-free orbital from another suitable particle. As a result, a new particle with an unpaired electron is formed. Generally, the reactions of radicals are chain reactions, causing a sequential loss of an electron (oxidation) of every particle included in the chain.

*Most of the oxidants are oxygen-based reactive species* (ROS), but they can also be based on atoms of carbon, nitrogen, sulfur, or phosphorus. Examples of potent (pro)oxidants, capable of interacting with cellular components are oxygen-centered superoxide anion ($\cdot O_2^-$), hydroxyl radical ($\cdot OH$), ozone ($O_3$), hydrogen peroxide ($H_2O_2$), hypochlorous acid ($HClO$), nitric acid ($HNO_3$), and also transition metals like iron or copper. In principle, similar oxidation processes occur at rancidification of vegetable oils, during browning of peeled apples and rusting of iron.

*Oxidative stress is nowadays regarded more and more as one of the potential inducers of different pathologies*. For example, it is believed that many teratogenic, mutagenic, and carcinogenic changes initiated by the endocrine disrupters are also connected with the generation and action of free radicals.

It is impossible to completely avoid the injuries caused to cells and organisms by the oxidants. It is not even essential for the free exogenous radicals to enter the cell. *Radicals*, necessary for a normal functioning of a cell, *are formed intracellularly* in the course of aerobic respiration, metabolism, and inflammations. *The exogenous sources of adverse radicals are* environmental pollution, solar radiation, x-rays, smoking, and consumption of alcohol. Fortunately, every cell of every organism has its own system of antioxidants comprising of synchronously acting components that guarantee maintenance of oxidant–antioxidant balance, both in cellular membranes (by fat-soluble antioxidants) and in cytosol (by water-soluble antioxidants). Unfortunately, these antioxidant systems are not perfect. Aging of an organism may cause an increase of this imperfection. This is the reason why the probability of oxidative stress-related diseases also rises with age. Therefore, elderly people must consume more antioxidants, mainly of vegetable origin, with food. But certainly, it is reasonable to consider taking the respective drug (e.g., vitamin) preparations as a very last possibility, since consumption of vitamin preparations can be connected with serious toxicological issues (see Chapter 18).

*Antioxidants* that antagonize oxidative stress can be

- Low-molecular compounds forming new radicals of lower activity (vitamins, minerals, polyphenols)
- Enzymes such as superoxide dismutase (SOD), catalase (CAT), or glutathione peroxidase (GPx) hindering the formation of radical chain reactions

Action mechanisms of both endogenous and exogenous food antioxidants are

- Retardation of superoxide formation in mitochondria
- Scavenging of ROS with the formation of more stable radicals
- Chelation or removal of the transition metals (Cu, Fe, Co, Ni, Zn, Mn, Cd, Cr, etc.) from the radical formation site
- Reduction of the already formed hydroperoxides
- Repair of the injured molecules

Well-known food antioxidants are vitamins A, D, E, and C (see Chapter 18), coenzyme $Q_{10}$ ($Co_{10}$ or ubiquione), plant flavonoids, especially fruit and berry pigments such as anthocyanins, red wine constituent *trans* resveratrol, glutathione, indoles, isothiocyanates, monoterpenes, saponins, and so forth.

*How much of food antioxidant additives can be reasonable consumed?* It can be considered that *the more the merrier* for one's health. But this is not always true. Respective studies with volunteers have yielded controversial results. Epidemiological studies have shown that antioxidant-rich diets that contain ample fruits and vegetables reduce the risk of pathologies like cancer, CHD, cerebral apoplexy (brain stroke), cataract, Parkinson and Alzheimer diseases, and arthritis. But at the same time, data exist, indicating that *the potent antioxidants may turn into prooxidants in the case of very high doses*. It is hardly possible to achieve such intakes by eating vegetables, but they can be achieved by consuming special drugs or medicine-like preparations.

### 3.3.7 *Inhibition of DNA repair*

It has been shown that metals such as *nickel, cadmium, cobalt, and arsenic* have a carcinogenic effect on humans and animals. The reason is the high sensitivity of DNA repair systems to these metals. Contacts with the metals that hinder the removal of the errors produced by endogenous or environmental toxicants inevitably increase the probability of tumor formation. The mechanism of the adverse effect depends on the metal. For example, divalent $Cd^{2+}$, $Ni^{2+}$, and $Zn^{2+}$ ions can inhibit the activity of human *N*-methylpurine-DNA glycosylase (MPG) toward a deoxyoligonucleotide with ethenoadenine (varepsilon A). MPG removes a variety of toxic/mutagenic alkylated bases and does not require any metal for its catalytic activity or structural integrity. Inhibition of MPG activity may contribute to the metal genotoxicity and depressed repair of alkylation damage by metals *in vivo* (Wang et al., 2006).

### 3.3.8 *Multiple interorgan effects*

A certain chemical compound can be primarily toxic to one organ, which in turn causes alterations of the functioning of other organs. Such chains of the effect have been observed, for example, in the case of polyols like *sorbitol* or *xylitol*. If rats were to be fed with a diet containing these polyols in elevated concentrations, appendiceal enlargement and increase of calcium absorption from the intestines will occur. In turn it causes an increase of calcium excretion with urine and formation of corticomedullar and renal pelvis nefrocalcinosis, acute tubular nephropathy, and renal calculus. Formation of medullar hyperplasia and neoplasia of adrenal glands have also been observed. Simultaneously multiple endocrine neoplasia is formed, especially in the pituitary gland, in pancreatic insula, and in thyroid C-cells.

## 3.4 *Biomarkers of toxic effect*

*To prove the exposure to a toxic substance*, and to assess the *response* and *sensitivity* of the organism, respective biomarkers are needed. Such biomarkers can be divided into three types (Timbrell, 2002, p. 67):

1. *Biomarkers of exposure* indicating that the exposure of an organism to a toxic substance has occurred
2. *Biomarkers of response* indicating that the organism has somehow responded to the effect of a toxic substance
3. *Biomarkers of susceptibility* showing a presumed susceptibility to the effect of the toxic substance

*These three types of biomarkers are interconnected*. The degree of the exposure of an organism to a xenobiotic can be roughly estimated by measuring the external dose of the substance. But you can never be sure that the entire dose was converted into the internal dose, that is, was assimilated by the organism (see Section 1.3). More precise results are obtained by measuring the concentration of the substance in the blood. The last number is much closer to the concentration of the substance in the organs that the blood passes, whereby one of these organs can be the target of the toxicant in the organism. The toxic effect can be possessed not by the xenobiotic that initially entered the organism, but by its metabolite(s). In that case, the metabolites are the correct biomarkers. In case a metabolite is of higher reactivity than the parent molecule, urine must be searched just for these reaction products or for example, their conjugates with glutathione.

Unlike probably not very abundant biomarkers of exposure, much *more biomarkers of response can be usually found*. These are, for example, the enzymes appearing in the blood after an organ injury, or special stress proteins, constituents of urine, or enzymatic activity and pathological changes at the general, microscopic, or subcellular level.

*The biomarkers of a presumed susceptibility* can be determined for every single member of a population. For example, such a marker can be

- Genetic deficiency of an enzyme (e.g., CYP2D6 or *N*-acetyltransferase) participating in detoxification or metabolism of a xenobiotic
- A certain substance reflecting an elevated reactivity of a receptor

- A certain substance appearing as a result of a metabolic disorder such as deficiency in glucose 6-phosphate dehydrogenase

Nowadays, more and more investigations are studying the alterations:

- In the genes of the organism exposed to a toxicant (*toxicogenomics*)
- In molecules of RNA, synthesized on the basis of DNA (*toxicotranscriptomics*)
- In proteins, synthesized on the basis of RNA (*toxicoproteomics*)
- In metabolites connected with the effect of these proteins (*toxicometabonomics*)

*Technologies of molecular biology are developing rapidly,* enabling quite a complete and fast analysis of complexes of important cell constituents like DNA, RNA, proteins, metabolites, and so forth. These technologies are extraordinarily powerful in the study of effects such as disorders in cellular homeostasis or structural integrities at the molecular level. They are being expanded to the study of lipids, lipoproteins, carbohydrates, and other components of the cell. Toxicological science as well as industry has quickly adopted these innovative technologies and they have already been accepted by the legislation. The whole of toxicological science has started to change. A new branch—*toxicogenomics*—studying the dependence of the structure and activity of the genome on adverse biological effects of foreign compounds is emerging (Aardema and MacGregor, 2002).

Most of the cases of exposure to toxic foreign compounds, except for a direct damage of cells or tissues by action of corrosive substances, are connected with smaller or bigger alterations in gene expression. Most pathological processes actually develop under genetic control. That is the reason why, a just analysis of gene expression provides us with powerful means for study of the toxicological processes. *The alterations in gene expression caused by toxic substances are often much more selective and sensitive than classical symptoms of pathologies.* Toxicogenomics allows the study of the defense and compensatory responses of the cell to the damaging effect of a xenobiotic, so preceding the formation of an actual pathological situation. It is also possible to establish the effects of low doses of xenobiotics, insufficient to elicit real pathologies. Since the doses used are superlow and in many cases closer to reality, the novel biomarkers sufficiently increase the possibility of human toxicological tests.

*Extrapolation of the results of the animal experiments to the humans is quite a complicated* and usually rather incorrect process. To raise the quality of risk assessment of a toxicant, it is necessary to find out and study *new biomarkers that are common to the test animals and humans*. It is just that the novel methods must enable to perform tests at different toxicant doses in order to allow a better comparison of the response of different species including *Homo sapiens*. Known patterns (fingerprints) of gene expression, which are specific for a particular xenobiotic and mechanism of toxic action that can be used as the biomarkers of a toxic effect, make it possible to reveal the mechanisms of toxic effects. *There are already gene expression chips* that enable simultaneous monitoring of thousands of genes. Probably the chips of the next generation will enable a parallel monitoring of all expressed human genes and analogical chips will soon be available for all important laboratory animals.

On exposure of an organism to alkylating agents like formaldehyde or epichlorohydrin, genes that participate in the repair of DNA and in the elimination and replacement of the alkylated proteins are expressed. It means a wide compensation of the cytotoxic effect of potent mutagens. If we had collections of the "fingerprints" for different toxic effect mechanisms, we could, by comparison with an unknown toxicity, quickly identify its mechanism.

*New developments in gene technology and genomics* have also stimulated and encouraged an enlargement of the study of nutrition from epidemiology and physiology to molecular biology and genetics (Liu-Stratton et al., 2004). As a result, *a new study area—nutrigenomics*—elucidating the effect of foodstuffs on the genome on the transcriptional level, has appeared. Earlier, for the study of mechanisms of action of different food components on health or diseases, methods based on either a single gene or physiological response were used. Regulation of all biological functions of an organism is a concert of thousands of genes, proteins, metabolites, and so forth. For example, the lipid homeostasis is achieved by a cooperation of several organs, hundreds of genes, various signal transduction pathways, and a large number of different biomolecules like receptors, hormones, transcription factors, enzymes, and apolipoproteins. Estimations based on single genes or physiological responses do not provide us with sufficiently exhaustive information to understand the mechanisms of both beneficial and adverse (toxic) effects of different food constituents. Novel methods of genomics such as *DNA microarray* are able to include a major part of the genome into one analysis and from this point, methods of proteomics and metabolonomics branch off (Shioda, 2004). Only a combination of the results obtained by all these methods enables a profound understanding of homeostasis and toxicology modulated by nutrition. This understanding allows setting up a system of biomarkers, which in turn makes it possible to prognosticate both the adverse and beneficial effects of novel foods.

For decades, the so-called *northern blotting* method has been successfully used for the determination of the expression of a single gene or quantification of mRNA, synthesized on the basis of this gene. A new method that is slowly relegating the older one is the *real time PCR* (polymerase chain reaction). This method enables determination of up to ten synthesized RNA molecules or about 0.1 pg ($10^{-13}$ g) of RNA within 2–3 h. Here, the number of simultaneously analyzable genes is limited. A significant development of the real time PCR is several types of DNA microarray, which make it possible to analyze transcripts all over the genome. *Here belong macroarray, cDNA microarray, high-density oligonucleotide microarray, and microelectronic array methods.* The second and third of these methods are most widely used (de Longueville et al., 2004).

## References

Altuğ, T. (2003). *Introduction to Toxicology and Food*, CRC Press, Boca Raton, FL.

Baken, K.A., et al. (2007). Toxicogenomics in the assessment of immunotoxicity, *Methods*, **41**, 132–141.

Bock, K.W. and Köhle, C. (2006). Ah receptor: Dioxin-mediated toxic responses as hints to deregulated physiologic functions, *Biochem. Pharmacol.*, **72**, 393–404.

Bosgra, S., Mennes, W. and Seinen, W. (2005). Proceedings in uncovering the mechanism behind peroxisome proliferator-induced hepatocarcinogenesis, *Toxicology*, **206**, 309–323.

Culp, S.J., et al. (2002). Carcinogenicity of malachite green chloride and leucomalachite green in B6C3F1 mice and F344 rats, *Mutat. Res.*, **506–507**, 55–63.

Fulda, S. and Debatin, K.M. (2006). Resveratrol modulation of signal transduction in apoptosis and cell survival: a mini-review, *Cancer Detect. Prev.*, **30**, 217–223.

Goldman, R. and Shields, P.G. (2003). Food mutagens, *J. Nutr.*, 133, Suppl. 3, 965S–973S.

IARC (2006). *Monographs on the Evaluation of Carcinogenic Risks to Humans*, Preamble, Lyon.

Jakszyn, P., et al. (2004). Development of a food database of nitrosamines, heterocyclic amines, and polycyclic aromatic hydrocarbons, *J. Nutr.*, **134**, 2011–2014, http://epic-spain.com/libro.html

Liu-Stratton, Y., Roy, S. and Sen, C.K. (2004). DNA microarray technology in nutraceutical and food safety, *Toxicol. Lett.*, **150**, 29–42.

Loit, A. and Jänes, H. (1984). *Toksikoloogia* [*Toxicology*], Valgus, Tallinn, Estonia (in Estonian).

de Longueville, F., Bertholet, V. and Remacle, J. (2004). DNA microarrays as a tool in toxicogenomics, *Comb. Chem. High Throughput Screen*, **7**, 207–211.

Lutz, W.K. (1986). Quantitative evaluation of DNA binding data for risk estimation and for classification of direct and indirect carcinogens, *J. Cancer Res. Clin. Oncol.*, **112**, 85–91.

Mocarelli, P., et al. (1996). Change in sex ratio with exposure to dioxin, *Lancet*, **348**, 409.

NTP, National Toxicology Program, Reports on Carcinogens, 11th edn., http://ntp.niehs.nih.gov/ntp/roc/toc11.html

Navas, J.M. and Segner, H. (2006). Vitellogenin synthesis in primary cultures of fish liver cells as endpoint for in vitro screening of the (anti)estrogenic activity of chemical substances, *Aquat. Toxicol.*, **80**, 1–22.

Newby, J.A. and Howard, C.V. (2006). Environmental influences in cancer aetiology, *J. Nutr. Environ. Med.*, 1–59.

Nyman, M., et al. (2002). Current levels of DDT, PCB and trace elements in the Baltic ringed seals (*Phoca hispida baltica*) and grey seals (*Halichoerus grypus*), *Environ. Pollut.*, **119**, 399–412.

Reasor, M.J., Hastings, K.L. and Ulrich, R.G. (2006). Drug-induced phospholipidosis: issues and future directions, *Expert Opin. Drug Saf.*, **5**, 567–583.

Shioda, T. (2004). Application of DNA microarray to toxicological research, *J. Environ. Pathol. Toxicol. Oncol.*, **23**, 13–31.

Sikkema, J., de Bont, J.A. and Boolman, B. (1995). Mechanisms of membrane toxicity of hydrocarbons, *Microbiol. Rev.*, **59**, 201–222.

Teo, S.K., Stirling, D.I. and Zeldis, J.B. (2005). Thalidomide as a novel therapeutic agent: new uses for an old product, *Drug Discov. Today*, **10**, 107–114.

Timbrell, J. (2002). *Introduction to Toxicology*, 3rd edn., Taylor & Francis, London, New York.

Vokes, E.E., Liebowitz, D.N. and Weichselbaum, R.R. (1997). Nasopharyngeal carcinoma, *Lancet*, **350**, 1087–1091.

Wallace Hayes, A., Ed. (2001). *Principles and Methods of Toxicology*, 4th edn., Taylor & Francis, Philadelphia, London.

Wang, P., Guliaev, A.B., and Hang, B. (2006). Metal inhibition of human *N*-methylpurine-DNA glycosylase activity in base excision repair, *Toxicol. Lett.*, **166**, 237–247.

Wu, K., et al. (2006). Meat mutagens and risk of distal colon adenoma in a cohort of U.S. men, *Cancer Epidemiol. Biomarkers Prev.*, **15**, 1120–1125.

Zheng, W., et al. (1998). Well-done meat intake and the risk of breast cancer, *J. Natl. Cancer*, **90**, 1724–1729.

# chapter four

# Analytical toxicology: Determination of foreign compounds

## 4.1 General principles

Analytical toxicology is an integral part of toxicology that allows one to

- Discover, identify, and quantify potentially harmful substances in the environment (soil, water, air), in feed, in the raw materials of food of plant origin or animal origin, and in food.
- Identify foreign substances and their metabolites *in tissues and body liquids* of experimental animals and victims of intoxications (blood, liver, kidneys, urine, stomach contents, etc.).

In earlier times, analytical methods were used mainly in clinical and forensic toxicology. Over the last years, their sphere of usage has significantly enlarged. A number of applications of analytical toxicology in the other branches of toxicology are enumerated in Table 4.1. Most of the applications represent determination of xenobiotics and their metabolites in

***Table 4.1*** Applications of Analytical Toxicology Methods in the Other Branches of Toxicology

| Branch of toxicology | Object of analytical toxicology |
|---|---|
| Food toxicology | Environmental, drug, pesticide residues in food, processing-borne toxicants |
| Clinical toxicology | Victims of food-borne and other intoxications |
| Forensic toxicology | Postmortem investigations |
| Analysis of drugs and narcotics in urine | Abuse of drugs, narcomania, doping in sports |
| Occupational toxicology | Exposure to toxicants at the working place |
| Veterinary toxicology | Intoxications of cattle, game fish, pets |
| Environmental toxicology | Environmental pollution and exposure to it |

live individuals, usually the determination of ultralow concentrations on the ppb or even ppt level.

The main methods used by analytical toxicology are methods of chemical instrumental analysis such as chromatography, electrophoresis, spectrophotometry, and enzyme immunoanalysis. Systematic toxicological analysis (STA) is making an increasing use of hyphenated methods, allowing to increase the volume and quality of the information received at the analysis of biological samples as well as to increase the probability of discovery and identification of formerly unknown toxic substances (Polettini, 1999). The STA approach is expanding to new areas that need chemical analysis such as drug addiction, drugs and car driving, and doping control. Extensive agricultural use of drugs and growth accelerators has generated the problem of residues of these substances in foodstuffs. More and more new biologically active anthropogenic chemicals or their degradation products are reaching the environment. To mitigate and regulate this matter, both regular and irregular random analytical control of food raw material of animal origin is necessary.

Necessity of determination of the ultralow concentrations of these substances in the presence of other interfering compounds has caused an increase of selectivity as well as sensitivity of chemical analysis methods, including that of the respective equipment. Owing to the recent developments in analytical toxicology, new contaminants such as acrylamide or the migration residues are being discovered. For example, in food, the real toxicity of ultralow doses of these contaminants is still to be approved.

The STA methods are also being more often used in the case of investigation of routine intoxications, notwithstanding that the causing agent may be known or at least quite clear. Participation of other compounds in the formation of the final toxic effect can never be excluded. For example, pure heroin poisonings are quite rare. The neurodepressing effect of heroin is often modified by other simultaneously administered substances (ethanol, cocaine, benzodiazepins, barbiturates, etc.).

## 4.2 *Hyphenated chromatographic and spectrophotometric methods*

Chromatography is a widespread method of separation of compounds, based on different affinities and hence partition of substances between two phases—stationary and mobile. *The stationary phase* is a spatially immobilized, mostly solid material, but it can be also a liquid either physically or chemically fixed on the surface or in the pores of a solid carrier. As an example of the latter, a pseudoliquid hydrophobic stationary phase consisting of long aliphatic (most usually octadecyl—$C_{18}$) hydrocarbon chains chemically fixed to porous silica gel can be given. Depending on the state of aggregation of the *mobile phase* or eluant *gas* (GC) and *liquid chromatography* (LC) are known.

The chromatographic stationary phase can be placed either in a tube or on a plate. In the first case, we have *column chromatography*; in the second case, with *thin layer* or planar *chromatograpy* (TLC). If one were to load one end of a tube, now called a chromatographic column, filled with a stationary phase or on one edge of the plate with a solution of the compounds mixture, and start to pass an appropriate liquid or gas through the tube or along the plate, the test mixture gets separated step-by-step, into components. The substances with higher affinity toward the mobile phase move faster along the tube or plate than the substances that have higher affinity to the stationary phase. By continuously changing the composition of the mobile phase in the case of LC or by raising the column temperature in the case of GC, it is possible to achieve a situation when practically all substances loaded simultaneously at one end of the tube or plate elute from the other end being separated from each other. Detection and quantification of the eluted substances by suitable detector(s) will follow. The GC columns are fine and long, with an inner diameter (ID) of tens of micrometers and lengths of several tens of meters; contemporary high-performance liquid chromatographic (HPLC) columns are 5–25 cm long, with ID of 1–10 mm and solid particle size of 3–10 $\mu$m.

*Earlier, GC was preferred in toxicological chemical analysis*. Even now, it is sometimes the method of choice in the analyses of clinical and forensic materials. *LC is gaining more and more popularity*. The reasons for such a change of mind will become evident from the following comparison of the two methods.

*GC* is an appropriate method for separation and determination of nonpolar, relatively volatile, and thermally stable compounds like aliphatic and aromatic carbohydrates and their halogenides. The volatility of substances can be often increased by derivatization, that is, by binding another molecule to the initial molecule. For detection of compounds separated by GC, the following nonselective detectors are being used: universal flame ionization detector (FID); electron capture detector (ECD), used mainly in the case of halogenides; photoionization detector (PID) for aldehydes, ketones, ethers, esters, and so forth. Separated compounds can be identified also by a selective mass-selective detector (MSD) that determines the ionic masses of the substances eluted from a column.

*GC with mass-selective detection* (GC-MS) is considered to be the *gold standard technique* in analytical toxicology. The strong advantages of GC-MS are high specificity, existence of big spectral libraries for compound identification, and a relatively low cost of the equipment. Disadvantages, stemming from the peculiarities of the gas phase, are a time-consuming sample preparation process, which very often includes derivatization. Drugs and toxicants as well as metabolites can be chromatographed in their original molecular shape only if they are volatile, thermally stable, and nonpolar or less polar. Artefacts that are quite often observed in GC-MS somewhat complicate the identification of compounds by this method.

*HPLC* is a good method for a simultaneous analysis of organic compounds of rather different polarity, molecular weight, and thermal stability. Its most used form is *reversed phase high performance liquid chromatography* (rpHPLC), where the stationary phase (mostly $C_{18}$) is less polar than the mobile phase, which is usually a mixture of water that is miscible in every ratio with organic solvents such as methanol, propanol, acetonitrile, and so forth. Raising the time gradient of an organic component in these mixtures is usually needed. *Normal phase chromatography* with a nonpolar mobile phase such as hexane and more polar supplements (e.g., propanol) on nonderivatized silicagel is also used.

LC can be easily coupled with the following sensitive detectors:

- *Absorbance detector*, working in the ultraviolet and visible region (UV-Vis detector). Contemporary UV-Vis photodiode array detectors UV-DAD continuously register the UV-Vis spectra of compounds exiting the chromatographic column. Unfortunately, these spectra are not very specific, but some of the regularities provide an opportunity to elucidate at least the compound's class. Since the optical density ($D_\lambda$) of the solution of an individual compound at definite wavelength ($\lambda$) is linearly related to the concentration of the compound in a rather wide interval of concentrations, UV-Vis detection enables a quantitative analysis.
- *Fluorescence detector* (Fl) uses the property of a part of a substance to emit light at higher wavelength $\lambda$ (lower energy) inherent for the particular compound, shortly after it has absorbed light at lower $\lambda$. This method is much more (about 5–10 times) sensitive and selective than UV-DAD, but its field of application is narrower. The reason is that the intensity (quantum yield) of the fluorescence is sufficiently high only in the case of compounds like polycyclic aromatic hydrocarbons (PAHs), containing condensed planar aromatic groups. Fortunately, it is possible to enlarge the field of application of Fl by derivatization of molecules with fluorescing groups like a phthalaldehyde or dansyl group. In this way, it is possible to use fluorescence for detection and also quantification of amines, alcohols, and carboxylic acids. In Figure 4.1, a two-step chemical scheme of sample preparation for fluorescence analysis of a potent carcinogen *N,N*-dimethylnitrosamine (NDMA) is illustrated (Cha et al., 2006). Step 1 comprises a reduction of nitrosamine to dimethylamine, which is then derivatized to dansyl amine by reaction with dansyl chloride (Step 2).

*Mass-selective detector* (MSD). This detection is highly selective and sensitive, but calls for a relatively expensive MS-(usually MS*n*, where *n* is 2–10) equipment. The method can be used for the detection and identification of all molecules that give either positive or negative ions. *MSD is generally not the best choice for quantification of substances.* MSD can be used by direct injection due to its high selectivity for semi-quantitative estimation of a compound concentration in a sample even without a preceding chromatographic separation. Two main types—*quadrupole and ion trap*—MSDs are in use. The former is more suitable for determination of very low concentrations

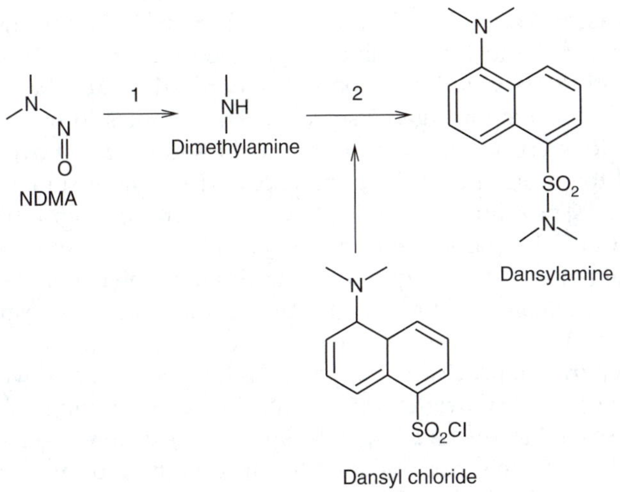

***Figure 4.1*** Denitrosation and dansylation of NDMA to obtain a fluorescent derivative.

of a single ion, whereas the second enables simultaneous determination of a bigger number of different ions, although with a slightly lower sensitivity. The ion trap method also permits multiple step fragmentation of initial or parent ion (tandem-MS or MSn) and the study of spectra of appearing fragments or daughter ions to carry out identification of substances. Several types of ionization methods are at the disposal of analysts. Nowadays, the most popular in analytical toxicology are *atmospheric pressure electrospray ionization* (APESI) as a milder, and *chemical ionization* (APCI) as a more powerful method of ionization.

Very often nowadays, combinations of two consecutively arranged different detectors such as HPLC-DAD-MSn or HPLC-Fl-MSn are used in LC. Such sequences allow simultaneous estimation of ultralow concentrations of known compounds, identification of different compounds, and a rather thorough investigation of the chemical composition of complex mixtures. Often it is not necessary to perform time-consuming sample preparation steps including the concentration of substance(s) to be estimated and removal from the sample of substances disturbing the analysis, by making use of different types of extraction like liquid/liquid or more modern solid phase extraction (SPE). Of course, such sample preparation steps like centrifugation and/or filtration still remain for removal of particles that can clog the chromatographic column and reduce its performance.

*TLC* is a simple, cheap, fast, and high-capacity chromatographic method for separation, purity assessment, and identification of different organic compounds. It has been applied for determination of toxicants like *mycotoxins*. Although nowadays, TLC is often superseded by HPLC and GC that usually guarantee a higher efficiency of the separation process, TLC is still the method of choice for a preliminary screening of suspicious samples, which need further application of more exact and also more expensive methods of analysis.

In the case of TLC, 0.2–0.25 mm thick, either hydrophilic or hydrophobic porous solid or liquid (water vapor condensed on the surface of the solid material) stationary phase is fixed on a glass or plastic plate. The mobile phase is a mixture of different organic solvents. The samples to be separated are dropped onto the plate close to its one end, the solvent is removed by evaporation, and the same end of the plate is placed into a vessel containing the mobile phase. The whole system is placed into a chromatographic container that is saturated by the solvent vapors. Capillary forces start to move the solvent together with the sample compounds along the plate. The higher the affinity of a substance to the moving phase, the faster the respective spot moves along with the solvent. When the solvent reaches the other edge of the plate, the separation process is stopped and the plate will be developed with a suitable method (UV-irradiation, chemical reactions giving colorful products, etc.) to establish the location of the spots of the different substances. The device used for the detection and quantification of the substances by registration of the light diffusively reflected from the plate is called densitometer.

It is possible to run several samples as well as standard compounds and their mixtures simultaneously on one TLC-plate on parallel lines to enable identification and quantification of the mixture components. For every spot, the respective $R_F$ value will be calculated:

$$R_F = \frac{\text{Distance from the center of the compound spot to the start line}}{\text{Distance from the solvent border to the start line}}$$

By comparing the $R_F$ values of sample compounds and standard compounds, it is possible to determine the sample components. In addition to normal- and reverse-phase plates, for example, ion-exchange and gel filtration plates are used in TLC. Separation is based on the differences in ion charges and dimensions, respectively.

To raise the separation ability of the method, two-dimensional TLC is used. After separation of a sample into components in the first dimension, the plate will be turned 90 degrees and eluted once more with another suitable mobile phase. *High-performance thin layer chromatography* (HPTLC) makes use of a thinner stationary phase (sorbent) layer and smaller sorbent particles than classical TLC. HPTLC reduces the spot areas, thus increasing the resolution. HPTLC is a very fast method; one separation process lasts about 10 min.

## 4.3 *Immunological methods for sample preparation and analysis*

### 4.3.1 *Immunoaffinity columns*

Immunoaffinity chromatography is used for sample pretreatment (purification and concentration) in the analysis of food toxicants (toxins) by analytical methods like HPLC, GC, TLC, UV-Vis, or fluorescence spectrophotometry, and so forth. A food sample extract is taken to a small column filled with immunoaffinity sorbent, usually respective antibodies linked to agarose gel beads. This binding step is followed by washing out the nonbound components of the sample and elution of the bound toxicant with a suitable solvent. The solution obtained in this way has a much higher purity and is more concentrated in regard to the toxicant than the initial extract. It is followed by a detection and quantification of the toxicant by one of the aforementioned analytical methods. There are commercially available immunoaffinity columns, for example, for the determination of aflatoxins, ochratoxins, and other mycotoxins as well as numerous other toxicants.

### *4.3.2 Enzyme-linked immunosorbent assay*

Although the enzyme-linked immunosorbent assay (ELISA) techniques have been elaborated for qualitative, semiquantitative, as well as quantitative determination of toxic compounds, they are used mostly for screening out the positive, in regards to the toxicant, samples from a great number of samples. These samples that presumably contain a measurable amount of a toxicant will be afterward analyzed by classical analytical (mainly chromatographic) methods to confirm the presence of the toxicant and to measure its concentration. ELISA methods have many different forms. Typically, the analysis is carried out on a microtiter plate using the competition method. An overview of the usage of immunoassays in analytical toxicology has been published by Sherry (1997).

## *References*

Cha, W., Fox, P. and Nalinakumari, B. (2006). High-performance liquid chromatography with fluorescence detection for aqueous analysis of nanogram-level *N*-nitrosodimethylamine, *Anal. Chim. Acta*, **556**, 109–116.

de Zeeuw, R.A. (2004). Substance identification: the weak link in analytical toxicology, *J. Chromatogr. B: Analyt. Technol. Biomed. Life Sci.*, **811**, 3–12.

Polettini, A. (1999). Systematic toxicological analysis of drugs and poisons in biosamples by hyphenated chromatographic and spectroscopic techniques, *J. Chromatogr. B: Biomed. Sci. Appl.*, **733**, 47–63.

Sherry, J. (1997). Environmental immunoassays and other bioanalytical methods: overview and update, *Chemosphere*, **34**, 1011–1025.

*chapter five*

# *Evaluation of toxicity of substances*

Acute, subchronic, and chronic toxicity of a chemical compound can be evaluated by five basic approaches:

- *Epidemiological studies* of human or other populations exposed to the toxic compound
- *Animal tests with higher organisms* (in vivo)
- *Tests with lower organisms* (in vivo)
- *Tests with cell cultures* (in vitro)
- *Computer calculations* (in silico)

## *5.1 Epidemiological studies*

Human exposure to a chemical compound can be

- *Incidental* via environment, occupation, or diet
- *Intentional*, for example, with a drug or a food additive

*Well-documented accidents with chemicals* can provide essential information about the human toxicity of the respective compound(s). The same kind of information can be obtained from the *occupational exposure* of workers to a substance, in case monitoring as well as recording has been on a proper level. Monitoring must contain an estimation of the concentration of the potentially toxic substance and its metabolites in the body liquids and use of biochemical indicators of pathological alterations. As an example, the inhibition rate of cholinesterase in the blood samples of agricultural workers dealing with organophosphorous pesticides can be presented. This data is rather complicated to obtain and is, unfortunately, seldom complete.

In principle, it is possible to carry out a toxicity evaluation with *volunteers*. In this case, the substances cannot, of course, be highly toxic. The exposure rate must be low, for example, such that it enables the study of metabolism and accumulation. Here, it is of the highest importance to find out the most sensitive biomarkers of exposure.

*A new drug* is administered, prior to its marketing, to a small number of volunteers. Thereafter, it is administered to a limited number of patients (phase I of clinical studies), next to a larger number of patients (phase II), and then to an even large number of patients (phase III). During the clinical studies as well as later, all of the adverse effects revealed should be documented. Study phase I provides information about the metabolism and deposition of the remedy, phases II and III about its side effects and the efficiency of the principal compound.

Data obtained from either human exposure or clinical studies are analyzed by the methods of epidemiology, although there are differences between incidental, occupational, and intentional exposures. To find out the relations between diseases or adverse effects forming at exposure to a toxic chemical, the data obtained are usually compared with analogical data of a nonexposed control group.

Epidemiological studies can be classified into four groups (Timbrell, 2000, p. 164):

1. *Cohort studies,* during which the exposed individuals are prospectively invigilated
2. *Case-control studies* of an incident, where the exposed individuals who have an established disease are retrospectively compared with control subjects not having this disease
3. *Cross-sectional studies,* where the spread of the disease in the exposed group is studied
4. *Ecological studies,* where the incidence of a disease in a certain geographic area in which the exposure to a dangerous chemical is highly probable is compared with the incidence in another area positively free of this chemical

*Cohort studies* are used for the clinical studies of medicines. The control group consists of individuals who have a disease curable with the same drug but to whom placebo is being administered instead of the real medicine.

*In the case of incidental exposure,* the analysis is mostly a retrospective study of *case control.* The control group is usually formed of individuals similar (in gender, age, and other parameters) to the exposed people. This method can be used, for example, for revealing the relationship between occupational exposure to a volatile chemical and formation of lung cancer. Of course, some nonexposed persons will also develop cancer, but in the case of a carcinogenic compound, the incidence of cancer must be statistically higher in the group of the exposed people. The data is processed analogically in the case of occurrence of adverse effects of a drug that is already in use.

*Food toxicology has close relations with ecological studies.* As an example, consider the relationship between the content of arsenic in the plants grown in arsenic-rich soil and the effect on the health of the people living in such areas (see Section 9.1).

To measure a toxic effect, the data obtained by epidemiological studies can be processed in different ways. For example, it can be presented as the ratio of the risks between the exposed and control groups—the *odds ratio*. In epidemiology, *relative risk* is the ratio of disease incidences between exposed and nonexposed populations. For example, if the probability of development of lung cancer among smokers is 20% and among nonsmokers 10%, then the relative risk of cancer associated with smoking would be 2. Smokers would be twice as likely as nonsmokers develop lung cancer.

An alternative measure is the *absolute excess risk*, which can be calculated by the formula $A \times B/C \times D$, where $A$ is the number of the cases of an illness in an exposed population, $B$ is the number of nonaffected individuals in a control group, $C$ is the number of exposed nonaffected individuals, and $D$ is the number of the individuals who have fallen ill without any exposure to the toxicant.

## 5.2 Animal tests

### 5.2.1 General principles

Although the information gained from epidemiological studies is very important for toxicology, most of the knowledge concerning the toxicity of a substance is still obtained from animal studies. This information is used, for example,

- For the risk assessment and safety evaluation of a new drug before the human tests
- For estimation of no-observed adverse effect level (NOAEL) values of food additives before taking into use
- For risk assessment of various chemicals used in the industry and in the environment

Nowadays, there is a strong worldwide endeavor to minimize the number of the toxicity tests carried out on higher animals. More and more toxicity tests with lower animals or cell cultures or principles of *quantitative structure–activity relationship* (QSAR) are used. This abandonment has been a very positive stimulus for the elaboration and validation of novel *in vitro* methods for toxicity assessment of compounds, partly replacing the animal experiments. However, it is not possible to completely abandon animal testing. At least in the near future, higher animals will still serve as the key component in the evaluation process of human risk of potentially toxic compounds. It takes time to elaborate equivalent test batteries with cell cultures and often it is, in principle, impossible. It is not possible to develop a complete substitute for an intact living organism that has metabolism and distribution of xenobiotics (Coecke et al., 2006). But certainly, it is both scientifically and ethically necessary to follow the principles of animal care and welfare, which are controlled by several legislative mechanisms. These mechanisms can be divided into two categories:

1. Guidelines and recommendations
2. Laws and government regulations

The second category requires also legislative mechanisms of mandatory compliance. The respective legislation depends on the country. For example in the United States, the penalties, annulation of licenses for animal tests, and even imprisonments are administered by the U.S. Department of Agriculture (USDA) in concordance with the Animal Welfare Act. The European Union (EU) uses certain minimal standards developed through the Council of Europe, which are used as the basis for national laws and regulations (Wallace Hayes, 2001, p. 774).

Since animal studies can be designed and controlled with great precision, they may provide toxicologists with results of high quality. The number of test animals must be sufficient to enable correct statistical processing of the toxicological data.

The two basic axioms in the design of the animal studies are

1. *The adverse effect of a toxicant can in "principle" be extrapolated from animals to humans:*

   The same dosage in mg/kg bw serves as the basis for this extrapolation. For human safety reasons, it is considered that humans are 10 times more sensitive than animals. Another basis may be the body area and the unit $mg/m^2$. For calculation of the body area, the following equation is used:

   $$m^2 = K \times w/100,$$

   where $K$ is a species-specific factor (9 in the case of rat and 10.6 in the case of humans), $w$ is the body weight of the laboratory animal in kg.
2. *To minimize the number of necessary laboratory animals,* it is reasonable to remember that the exposure of a small number of animals to high doses of the toxicant may replace the exposure of a large number of animals to low doses of the toxic substance as below:
   - It is necessary to have 30,000 laboratory animals to detect cancer with an incidence of 0.01% (20,000 out of 200,000,000).
   - Therefore, high toxicant doses and extrapolation to small doses are used for the human risk assessment of the toxic substance in small animal groups (see Table 5.1).

*It is necessary to be extremely careful* with the interpretation of the results of toxicity studies on laboratory animals and extrapolation to humans. Many toxicants are capable of causing serious harm to an organism, not producing any adverse effect in another physiologically sufficiently different organism.

***Table 5.1*** Equivalent Toxic Doses for Four Animal Species

| Species | Weight (g) | Dosage mg/kg bw | Dose mg/ animal | Body area $cm^2$ | Dosage $mg/cm^2$ |
|---|---|---|---|---|---|
| Mouse | 20 | 100 | 2 | 46 | 0.043 |
| Rat | 200 | 100 | 20 | 325 | 0.061 |
| Dog | 12,000 | 100 | 1,200 | 5,770 | 0.207 |
| Humans | 70,000 | 100 | 7,000 | 18,000 | 0.388 |

*Let us compare animals and plants*. Plants lack a nervous system, an efficient circulation system, and blood and muscles, but, unlike animals, they have photosynthesis apparatus and rigid cellular walls. The toxicity of many insecticides originates from their effects on the nervous system, plants are less sensitive here. And oppositely—most animals are less sensitive to most herbicides (see also Chapter 14).

*The variability of toxic response can be caused by differences in the functioning of the main biochemical pathways of the organisms*. For example, bacteria do not absorb folic acid, but they synthesize it from *p*-aminobenzoic acid, glutamic acid, and pteridine, whereas mammals are not capable of synthesizing folic acid and hence must obtain it with food. Sulfonamides, which are toxic to bacteria just due to a resemblance of their molecules to *p*-aminobenzoic acid both in dimensions and in charge distribution, antagonize the synthesis of folic acid from *p*-aminobenzoic acid. Humans do not exhibit this type of biochemical conversion at all.

Fortunately, both anatomical and physiological similarities outnumber the dissimilarities between humans and laboratory animals, thus justifying the use of animals in toxicological studies. At the same time, there are numerous qualitative as well as quantitative interspecies differences that cannot be ignored when extrapolating the results of toxicological tests (see Table 5.2).

The structure of animal toxicological tests depends on the (bio)chemical properties of the particular compound, on its supposed end use, and on legislation. Relatively few data are needed for the toxicological assessment of the industrial chemicals produced and used in small amounts. At the same time, especially thorough toxicological studies are to be performed with compounds considered for drug use. Pesticides should be tested on various animal as well as plant species; also, knowledge of their stability and behavior both in the environment and in the food chain should be obtained. Ecotoxicology comprises much wider studies of residues than drug toxicology. These include tests with invertebrates like daphnia (*Daphnia magna*), earthworms, fish, phytoplankton, and higher plants. *Nowadays, toxicologists have started to recognize that persistent drug residues likewise call for ecotoxicological tests*.

Toxicologists very often face the problem of selection of the appropriate *laboratory animal(s)* for their testing. The species of choice depends on

***Table 5.2*** Most Important Differences between Humans and the Widely Used Laboratory Animal Rat

| | |
|---|---|
| General | Anatomical |
| Body dimensions | Absence of gall bladder |
| Fast multiplication | Yolk-sac placenta |
| Short gestation/lactation | Presence of udder |
| Short lifespan | Forestomach |
| Dry diet applicable | Fur |
| Physiological | Biochemical |
| Multiparous | Formation of $\alpha_2\gamma$-globulinin |
| Absence of the vomiting reflex | the case of males |
| Nutritional | Behavioral |
| Different needs for vitamins and minerals | Nocturnal way of life |
| Independence of ascorbic acid in food | Coprophagy |
| | Cannibalism |
| Living conditions | Genetic variability |
| Controlled illumination/temperature/ | Interstrain differences in genesis |
| humidity in the case of test animals | of spontaneous tumors |

*Source*: Adapted from Dybing, E., *Food Chem. Toxicol.*, **40**, 237–282, 2002.

the type of test, on the preliminary information about the toxicity of the compound to be tested, and also on ethical and financial considerations. The last two parameters are often the decisive factors in the exclusion of nonhuman primates from the list of test animals. In the majority of cases, rats and mice, about which much background data are available, are used for the toxicity tests. Such background data are much scarcer in the case of less-used species like guinea pigs, hamsters, or jerboa. Genetically, homogenous (inbred) as well as genetically modified mice strains are commercially available. The degree of physiological similarity between human and laboratory animal organisms with respect to the action of the substance under study and simplicity and/or cheapness of their usage are also essential. For testing of veterinary drugs or environmental poisons, mostly younger adult individuals of both genders are used.

*In the starting phase of an animal test*, the approximate toxicity level of the toxic dose of a compound should be determined. While in the case of a drug, the necessary toxicological information is already included in the results of the preceding pharmacological studies; for new industrial chemicals, in most cases, no reliable information about their biological activity is available beforehand. In this case, preliminary tests using a logarithmic scale change of the compound dose should be carried out with a careful observance of the animal behavior. When the region of toxicity has been already roughly estimated, the studies will continue with different, more detailed toxicity tests. Usually acute, subchronic (28- or 96-day), or chronic (lifelong) test and mutagenicity, carcinogenicity, teratogenicity, reproduction and *in vitro* tests are used. In the case of some compounds, special studies like irritability or skin sensitivity tests must be performed.

Obligatorily, the test animals must be healthy, without any contagious disease, without parasites, and so forth. According to their source, the test animals can be divided into three classes:

1. *Animals obtained from wild populations or open colonies.* Their health conditions may be very variable.
2. *Animals derived from closed colonies* where new animals are either not introduced or introduced in a very small number. These animals are vaccinated and cured of microbes and other foreign organisms. Although interindividual health differences can occur in this group as well, the control measures exclude the existence of foreign organisms.
3. *Animals stemming from colonies obtained by Caesarean section or embryo transplantation.* These measures fully eliminate any unwanted microorganisms and parasites. These animals are kept in the bioelimination systems, protecting them against contamination with species-specific pathogens or in some cases, with any organisms capable of influencing the results of the studies. In the case of big test animals (cats, dogs, etc.), this group is very difficult to form and the animals are extremely expensive.

### 5.2.2 *Organism-independent factors influencing compound toxicity*

While planning and conducting animal tests and interpreting the results, one must certainly take into consideration the dependence of compound toxicity on animal species as well as the intraspecies and intrapopulation variabilities of the character and strength of the toxic response (Dybing et al., 2002).

Much of the following material is in some way also usable for estimation of the human toxicity of a particular compound in a particular situation.

#### *5.2.2.1 Dependence on species*

The sensitivity of different animal species including *Homo sapiens* to one and the same toxicant can be very different. Such variability can be caused by differences in absorption, metabolism, and excretion of the compound and its metabolites. The response of humans to a chemical compound can differ from the response of an animal both qualitatively and quantitatively. For example, humans and rats synthesize methemoglobin. But in the organism of guinea pigs or rabbits, this conjugate of nitrite ion with hemoglobin is practically not formed. A dose of hydrogen cyanide causing a heavy intoxication of dogs has a very weak effect on humans. On the contrary, dogs and rabbits can endure a hundred times higher dose of atropine than man.

Although the general direction of metabolic reactions (phases I and II) is common for different mammals, the basic reactions can differ from each other. For example, *aromatic amines are acetylated* by humans, rats, and rabbits but not by dogs. *Glucuronides are formed* in most mammals, except for cats. *Glutamine conjugates have been discovered* only in the case of humans and chimpanzees, conjugation with ornithine (2,5-aminopentanoic acid) occurs only in birds.

There are still many other interspecies differences; some of them will be described in the discussion of particular food-borne toxic substances (see Section two).

#### 5.2.2.2 Genetic variabilities

It is clear that there are genetically caused interindividual variabilities in the activities of enzymes participating in the activation and/or detoxification of xenobiotics. This in turn causes interindividual variabilities in the responses to a toxicant effect. In the case of absence of such variabilities, the slope of the dose–response curve (Figure 1.1) would have been infinite. The hereditary genetic differences that occur in the case of more than 1% of a population are called *genetic polymorphism*. The latter can be caused by one of the following factors:

- Mutations in the coding domains of the structural genes
- Mutations in transcriptional factor binding sites
- Mutations in the coding sites of genes that are working as transregulatory elements
- Base-pair mutations that increase abnormal mRNA splicing and gene duplication
- Absence of whole genes

For example, *favism*, a pathological condition occurring after intake of broad bean (*Vicia faba*) causes hemolytic anemia, hematuria, and jaundice in the case of some individuals. The susceptible subpopulation is enzyme glucose-6-phosphate dehydrogenase (G6PD)-deficient (see also Section 8.10).

#### 5.2.2.3 Generic variabilities

Generic differences in the CYP-mediated metabolism have been observed in the case of rodents. These differences are controlled by the hypothalamus and pituitary gland. As a result, a number of compounds are carcinogenic only to one sex. Less studied but perhaps not as serious differences in metabolism have been found in the case of other species including humans.

*There is a general opinion that males are more susceptible to any adverse effects.* The male sex hormones weaken and female sex hormones strengthen the resistance of an organism. As a general biological peculiarity, females have a longer lifespan. The most important difference between males and females consists in the difference of metabolic rates of organic compounds catalyzed by hepatic microsome enzymes. Many substances are oxidized remarkably faster in the liver of males. Before sexual maturity, the susceptibility to the effect of toxicants is about the same. For example, small concentrations (about 3–4 mg/L) of chloroform produce hepatic damage only in the case of male mice and many of them die over some days after exposure. Analogical difference exists between the sensitivities of different animals to the effect of trichloroethane. Benzene, morphine, and adrenalin appeared to be more toxic to male rats, nicotine to female rats, and folic acid to female mice. Female rats are two times more susceptible to the effect of organophosphorous insecticide parathion than male rats. This difference can be explained by the fact that male rats have a higher activity of CYP, metabolizing parathion into harmless compounds. Men have 3–6 times more frequent liver tumors, induced by chemical compounds, than do women.

The case of monoterpenic hydrocarbon *d-limonene* that is found in many citrus fruits and used as an aroma and flavoring substance and industrial solvent can serve as an example of nonendocrine sexual differences. *d*-Limonene is carcinogenic only to male rats, which, after contact with this compound, start accumulating $\alpha_2\mu$-globulin in the renal proximal tubules. This is followed by an overload of protein, progressing renal injury, and compensatory proliferation of cells. The process can end with the formation of a tumor of the renal tubules. Female rats do not synthesize $\alpha_2\mu$-globulin.

Women have a much higher toxicity of ethanol and probability of developing of alcoholism than men (see Section 16.3).

#### 5.2.2.4 Dependence on age

The age of an individual is a very important factor in the susceptibility to the effect of toxicants—in the case of both test animals and humans. The reasons are the age-caused differences in the functioning of the nervous, respiratory, endocrinal, and cardiovascular systems, gastrointestinal tract, and different membranous barriers. The aging of an organism is accompanied by an alteration of the rates of substance distribution, metabolism, detoxification, and excretion.

*In general, young individuals are more sensitive to toxic effects than adults are.* Newborn animals and children lack phase I enzyme CYP. This enzyme family reaches normal activity by the 30th or 60th day of life in the case of rats and humans, respectively. The disturbances in phase II conjugation reactions of newborns are caused by the deficiency of the enzyme catalyzing formation of uridine phosphate glucuronic acid. *Different enzyme deficiencies render newborns very susceptible to the effect of carcinogens.* That is why many drugs and food additives are toxic to the newborns, who are more sensitive to the depressants of the central nervous system (CNS), analgetics, and cholinolytics, but less sensitive to the effect of CNS stimulators. Susceptibility to heavy metals is relatively independent of age. Toxicity studies of benzene on rats and mice have shown that the most susceptible are young animals and the most resistant are, in the case of both intrapulmonary and oral administration, adults, whereby the sensitivity increases again with age. After starvation for some days, the overall absorption ability of the digestive tract of young animals increases and that of adults decreases. Absorption is more complete in the case of young animals. *Elderly people have usually an especially severe progress of intoxication.*

#### 5.2.2.5 *Dietary conditions*

Dietary conditions may have a remarkable effect on bioavailability and, hence, toxicity of many chemicals. For example, absorption of toxic cadmium from the gastrointestinal tract increases in the case of iron deficiency in food. As the result, the absorption of cadmium is twice as normal in the case of women with a low serum ferritin level.

*Deficiency of mineral substances* (calcium, copper, magnesium, zinc, selenium) in food reduces both CYP-catalyzed oxidation and reduction. It leads to the decrease of the summary biotransformation of xenobiotics. Restoration of the normal dietary level of these minerals returns the enzymatic activity to its physiologically normal level.

*Deficiency of the vitamins C-, E-, and B-complex* reduces the biotransformation rate of xenobiotics. All these vitamins are either directly or indirectly bound to the system of CYP. Their deficiency can alter the redox—conditions of the cell, thus, inhibiting the synthesis of the macroergic substances necessary for running of the biotransformation phase II. Again, the restoration of a normal level of vitamins in the diet restores the normal enzyme activity. On the other hand, some food components can influence the endogenous activity of different vitamins. For example, rats that were fed with antioxidants like hydroxytoluene suffered from a fall of vitamin K-dependent blood coagulation activity, resulting in the death of the animals via bleeding. These effects can be avoided by addition of vitamin K to food.

*Low-protein diets* remarkably increase the toxicity of xenobiotics, which are active in their original form, but reduce the toxicity of the compounds needing preliminary biotransformation. For example, the lethality and hepatotoxicity of *N,N*-dimethylnitrosoamine (NDMA) decrease remarkably in the case of rats that are kept at a low-protein diet. A reduction of toxicity correlates with CYP-mediated *N*-demethylation of NDMA, the first step of biosynthesis of a reactive alkylating agent. Several experiments have shown that alterations in the activities of the CYP isoforms are also connected with a shortage of dietary protein.

Protein deficiency can enhance the toxicity of arsenic compounds. In most animals, methylation of very toxic inorganic arsenic to less toxic methylated metabolites is the main way of its detoxication. Protein deficiency reduces the velocity of biomethylation.

*The carbohydrate-rich diet* decreases the activity of the enzymes involved in the foreign compound metabolism.

The activity of biotransformation enzymes, especially the membrane-bound ones, depends on the types of *dietary lipid*. Diets containing much of the polyunsaturated fatty acids (PUFAs) reduce the CYP content in rat liver. The main reason is the higher susceptibility of PUFAs to peroxidation, causing a degradation of the microsomal membranes with a simultaneous disappearance of CYP. The nature of the membranous fatty acids also influences the membrane fluidity and thus, the toxicity of the agents that are perturbating the membranes.

*Caloric deficiency*, on the one hand, increases the longevity of laboratory animals. On the other hand, it reduces the incidence of tumor formation in the case of both spontaneous and induced-by-agents tumors. And on the contrary, connections between overfeeding, overweight, and high mortality have been proven in the case of rodents.

Many fruits and vegetables contain natural compounds—inducers of phase II enzymes that are capable of accelerating the inactivation of the food-borne toxicants. For example, 1,2-dithiol-3-thions, potent inducers of glutathione S-transferases (GSTs) (see Section 2.3.3) found in cruciferous plants like cabbage or Brussels sprouts, defend rat liver against carcinogenic aflatoxin $B_1$ by "trapping" its very carcinogenic metabolite aflatoxin $B_1$-8,9-epoxide (see Chapter 11).

*Stimulation of a limited number of colonic bacterial strains* is an alternative mechanism, which natural compounds may use to facilitate the defense of an organism against food-borne toxicants. *Inulins*, a widespread group of the natural polymers of fructose, are resistant to the action of pancreatic as well as enteric hydrolases. Inulin gets fully fermented only in the colon, where it promotes a preferred growth of beneficial *Bifidobacteria* and *Lactobacillus* and accelerated production of lactic acid (butyrate). These bacteria have elevated, in comparison with a number of strains of *Bacteroides*, *Clostridium*, and *Enterobacterium*, activity of such enzymes as azoreductase, nitroreductase, $\beta$-glucosidase, $\beta$-glucuronidase and $7\alpha$-hydroxylase. Since the enumerated enzymes participate in the activation of a number of food toxicants, a prevalence of *Bifidobacteria* and *Lactobacillus* can reduce the risk of colon cancer (Geier et al., 2006). The last effect has been demonstrated in several chemically induced colon carcinogenesis models. Furthermore, both inulin and butyrate induce apoptosis of the colonic cancer cells.

#### *5.2.2.6 Health conditions*

*Pathological conditions of the liver* have a strong influence on the metabolism of xenobiotics through the following three factors:

1. Alterations in the blood supply of the liver that influence the arrival of xenobiotics to the site where metabolic reactions will occur.
2. Reduction of the number of viable hepatocytes.
3. Reduction of production of albumin. It leads to an increase of both the content and potential toxicity of nonbound foreign compounds in the blood as well as in tissues.

The metabolism and immunotoxicity of xenobiotics are altered by diseases like diabetes and hypertension. In a number of human populations that are characterized by a high incidence of hepatic cancer, a tumor is often accompanied by hepatitis B virus infection and there must be a high level of exposure to the dietary aflatoxins. There is likely to be a synergism of action between the virus and aflatoxin $B_1$, caused by the acceleration of cell proliferation in the case of hepatitis.

Glomelural filtration and tubular secretion of the xenobiotics decrease in line with deepening of tissue damage in the case of renal diseases. As a result, inhibition of elimination of the xenobiotics from the organism takes place.

#### 5.2.2.7 Simultaneous contact with several xenobiotics

The total effect of several foreign compounds can be (Dybing et al., 2002) as follows:

1. *Summation of doses*—the action mechanisms of different chemicals are the same, but their potencies are different. Such a simple joint action enables a mathematical summation of the effects after taking into account the potencies.
2. *Summation of response/effect*—the action mechanisms and probably the site are different. The effects do not combine and do not modify each other. Mathematically, the sum of the responses is the sum of the effects of all active compounds.
3. *A more complicated combination of responses*—the summary effect is either bigger or smaller than the simple sum.

## 5.3 Cell culture tests

*Human society demands minimization of animal use in various medicinal experiments* including toxicity testing. Therefore, replacement of animal tests with various *in vitro* experiments is on the agenda. In some cases, this replacement has been successful. *In vitro* systems are a good tool, for example, for searching of biomarkers of toxicity; they estimate the cytotoxicity of compounds and cellular response and enable toxicokinetic modeling. The results obtained are valuable for the assessment of risk of compounds to humans via extrapolation of the effects to the *in vivo* situations and from animals to humans.

*There are many end points of* in vitro *toxicology*. One widely used approach is the estimation of the genotoxicity of compounds for a preliminary assessment of its carcinogenicity. The methods used here are fairly well designed.

Since different genotoxicity tests detect different genetic processes, it is reasonable to use a set of tests for a complete characterization of a substance:

1. *Unscheduled synthesis of DNA (USD)*. In this method, the intensity of DNA repair is assessed by measuring the amount of thymidin, labeled with radioactive tritium ($^3$H-TdR—one of DNA monomers), entering the cell. Cell culture is exposed to the chemical studied during a definite period (from 2 h to some days), labeled thymidin is added and the mixture incubated. DNA repair is quantitated by radioautography. The cells must be kept under conditions preventing

normal semiconservative replication but allowing DNA repair (Madle et al., 1994).

2. *Salmonella/mammal microsome* (*Ames*) *test*. This test enables detection of the reversed mutations formed in the DNA of Gram-negative enterobacterium *Salmonella typhimurium* after exposure to a toxic substance and to the rat liver homogenate. The number of mutant colonies that are able to grow in the histidine-free environment is estimated. The Ames test indicates the number of bacterial cells, which after an exposure to the chemical, have mutated from $His^-$ genotype to $His^+$ genotype. The test bacterium must be an auxotrophic mutant ($His^-$) that unlike the "wild" bacterium requires histidine for its growth. Bacterium must also have plasmid pkm101 that codes error-proof repair of DNA. Furthermore, this bacterium must have rfa mutation, that is, a defect in the cell wall that enables the substance to be tested to make diffusion easier into the cell. Usually several *S. typhimurium* strains are used in the test, since various strains help detect various changes in the DNA molecule. For example, the strain TA 98 detects frameshift mutation and the strain TA 100 bp detects replacements in a DNA molecule. Since part of the compounds induce the first, and a part the second mutation, it is reasonable to use both strains in parallel. The rat liver homogenate contains CYP enzymes and cofactors. In addition, histidine-free minimal glucose agar plates are necessary for observation of the bacterial growth. The reliability of the test is 80–85%; both false-negative and false-positive results can occur. The method is relatively simple and cheap (Ames et al., 1975).
3. *Sister-chromatid exchange* (*SCE*) *test*. Genotoxic agents are often able to induce aberrant staining patterns in the leukocytary chromosomes, which can be observed under a microscope. SCEs involve a breakage of both DNA strands, followed by an exchange of whole DNA duplexes. This occurs during the S phase and is efficiently induced by mutagens that form DNA adducts or that interfere with DNA replication. The formation of SCEs has been correlated with recombinational repair and the induction of point mutations, gene amplification, and cytotoxicity. The SCE test is usually performed on human peripheral blood lymphocytes. The test is reliable and cheap (Wolff, 1983).

*For determination of genotoxicities,* other microorganisms such as bacteria *Escherichia coli* or yeast cells, and also mammal cells can be used. Of the mammal cells, murine lymphome and Chinese hamster ovary cell lines have found the most extensive use. Human lymphocytes can also be used for testing of the chromosome destruction. It has to be stressed that the correlation between the positive results of bacterial mutagenicity tests and real carcinogenicity of a compound to laboratory animals is by far not 100%. It means that the well-known animal carcinogens are not always mutagenic in procaryotic cell systems and, vice versa, some compounds, mutagenic to bacterial cells, are not animal carcinogens. For example, a food additive such as sulfur dioxide ($SO_2$) is mutagenic in *in vitro* but not *in vivo* test systems. In general, compounds that are regarded as suspicious but are not mutagenic should be tested for their carcinogenicity *in vivo*.

The *in vitro* studies have generally been successful for testing new ingredients of *cosmetics*. Just the usage of human skin cells and simple *in vitro* systems have caused the cosmetics industry to reduce *in vivo* testing of *skin irritability*.

Recently, an *in vitro* alternative to the *in vivo allergenicity/sensibilization test* has been offered. Often, the *in vitro* tests need only supplementary data from *in vivo* tests for decision making. For example, a bacterial mutagenicity test can show a substance as a potential genotoxic carcinogen, but real carcinogenicity can be demonstrated only on laboratory animals. The result of a bacterial test can be sufficient to stop the development of a substance for drug or component of a cosmetic, but in the case of industrial chemicals, demonstration of real carcinogenicity is necessary.

One of the most widely used *in vitro* test-systems is the isolated, either human or animal, hepatocyte or hepatic tumor cell line HepG2.

A number of problems like an insufficient viability of the primary cells during long experiments are connected with *in vitro* cell tests, for the reason that they can be used only in the short-term tests. The results of an experiment can be influenced by biochemical alterations like changes in the level of CYP and isoenzyme ratios that may occur already during tissue preparation. Alternatively, freezable immortal cell cultures can be used just when it is necessary. Unfortunately, since such cells originate from tumors, they are not identical to the cells of a normal tissue. *Results of the* in vitro *tests generally tend to underestimate the real* in vivo *toxicity of the substance.*

## 5.4 Computer calculations

The chemical structure and especially the presence of specific atomic groups provide a definite chemical compound with a possibility of particular toxic action as well as particular metabolism route. For example, epoxide, carbamate, or nitrosated amino groups indicate a mutagenic and carcinogenic potential. When coupled with a prediction of a possible intake of this chemical by humans, this qualitative information can be developed by the use of more precise and informative quantitative approaches such as QSAR (Barlow et al., 2002). The parameters that can be measured or estimated using different molecular modeling software and can be used in the QSAR studies are atomic charges and partial charges, electron densities, polarizabilities, van der Waals areas, hydration energies, and so forth. QSAR studies can be used for screening, understanding, and predicting the reactivities of chemicals, for estimation of their toxicities, and hazard-ranking for further testings. QSAR studies have also a number of weaknesses. For further acquaintance with the methods of *in silico* calculations, we refer to the recent review by Simon-Hettich et al. (2006).

## 5.5 Acute toxicity tests

Acute toxicity tests are designed for the study of toxic responses manifesting in an organism during a short period (up to 24 h) after mostly a single (acute) contact with the toxicant. For example, a single dose of HCN (50–60 mg) induces death within minutes. In the case of death that follows the contact with a toxicant after 24 h, we have to contend with the *delayed effect*.

*Results of the classical acute toxicity test enable* to construct a dose–response curve one and to estimate the $LD_{50}$ or $ED_{50}$ values (see Section 1.3). Classically, at the first stage of studies when almost nothing is known about the toxicity of compounds, four different doses in the logarithmic progression are used. The most used animal species are the mouse, rat, and dog. Since mouse and rat differ from each other rather substantially, these species are often used in parallel. As a rule, 8–10 animals (4–6 of both sexes) are used. The route of administration is usually the one that is most effective in the case of humans. If this is the oral route, then the substance is mostly taken directly to the stomach through a pipe, but it can be administered with food as well. Direct administration guarantees, in most cases, a higher toxicity of the substance, although opposite cases have also been observed. During *postmortem* studies, samples are taken by autopsy from all organs of all animals—those that died during the test and those that survived. Profound general and histopathological studies of the samples are thereafter carried out.

*The conventional $LD_{50}$ test was finally cancelled by the end of 2002*. The following *alternative animal tests* of acute oral toxicity have been developed:

- Fixed dose procedure (FDP)
- Acute toxic class (ATC) method
- Up-and-down procedure (UDP)

These methods guarantee significant improvements in animal welfare and permit to reduce dramatically the number of test animals needed. They have recently undergone a revision to improve their scientific performance but more importantly to increase their regulatory acceptance. They can now be used for all types of test substances and for all regulatory and in-house purposes. *In vitro* cytotoxicity tests could be used as adjuncts to these alternative animal tests within some years to improve the dose level selection and thus give further modest improvements in the numbers of animals used. However, the total replacement of animal tests, if ever, may happen not earlier than in about 10 years after considerable development and validation. (Botham, 2004).

*FDP* was first proposed in 1984 by the British Toxicology Society, as an alternative to the conventional $LD_{50}$ test (Organization for Economic Cooperation and Development (OECD) Test Guideline 401), for the determination of acute oral toxicity. The FDP uses fewer animals and causes less suffering than the $LD_{50}$ test, and has provided information on acute toxicity, which allows substances to be classified according to the EU hazard classification system. In 1992, the FDP was introduced as OECD Test Guideline 420; the revised FDP was adopted by the OECD in 2001 (Stallard and Whitehead, 2004).

*The FDP method is briefly as follows.* A small number of laboratory animals (e.g., five of both genders) is exposed to a dose of a chemical of 5 mg/kg and the occurrence of toxicity signs is monitored. If 90% or more of the animals die, the substance will be classified as *very toxic,* but if 90% or more or test animals survive without any sign of intoxication, a 10 times higher dose (50 mg/kg) will be administered to the next group of animals. If again, 90% or more die, the substance is *toxic,* but if again 90% or more of the animals do not respond, the dose will be increased again by 10 times (to 500 mg/kg) for the third group of animals. If 90% or more die, the substance will be classified as *harmful,* but if once more 90% or more of the animals stay indifferent, the substance will be assessed as *unclassified* (safe) (Timbrell, 2002, p. 168). After the test, all the animals will be sacrificed and their tissues studied profoundly.

*The oral ATC method.* The ATC method is also a sequential testing procedure using, depending on the mortality rate, from three to six animals of one sex per step at any of the defined dose levels. The number of test animals is reduced by 40%–70% in comparison to the $LD_{50}$ test. The method is based on the Probit model. The results of both national and international ring tests demonstrated an excellent agreement between the toxicity and the animal numbers predicted biometrically and observed in the validation studies. The oral ATC method was adopted as an official test guideline by OECD in 1996 and was slightly amended in 2001. The ATC method has been widely and successfully used in Germany. In the EU Member States, the ATC method is used in the range of 50% of all tests conducted (Schlede et al., 2005).

*Up-and-down procedure.* UDP also uses sequential dosing, combined with computer-assisted computational methods during the execution and calculation phases of the test. Staircase design is applied to the acute toxicity testing with its binary experimental end points (yes/no outcomes). UDP provides a point estimate of the $LD_{50}$ with approximate confidence intervals in addition to the observed toxic signs for the substance tested. UDP does not provide information about the dose–response curve. Computer simulation is used to test the performance of the UDP without the need for an additional laboratory validation (Rispin et al., 2002).

## 5.6 Subacute/subchronic toxicity tests

*Acute toxicity tests are by far not sufficient,* for example, for the assessment of the toxicity of *food additives.* They also have only a limited value in the studies of both industrial and agricultural exposure to pesticides. Since the same food can be consumed during a longer period, it is essential to study the effect of relatively low but repeated doses of the food additive. Acute toxicity tests are usually followed by *subacute/subchronic tests.*

In these tests, the chemical to be studied is administered to laboratory animals either daily or a little more seldom during 90 days or 3 months. In the case of rats, the subacute tests last for 14, 21, or 28 days, and the subchronic tests for 90 days. These studies are to be carried out with two different animal species, one of which is a rodent. Most often, these two species are the rat and dog, since those are easily available and a lot of background information exists concerning these species. In the case of rats, the tests are started immediately after weaning when their growth rate is the highest. Every dose group (not less than three) contains at least ten animals of both sexes and the experiment lasts for a minimum of 10% (three months) of the animal lifespan. If there is an intention to study the pathogenesis and reversibility of the injuries, organism adaptation or biochemical kinetics, observations are carried out after every 3 weeks up to 3 months. Simultaneously, the concentration of the toxicant and its metabolites in blood, urine, and tissues can be estimated and other clinical–chemical analyses carried out. The information obtained can be correlated with the intensities of the toxic effects manifesting themselves during the test.

*In food toxicology*, the subacute/subchronic tests are often used for *determination of no-observed-effect-level* (NOEL) or NOAEL *values* for noncarcinogens and the *maximum tolerated dose* (MTD) values for carcinogens. MTD is the highest level of a substance in feed that can be administered to an animal without manifestation of other signs of toxicity, but the toxicity caused by the developing tumor. The data obtained from subchronic studies helps in turn to design chronic toxicological studies.

## 5.7 Chronic toxicity tests: Acceptable daily intake

These tests last for 12–24 months in the case of rodents (for about 50% of their lifespan) and 6–12 months in the case of other animals, and are necessary for the estimation of MTD values (see Section 5.6) of carcinogens requiring a longer developing time of the tumor. Chronic toxicity tests can be combined with *in vivo* carcinogenicity tests. Like subchronic toxicity tests, they are wound up with *postmortem* pathological and microscopic histological studies. During a chronic toxicity test, it is possible to perform a clinical–chemical analysis for the evaluation of development of pathological alterations. Among the indicators of adverse effects, there can be an alteration of simple parameters such as animal body weight, feed, or water consumption. Special attention is paid to the target organs of the toxic effect and to the animals that die before the end of the experiment. *Chronic toxicity tests are very significant in regard to the following*:

- Prolonged drug use
- Lifetime exposure to food additives
- Long-term, low level exposure to industrial and environmental chemicals

In food toxicology, chronic toxicity tests are used for estimation of NOAELs, which are in turn necessary for derivation of the *acceptable daily intake* (*ADI*) *values of food additives and contaminants*. In the case of all these tests, the right choice of doses, animal species and strains, the type of exposure, and measurable parameters are very important.

Despite a rather wide choice of animal species for chronic toxicity tests, rodents are again mostly preferred. Bigger animals are used when larger amounts of blood are needed for the analyses. The chemical studied is administered daily during the whole study period, which covers usually most of the animal's lifespan. As in the case of acute toxicity tests, the route of administration must be the same that most probably is effective in the case of humans. In the case of gases or volatile industrial solvents, it is inhalation, and in the case of food additives such as pesticide residues and other contaminants, it is the peroral way. An appropriate method is the addition of the substance to either food or drinking water, whereby checking of its stability in this medium is obligatory. If the substance to be tested has a bad smell or taste, it should be either encapsulated or taken directly to the stomach by a pipe.

In the carcinogenicity and mutagenicity tests, also mainly rodents—rats, mice, golden Syrian hamsters, and so forth—are used. Preference of rodents is substantiated by their generally higher susceptibility to induction of tumor, relatively short lifespan, lower maintenance expenses, their former wide usage in toxicological as well pharmacological studies (much background data), and existence of inbred strains. Naturally, nonrodent animals like dogs and primates can also be used.

## 5.8 Other toxicity tests

*New Substances Notification Scheme*, a document that is in force in the EU, specifies the demands on physicochemical, toxicological, and ecotoxicological tests required for marketing of a new chemical (Ginzky, 2001). According to this scheme, the number of tests depends on the expected production volume of the substance. Absolutely necessary are acute toxicity tests using three different routes of exposure (peroral, inhalation, and percutaneous), irritability tests of skin and eyes (Draize test), skin sensibilization test, subacute toxicity test (28 days), and mutagenicity tests, both with bacteria and with a mammal cell culture. In addition, one may find it necessary to perform ecotoxicological tests, such as toxicity to fish and to daphnia (*Daphnia magna*) and degradation rate of the substance under question in the environment (*biochemical oxygen demand—BOD, chemical oxygen demand—COD*). Depending on the volume of use of the substance and results of the former experiments, it may also be necessary to perform teratogenicity, reproductivity, carcinogenicity, and other tests. Sometimes, it is necessary to repeat earlier tests using alternative ways of administration or another test animal or to carry out supplementary ecotoxicological tests like prolonged fish and daphnia tests, studies of effects on higher plants, and bioaccumulation studies in fish and probably also in other organisms.

Toxicity and ecotoxicity tests should be performed in accordance with internationally recognized directives like the guides issued by the *OECD* and in accordance with the system of *Good Laboratory Practice* (GLP).

## *References*

Ames, B.N., Mccann, J. and Yamasaki, E. (1975). Methods for detecting carcinogens and mutagens with the Salmonella/mammalian-microsome mutagenicity test, *Mutat. Res.*, **31**, 347–364.

Barlow, S., et al. (2002). Hazard identification by methods of animal-based toxicology, *Food Chem. Toxicol.*, **40**, 145–191.

Botham, P.A. (2004). Acute systemic toxicity—prospects for tiered testing strateies, *Toxicol. In Vitro*, **18**, 227–230.

Coecke, S., et al. (2006). Metabolism: a bottleneck in *in vitro* toxicological test development. The report and recommendations of ECVAM Workshop 54, *Altern. Lab. Anim.*, **34**, 49–84.

Dybing, E., et al. (2002). Hazard characterization of chemicals in food and diet: dose response, mechanims and extrapolation issues, *Food Chem. Toxicol.*, **40**, 237–282.

Geier, M.S., Butler, R.N. and Howarth, G.S. (2006). Probiotics, prebiotics, and synbiotics: a role in chemoprevention for colorectal cancer? *Cancer Biol. Ther.*, **5**, 1265–1269.

Ginzky, H. (2001). EU and US strategies under reform: A comparison of strategies and weaknesses, *RECIEL*, **10**, 199.

Madle, S., et al. (1994). Recommendations for the performance of UDS tests *in vitro* and *in vivo*, *Mutat. Res.*, **312**, 263–285.

Rispin, A., et al. (2002). Alternative methods for the median lethal dose LD(50) test: the up-and-down procedure for acute oral toxicity, *ILAR J.*, **43**, 233–243.

Schlede, E., et al. (2005). Oral acute toxic class method: a successful alternative to the oral $LD_{50}$ test, *Regul. Toxicol. Pharmacol.*, **42**, 15–23.

Simon-Hettich, B., Rothfuss, A. and Steger-Hartmann, T. (2006). Use of computer-assisted prediction of toxic effects of chemical substances, *Toxicology*, **224**, 156–162.

Stallard, N. and Whitehead, A. (2004). A statistical evaluation of the fixed dose procedure, *Altern. Lab. Anim.*, 32, Suppl. **2**, 13–21.

Wallace Hayes, A., Ed. (2001). *Principles and Methods of Toxicology*, 4th edn., Taylor & Francis, Philadelphia, London.

Wolff, S. (1983). Sister chromatid exchange as a test for mutagenic carcinogens, *Ann. NY Acad. Sci.*, **407**, 145–153.

# chapter six

# Toxicological safety and risk analysis

## 6.1 Toxicological safety

*Safety generally means a complete absence of danger*. There is not a single chemical compound, either natural or anthropogenic, in the world, for which it has been conclusively proved that it is absolutely harmless to living organisms. The more negative results from high-quality experiments, the more probable it is that the substance is safe. But from a statistical point of view, it is always possible that the subsequent experiment will already have a positive result. The term "safety" has been in continuous development alongside the growth of the medicinal and toxicological knowledge of mankind.

*In the past, a substance was considered to be safe* if it did not cause an immediate death or at least an acute serious injury in case of contact with a living organism.

*Nowadays, a substance is considered* to be *relatively harmless* if it does not elicit any adverse effects in definite biological systems either on the level of organs or physiological systems or on the level of macromolecules or single links of the metabolic system.

*A substance is declared as hazardous* (or at least suspicious) even if it does no harm other than alter the activity of one single enzyme. Nowadays the term 'dangerousness' includes also the blastomogenic, especially mutagenic, carcinogenic, embryotoxic, and teratogenic properties of a substance. Speaking about safety requires an indication of the conditions (test animals, experiment method, etc.), in which the safety of a compound has been proved. Much information can be obtained by acute toxicity tests, either by a conventional estimation of $LD_{50}$, which is very easy to perform and the results of which are well observable and can be easily quantitated, or that of its surrogates.

It is often advantageous and/or necessary to compare the doses of a substance like a drug or a food additive that cause desirable and adverse effects, respectively. By practical considerations, the *safety margin* is considered to be large if the ratio of these two doses is big. But if this ratio is small, then the use of the substance, for example, as a food additive can be connected with serious hazards to the health or even to the life of the eater. The concept of *therapeutic index* (TI) has been elaborated for quantitative assessment of the safety margin of drugs. TI is the ratio of *the toxic dose* (TD) of a substance to its *effective dose* (ED) that is necessary for achieving the expected therapeutic effect. $TI = TD_x/ED_y$. In this way, the same type of toxicities of different compounds or different toxic effects of the same compound can be compared with each other.

Many compounds have been labelled as *GRAS—generally regarded as safe.* These compounds have been used already for years (sometimes for centuries) without an emergence of any adverse effect on any organism. *But it is still reasonable to be a little cautious* since a long-known foodstuff occurring in a new human population or food composition may prove to be, if not a very strong poison, then at least problematic. A foodstuff can turn out to be toxic for some risk groups. A food that is quite normal and safe for adults may have a fatal effect in the case of babies. For example, honey has caused sudden death in some infants younger than six months. As it has been revealed, the reason of it lay in the contamination of honey with the spores of bacterium *Clostridium botulinum* causing botulism (see also Section 16.4).

The policy of food safety must be based on a versatile and integrated risk analysis along the whole food chain, whereby its three basic components are *risk assessment, risk management, and risk communication.*

## 6.2 Risk assessment

*Risk* is a mathematical concept meaning a probability of genesis of adverse effects after an exposure to a definite chemical compound under definite conditions. Risk can be defined as a product:

$$\text{Risk} = \text{hazard} \times \text{exposure}$$

*Hazard* is an intrinsic property of a substance to cause adverse effects, arising from its chemical structure. And vice versa, safety is, in this context, a practical certainty that no adverse effects will occur when a substance is used in the manner and quantity anticipated for its use. As the exposure increases, the risk also increases; and vice versa, when the exposure decreases both in time and in quantity, the risk of intoxication with that compound decreases as well.

*Risk assessment is necessary, if*

1. A human exposure to this compound either during preparation or usage is likely.
2. Occurrence of the substance in the environment is potentially hazardous to humans and/or animals.
3. An exposure of sensitive human and ecological populations to significant doses of this compound is very likely.

4. The compound is persistent in the environment and disposed to bioaccumulate.

*In the course of a risk assessment, the hazard, exposure, and risk are estimated.* It is based on the principle that in the case of most, but not all, substances, the adverse response depends on the dose. Consequently, safe doses must also exist and it should be possible to determine the level of exposure that is not connected with a remarkable risk for both humans and ecosystems. Risk assessment is a scientific process, followed by a risk–benefit analysis and risk management. *Risk management* is a process of consideration of different ways of action and choosing the most appropriate legislative action based on the results of risk assessment as well as on social, economical, and political considerations.

*Risk assessment consists of four basic stages:*

1. *Hazard identification.* During this stage, it is found out which toxic effects (if at all) the substance under question can elicit.
2. *Dose–response relationship demonstration.* Evaluation of the causal relationship between the exposure to a hazardous substance and the occurrence of adverse effects either on the individual or population level.
3. *Exposure assessment.* Determination of the level, incidence, and duration of human exposure to the toxic substance.
4. *Risk characterization.* Estimation of the incidence of adverse effects at various conditions of human exposure.

### 6.2.1 *Hazard identification: Principle of the three Rs*

The potency of a substance to cause various toxic effects is studied here. The necessary initial data is derived from

- Human epidemiology
- Animal toxicity tests
- *In vivo, in vitro,* and other studies

A substance may have an internal potency to cause toxic effects of various types and severity. *It is necessary to estimate the most important primary hazard.* For example, a compound can be reversibly hepatotoxic at high doses but may cause skin cancer at much lower doses. Carcinogenicity is certainly the primary hazard. Unfortunately, the picture is not always so clear.

Although human data is most ideal, it is not always available and the dose–response relationship should be established mainly on the basis of other available information. Even scarce epidemiological data can at least give hints about the existence of causal links between human exposure and toxic effect. Supplementary data from animal studies is usually needed. Necessary histopathological, clinical–chemical, and biochemical information is then obtained under controlled conditions. The interspecies variations should, of course, also be taken into account (see Section 5.2).

Nowadays, both designing and execution of animal tests are subjected to the principle of the *three Rs—replacement, reduction, and refinement*. The classical animal tests should be, as much as possible, replaced by *in vitro* tests, their number should be reduced, and their methodics refined.

Often it is possible to use results of the *in vitro* tests, carried out with human and animal cells and tissues, in decision-making on the toxicity of a compound and, hence, on its usability as a drug or a component of a cosmetic preparation. But it is clear that the *in vitro* data cannot entirely replace the results of animal tests and epidemiological data. Toxicokinetics (absorption, distribution, and metabolism), which is characteristic only of an organism, should be also taken into account.

More and more *quantitative structure–activity relationship* (QSAR) *in silico* method is used for predicting the toxicity of new compounds (see Section 5.4).

### 6.2.2 *Demonstration of a dose–response relationship*

*Here the quantification of the hazards, identified at the first stage, is performed* and the dependence of the intensity of the adverse effect on the dose is established. The latter procedure very often demands an extrapolation from high doses in animal tests to much lower doses of a potential human exposure. Such extrapolation depends on the type of the effect declared as the primary one. In the case of genotoxic carcinogenicity, no threshold dose is expected; and for the risk assessment at low doses, a suitable mathematical model can be used. When the primary toxic effect is nongenotoxic, the existence of a threshold dose should be expected.

*The risk assessment of potentially carcinogenic substances is also a two-stage process*. At the first stage, a qualitative assessment of the results of hazard identification is carried out and at the second stage, this risk is quantified for certain or probable human carcinogens (see Table 3.3).

For a quantitative assessment of the dose–response relationship in carcinogenesis, various models are being used:

1. *One-hit model*. This is an ultraconservative approach regarding the formation of cancer as a one-step process, induced by a single event or hit on the molecular level.
2. *Linearized multistage model*, used, for example, by the US EPA. The method allows estimating the cancer risk at low doses of a potential carcinogen by a linear extrapolation of the results in the coordinates, dose–response in percent to the threshold dose zero. This method provides a daily dose of the substance in mg/kg, which is necessary for eliciting cancer over 70 years (medium human lifespan).
3. *Multihit model*. This approach assumes that the transition of a normal cell into a tumor cell needs several events or hits on the molecular level.

4. *Probability unit (probit) model.* It assumes a normal-logarithmic or Gaussian distribution of the resistance against the toxic effect in a population exposed to the substance.

Another model that is increasingly being used for the assessment of tumor risk is the *physiologically based pharmacokinetic model* (PB-PK), which uses data of absorption, distribution, metabolism, accumulation, elimination, and toxicity mechanism for estimation of the target dose usable for extrapolation. This model needs a large number of various initial data (Nestorov, 2003).

The cancer risk figures, obtained by means of the different enumerated models, are rather variable. For example, the concentration of the chlororganic pesticide chlordane in drinking water must be 0.03 ppm (one-hit model), 0.07 ppm (linearized multistage model), or 50 ppm (probit model), respectively, to cause one human being out of a million to die of cancer.

Especially complicated is the evaluation of the results of animal tests, since here the increase in the tumor frequency as a response to the toxic action of a compound must be demonstrated in a small population, in which already, normally, some types of tumor can occur. A *practical statistical limit* exists that determines the minimal detectable incidence of tumor formation. For example, at least 6 animals out of 1000 must develop cancer during the experiment to prove the carcinogenicity of a compound with 99% reliability. Use of yet larger numbers of animals is impractical, expensive, and ethically very problematic. One practical solution in the case of infrequent processes could be an increase of the experimental doses of the toxicant, provided that the dose–response relationship is linear and hence, the respective back-extrapolation is justified. Various models have been elaborated, but the estimates obtained by them are variable. The accuracy of a mathematical model is of no importance if the initial toxicological data is poor, highly erroneous, or uncertain.

In the carcinogenicity tests, doses close to the *maximum tolerable dose* (MTD) are generally used. *Such an approach is rather disputable,* since metabolic routes and toxicokinetics may depend on the dose. It can be a real situation that a compound metabolizes at high doses, both qualitatively and quantitatively differently, from what happens at the realistic lower doses and exposure. The other pathologies elicited by the high doses of a xenobiotic may also influence the carcinogenicity of this compound.

For example, *hydrazine* is a weak carcinogen at its high doses. Hydrazine causes methylation of DNA, that is, probably reveals mutagenicity, which can lead to the formation of cancer. But this methylation takes place only at high hepatotoxic concentrations of hydrazine. In this case, the acute toxic effect of hydrazine, which is somehow related to the methylation of DNA of the substance, is probably necessary for cancer formation. Consequently, extrapolation of the carcinogenicity of hydrazine from high to low doses is not reasonable (Timbrell, 2002, p. 176).

In the case of nongenotoxic and noncarcinogenic substances, when it is reasonable to speak about a threshold dose of the toxic effect, it is also possible to estimate the highest dose—exposure to which does not cause any toxic response as yet. Such a dose or level is called *no observable adverse effect level*—NOAEL. But when the data is too inaccurate, the *lowest observable adverse effect level*—LOAEL—can be estimated. The use of the latter value is still less recommended.

### 6.2.3 *Assessment of exposure*

Exposure of an organism to a substance converts the latter from a hazard to a real source of risk. Therefore, determining the probability of the exposure is a very important step in the risk assessment process. It includes an assessment of the source, ways, and level of the exposure. Sometimes, as in the vicinity of chemical factories or landfills, we may have to do with a simultaneous exposure of the organisms to several or even to numerous potentially toxic substances. Different substances usually have different effect(s) on the organism, and their summary effect can be additive, synergistic, or antagonistic (see Section 1.4). Various ways of exposure—inhalation, contact through the skin, ingestion, and so forth can be realized with various types of organisms (humans and animals, children and adults, etc.). These and several other factors must be taken into account in a rather complicated occupational or environmental risk assessment. Often when the actual exposure level is not known, the auxiliary models involving parameters such as movement of air or ground waters must be used. Essential information for employment of these models is provided by the physico-chemical parameters of the potentially toxic substances like vapor pressure, solubilities in water and lipids (e.g., $K_{ow}$ values), and so forth. Important information concerning the exposure is obtained from the use of biomarkers (see Section 3.4).

### 6.2.4 *Risk characterization*

*In the last step of risk assessment, results of the preceding steps are integrated* to learn the probability of occurrence of the adverse effects of a chemical compound on humans, no matter whether these effects have threshold doses or not, whereby the biological, statistical, and other uncertainties have been taken into account. For carcinogens, this probability is calculated from the slope of the straight line considering a 70-year-long daily administration in units of mg/kg per body weight (bw) and is expressed as an increase of the risk of cancer formation (e.g., one incidence of cancer to $10^6$ exposures).

On the basis of NOAEL and LOAEL values obtained by the animal testing, the numerical values of several parameters, important for food, can be calculated. For food additives and veterinary drug residues, such parameters are *the maximum allowed daily intake* (ADI—see Chapter 17) and *reference dose* (RfD) used by US EPA. In the case of food contaminants, the analogical parameter is the *tolerable daily intake* (TDI) (see Chapter 17). To guarantee a maximum reasonable security of the calculated ADI or TDI values, *the following safety or uncertainty factors are used* in the calculations:

10 times—considering interindividual (generally intraspecies) variability of humans in response to a toxic effect
10 times—considering interspecies variability (for transition from animals to humans)
10 times—when lower than chronic doses are being used in the animal studies
10 times—when LOAEL instead of NOAEL is used
0.1 to 10 times—modification factor used by US EPA for calculation of RfD values

For calculation of ADI or TDI values, NOAEL (or LOAEL) is divided with the combination of these factors:

$$\text{ADI (TDI)} = \text{NOAEL/safety factor(s)}$$

The ADI or TDI is expressed in milligrams of the substance per kilogram body weight.

*Most often, the combined safety factor 10 × 10 = 100 is used in risk assessment.* This factor is supposed to take into account both interhuman and interspecies variabilities. This approach can be used in the case of both chronic and shorter studies. It is correct to perform studies lasting for an appropriate period. Doses are usually expressed either per body weight or area unit of the test animal and thereafter, extrapolated to various species. Such extrapolation presumes similar sensitivities of the respective species either per body weight or area unit. In the case of noncarcinogenic substances, during risk assessment, the actual exposure is compared either with the ADI or with another equivalent parameter such as simultaneous exposure to several substances that are considered to have an additive effect.

Interspecies variability in the toxic effect is a serious problem for both risk assessment and interpretation of the toxicological data. One should always carefully consider the type of animal to be used for the extrapolation, especially with regard to its sensitivity, or whether by character of the toxic response or by distribution of the substance in the body it is most closely related to humans. It may happen that some species or strains used for testing have an intrinsic elevated incidence of tumor formation. In this case, it is difficult to estimate the significance of the increase of the tumor incidence caused by the putative carcinogen. It can be concluded that the application of the results of the carcinogenicity tests in the risk assessment process is much more complicated than the application of other toxicity data. In the case of an acute toxicity, the picture is usually much clearer and single-valued, and a relatively accurate estimation of NOAEL is much easier. Certainly, the biochemical mechanisms of the toxic effect must be known and the statistics must not be trusted too much.

*In the case of food*, it is sometimes necessary to carry out the risk assessment of a substance or substance group *separately for individuals of a medium (normal) and high exposure*. The high degree of exposure can be connected with a certain stage of life or qualitative or quantitative dietary preferences of a person. A very important cause of interindividual variabilities in the toxic response is the differences in the toxicokinetics of the substances. These differences can be determined not only genetically, but also by dietary and pathophysiological conditions (see Section 5.2.2). The hazard identification as well as characterization procedures usually consider isolated compounds. But food is an extraordinarily complicated mixture of compounds that simultaneously influence the organism and also each other. During the risk characterization, special attention should be paid to substances with a similar action, the toxic effects of which can be summarized. Humans can be exposed to suspicious food constituents either chronically (usually to low doses), shortly (often to high doses) or chronically to low doses that shortly can increase. Hence, sometimes it is necessary, in addition to ADI values that are obtained in the course of long-term studies, to elaborate another guidance value—the *acute reference dose* (ARfD). ARfD values are obtained from short-term studies of acute exposure. In the case of a combination of long-term low-dose and short-term high-dose exposures, the results of acute and chronic studies should be averaged.

*The xenobiotic doses exceeding the ADI value constitute a special issue*. Since ADIs are obtained by chronic exposure tests and counted by means of safety factors, which are naturally rather approximate and probably overestimated, then such an excess need not always be connected with unavoidable and fast toxic responses of the organism. In these cases, very accurate studies of the mechanism of the particular toxic effect, the degree and duration of the in excess of the ADI values, and the half-life of the substance are necessary. Naturally, every such short-term excess reduces the *safety margin of the organism* against the toxic action of this as well as similar compounds.

*In certain cases, risk assessment is necessary for compounds spreading by air and/or water*. Respective instructions have been set up by the World Health Organization (WHO). Contaminants spreading by air can elicit both irritation and chronic effects. The respective guidance values, calculated on the basis of either NOAEL or LOAEL (in a poor case), provide exposure levels in combination with exposure periods, during which the adverse effects should not appear as yet. Analogical values are also available for the substance concentrations in water. Making use of NOAELs and the respective safety or uncertainty factors, the TDI values can be calculated also for the substances contaminating drinking water. Guidance values are determined on the basis of TDIs and the known daily consumption for an adult weighing 70 kg for a 70-year period. In the case of both airborne and waterborne carcinogenic contaminants, the absence of threshold doses is taken into account (Renwick et al., 2003).

It is essential to use, at different stages of risk assessment, biomarkers of the toxic effect indicating an existence of the internal dose, for example, for a correct description of the dose–response relationship. *Biomarkers of response* are also necessary for estimation of NOAEL. They help elucidate the important link between exposure and response.

A profound overview of all risk assessment procedures of foods and foodstuffs is given in the respective review papers (Barlow et al., 2002; Dybing et al., 2002; Renwick et al., 2003).

It should be recognized that despite the versatile and voluminous toxicological research carried out in recent years, the actual knowledge of the mechanisms of toxic effects is still rather incomplete, touching mainly a small number of both natural and anthropogenic chemicals. This is the reason why risk assessment is still a complicated and often rather incorrect process. Toxicologists are not able to answer all questions. It is never possible to speak about complete safety. Nowadays, when human exposure to various chemicals and their combinations is increasing, it is of growing importance to apply the *risk–benefit principle* in risk assessment. For example, it is not possible to feed mankind and keep tropical diseases under control without the use of pesticides, to avoid botulism without an addition of nitrites to food, and so forth. But one must always benefit from the use of a potentially toxic substance at the scientifically substantiated minimal level to overweigh the risks connected with the use of this substance.

## References

Barlow, S.M., et al. (2002). Hazard identification by methods of animal-based toxicology, *Food Chem. Toxicol.*, **40**, 154–191.

Dybing, E., et al. (2002). Hazard characterization of chemicals in food and diet: dose response, mechanisms and extrapolation issues, *Food Chem. Toxicol.*, **40**, 237–282.

Renwick, A.G., et al. (2003). Risk characterization of chemicals in food and diet, *Food Chem. Toxicol.*, **41**, 1211–1271.

Timbrell, J. (2002). *Introduction to Toxicology*, Taylor & Francis, London, U.K. and New York, U.S.A.

# chapter seven

# Internet sources of toxicological information

Nowadays, there are a number of sources of toxicological information. It is especially convenient to use the Internet. The most famous Internet-based databases for toxicological data are:

*ChemFinder* www.chemfinder.cambridgesoft.com

*Toxline* www.nlm.nih.gov/pubs/factsheets/toxlinfs.html

*TOXNET*®, developed by the National Library of Medicine of U.S.A (NLM), is a web-based system of databases, providing information on toxicology, hazardous chemicals, and the environment. Among TOXNET's preeminent databases are the Hazardous Substances Databank and TOXLINE file of bibliographic references (Wexler, 2001).

*Toxline* is a database of the NLM, which contains bibliographic information about biochemical, pharmacological, physiological, and toxicological properties of drugs and other chemicals—altogether, there are over 3 million bibliographic references, all equipped with summaries, indexes, and Chemical Abstracts Service (CAS) registration numbers.

Toxline references are grouped into two parts—Toxline Core and Toxline Special. Toxline Core covers the majority of the toxicological periodicals. Toxline is a separately treatable part of a large biomedical database called Medline® that can be reached through the PubMed® system by selecting "toxicology" as the subset limit. Toxline Core is also reachable through TOXNET.

Toxline Special amplifies Toxline Core with references from specialized journals and other sources.

*CANCERLIT*® is a bibliographic database concerning tumors, which contains about 2 million references and summaries from more than 4000 sources, including biomedical journals, proceedings, books, reports, and dissertations.

## *7.1 Pesticide residues*

U.S. EPA has created a comprehensive Internet database (www.epa.gov/pesticides), where one can get answers to various questions connected with pesticides. At an international level, it is possible to get data concerning the allowable content of pesticide residues in food at the web page of FAO/WHO (See www.codexalimentarius.net/mrls/pestdes/jsp/pest_q-e.jsp).

A very good university-maintained pesticide database is Extension Toxicology Network or EXTOXNET (http://www.ace.orst.edu/info/extoxnet), maintained jointly by five universities of the United States. See also its pesticide profile at: http://www.ace.orst.edu/info/extoxnet/faqs/pesticide/pesthome.htm.

### 7.2 *Food additives*

U.S. FDA Center for Food Safety and Applied Nutrition (U.S. FDA CFSAN) keeps a versatile database at http://www.cfsan.fda.gov/~lrd/foodadd.html. At an international level, the Joint FAO/WHO Expert Committee on Food Additives (JECFA) (http://www.who.int/pcs/jecfa/what_is_jecfa.htm) is the expert committee assessing the safety of food additives, veterinary drug residues, natural toxicants, and pollutants in food.

### 7.3 *Food allergens*

NLM also maintains a database of food allergens (http://www.nlm.nih.gov/medlineplus/foodallergy.html), which contains news, reviews (synopses), descriptions of avoiding/monitoring programs, research data, and statistical materials in this field. National Institute of Allergy and Infectious Diseases (NIAID) maintains its database at http://www.niaid.nih.gov, which contains data about ongoing scientific research on allergies. On the Internet, one can also find nongovernmental food allergen databases. For example: http://www.foodallergy.org/; http://allergysa.org; or http://www.farrp.unl.edu/.

## *Reference*

Wexler, P. (2001). TOXNET: An evolving Web resource for toxicology and environmental health information, *Toxicology*, **157**, 3–10.

*section two*

# *Main groups of food-borne toxicants*

*There is a widespread opinion* that the safety of traditional foods has been proven by long-term nonproblematic consumption. It is believed that a food is safe if no specific serious hazard has been identified by its consumption. But it has been revealed that *many foods used for centuries still contain toxicants and antinutritional substances*. Development of chemical analysis technique as well as an increase of knowledge of long-term and chronic effects enable one to be aware of the health hazards that were earlier either unknown or unrecognized. Even very small quantities of toxic substances occurring in prepared food can, due to continuous long-term accumulation in the adipose tissue, liver, kidneys, and other tissues, become dangerous for humans.

*Life standards, expectations, and conceptions of life and health* have substantially changed during the last decades. Diets have changed and there has been an expansion of vegetarianism and *consumption of exotic, minimally prepared and fast foods*. Against this background, new conceptions of food safety, concerning primarily green foods, are forming. *Acute intoxications* with food are nowadays actually relatively *rare*; much more frequently occurring and more significant are long-term and often hard-to-diagnose effects of plant and animal toxins that by the time of discovery have already developed extensively.

*Over 99% of all toxicants endangering humans are natural compounds*. Most of them are produced by plants and fungi for "chemical" defense against herbivorous animals and other plants or fungi (Rasmussen et al., 2003). *The idea that everything natural is automatically safe is very often wrong*.

Toxicants reach ready food primarily through contaminated plants and animals during their growth, and they enter food raw materials during storage and transportation. Chemical residues and contaminants are either a result of anthropogenic activities or originate from nature, that is, where the plant is cultivated or grows. According to their origin, food toxicants can be categorized into the following large groups:

1. Plant endogenous toxic metabolites that are synthesized for defense against animals, other plants, or microorganisms
2. Geochemical contaminants that a plant acquires from soil (fluorine, selenium, etc.)
3. Anthropogenic environmental pollutants that have formed from burning of fossil fuels, radionuclides, and components of industrial emissions. These include, for example, toxic elements such as mercury or lead, radionuclides, and so forth
4. Toxic metabolites of microorganisms (mycotoxins, enterotoxins)
5. Endogenous toxic compounds of animal origin (prions, fatty acids)
6. Phytotoxins, accumulated in fish and crustaceans (tetrodotoxin, saxitoxin)
7. Residues of compounds used for defense against agricultural pests
8. Residues of animal breeding chemicals (veterinary drugs and feed additives)
9. Toxic substances unintentionally added to food during preparation, storage, and transportation. These include components of plastic packaging and cleaning materials, disinfectants, toxic amines, and so forth
10. Toxic substances formed during the digestion of food
11. Food additives, including functional ingredients and vitamins.

Following is a short systematic overview of more or less toxic substances belonging to these groups.

# *chapter eight*

# *Endogenous plant toxicants*

Plants provide humans with over 70% of the necessary protein and a number of other useful compounds. Although over millenniums the acutely and severely toxic plants or their toxic parts have been excluded from the human diet, mostly by the method of trial and error, in some cases, we still have to deal with plant compounds that have an adverse effect on long-term consumption with the so-called antinutritional compounds, or with playing a risky game. A half-ripe or a wrongly stored part of an otherwise edible plant might turn out to be poisonous. Toxic components can also reach our dining table, for example, via the milk of the animals that have been fed toxic plants.

The plant toxicants involved in our further discussion is not by far exhaustive. Here you will find only the most important plant-borne toxicants from various parts of the world.

## *8.1 Lectins or hemagglutinins*

*Lectins are nonenzymatic termolabile proteins, glycoproteins, or lipoproteins that specifically bind to the saccharide* (glucose, galactose, etc.) *groups*. They have been found in more than 800 edible plants. Lectins are widely distributed in the leguminous plants (*Leguminosae*) such as various beans, peas, and so forth. In addition to plants, they have been found in animals like sponges, mollusks, fish (in blood), amphibians (in the eggs), and in mammal (including human) tissues. Since many, but not all, lectins are capable of agglutinating erythrocytes and other cells, they are often also called hemagglutinins. In practice, lectins are used for blood group typing, recognition of tumor cells, for investigation of signal transmission, mitogenesis, and cell death.

*Some plant lectins can exert adverse effects when one eats only the raw plant.* Their general antinutritional effect is caused by binding of the lectin molecules to the membranes of the intestinal cells. This binding is followed by a nonspecific inhibition of both active and passive transport through the cell wall of important nutrients like amino acids, fats, vitamins, minerals, and necrosis of the cells of the intestinal epithelium. For example, a long-term consumption of raw legumes may lead to growth retardation and even a goiter.

An acute systemic exposure to the lectins may cause fatal injury to the liver and other organs (Vasconselos and Oliveira, 2004).

Some lectins, like *ricin* from the seeds of the castor oil plant (*Ricinus communis*) and *abrin* from *Abrus precatorius* are really dangerous poisons. A ricin molecule consists of two polypeptide subunits (short A and long B chain) linked together with S–S bridges. In case of ingestion, ricin causes a severe necrosis of intestinal wall cells and death via multiorgan damage. Yet, absorption of ricin from the intestines is rather small and percutaneous absorption is negligible. Inhalation and injections are much more dangerous routes of administration. Ricin produces, for example, hepatotoxicity, nephrotoxicity, and oxidative damage at 24 h of post intraperitoneal (i.p.) treatment of rats. Hepatotoxicity is the most prominent type of ricin toxicity (Kumar et al., 2003). $LD_{50}$ of ricin to humans is 5–10 $\mu$g/kg or 350–700 $\mu$g for an adult weighing 70 kg in case of inhalation of solid particles or intravenous or intramuscular injection. *The toxicity of ricin is extremely high*—one molecule of ricin is capable of killing one cell.

It is supposed that the main general toxicity of ricin is caused by its destructive effect on the cells of the reticuloendothelial defense system (blood monocytes, macrophages of liver and lungs, the endothelial and reticular cells of the spleen, the lymph nodes and red bone marrow, glial cells of brain, the thymus, tonsils, and various types of the lymphoid tissues). It results in liquid and protein loss, bleeding, edema, and weakening of the defense system against endogenous toxins in every single cell as well as in the whole organism.

The mechanism of the toxic action of ricin is as follows:

1. The longer subunit B, with lectin properties, binds to the galactose group of a glucoprotein on the outer side of a mammal cell.
2. Then follows endocytosis, during which the whole ricin molecule is taken into the cell inside a vacuole formed.
3. The ricin molecule is released from the glucoprotein molecule and subunits A and B separate from each other by breaking the S–S bond.
4. The B-chain of ricin makes a channel through the vacuole wall; the A-chain enters the cytoplasm and on reaching the ribosomes, it blocks protein synthesis and kills the cell.

It has been shown that the thyroid gland damage of a rat, caused by ricin, can be associated with lipid peroxidation mediated by the reactive oxygen species (ROS) (see also Section 3.3.6) (Wallace Hayes, 2001, p. 495; Kumar et al., 2003).

A ricin-containing ampoule attached to the top of an umbrella was used to kill journalist Georgi Markov by the Bulgarian secret police in London, in 1978.

## 8.2 *Enzyme inhibitors*

Although many foods of plant and animal origin contain inhibitors of the proteases, amylases, and lipases, it is reasonable to touch on only the *proteases* here. These extremely proteolysis-resistant proteins are found in all plants, but in large quantities especially in the soybean (*Glycine max*—the Kunitz inhibitor) and in other beans, peas, beet, cereals (corn, wheat, rye, rice), clover, and potato. They are mostly trypsin inhibitors, but the inhibitors of chymotrypsin and carboxypeptidase B have also been found. It is not likely for anybody to eat such a large amount of one raw source of an inhibitor to cause a serious acute poisoning. But it has been shown that eating even small amounts of various protease inhibitors can increase the risk of growth retardation, pancreatic hypertrophy, and cancer. At binding of the inhibitor molecule to the enzyme molecule, the pancreas receives a signal for synthesis of new enzyme molecules, which may finally lead to the hypertrophy of the pancreatic tissue (Gumbmann et al., 1986).

Thermal processing usually denaturates proteinous inhibitor molecules. For example, in the case of potatoes, microwaving and boiling are more effective than cooking. Since many vegetables are eaten without any processing or after a short time of cooking, they can still exert at least a partial inhibition capability of the digestive enzymes.

## 8.3 Alkaloids

*Alkaloids are nitrogen-containing heterocyclic compounds* that exert a pharmacological effect in humans and animals. In plants, the alkaloids have a defense function against herbivorous animals, parasites, and insects. Alkaloids are also found in fungi and in animals. The name alkaloid originates from the Arabic word "al-qālī," meaning both alkali and the alkaline properties. Alkaloids, due to their alkaline properties, form water-soluble salts with acids.

A number of plant alkaloids are toxic for animals and some of them have a bitter taste. Alkaloids, like morphine or codeine, are used in medicine as analgetics. Humans may have a contact via food mostly with *pyrrolizidine, xanthine,* and *solanine—alkaloids,* less with *efedrine alkaloids.* The compounds that can be of some interest for food toxicologists are also the piperidines of tobacco (*Nicotiana tabaccum*) and hemlock (*Conium*), quinolizidines of lupine (*Lupinus*) and milk vetch (*Astragalus*), indolizidines of swainsona (*Swainsona canescens*) and red clover (*Trifolium arvense*), which can reach our table through the milk of the animals that have eaten these and several other plants. In fact, we routinely ingest alkaloids but fortunately, their level in the edible plants is too low to exert any serious acute toxic effects. It is extremely complicated to determine the toxic effect of the low doses of the alkaloids.

### 8.3.1 Pyrrolizidine alkaloids

*Pyrrolizidine alkaloids* (PA) is a group of substances, the molecules of which contain an N-atom in the *pyrrolizidine-cycle* (Figure 8.1). PAs have been found in more than 250 plant species and over a half of them are toxic. For humans, the most important toxic PA containing plants are ragwort (*Senecio*), crotalatia (*Crotalaria*), and heliotrope (*Heliotropium*). Humans can have a direct contact with plants like coltsfoot (*Tussilago farfara*) or compfrey (*Symphytum*

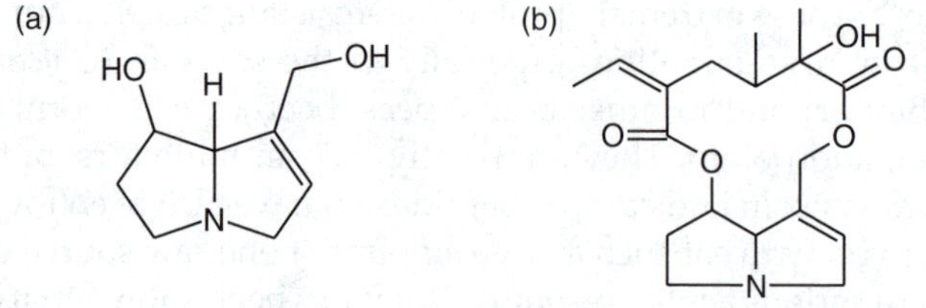

Figure 8.1 Structure of pyrrolizidine alkaloids: (a) retronecine and (b) senecionine.

*officinale*) via the vegetable salads and teas prepared from these plants, with viola (*Viola x wittrockiana*) or bugloss (*Echium*) via honey or milk or bird's eggs. In the honey of plants of the genus *Senecio,* the total concentration of PAs has been measured between 0.3 and 3.2 μg/kg, and extremely high PAs concentrations (30–70 μg/kg) have been registered in the honey collected in the foothills of the Swiss Alps in Europe. *Epidemic poisoning of humans with cereals* contaminated with seeds of the weeds *Heliotropium lasiocarpium* and *Trichodesma incanum* took place in Uzbekistan. After this incident, the official upper limits to the content of the seeds of these plants in grain (0.2% and 0% respectively) were established in Uzbekistan. Poisoning of humans with PAs and even deaths have been recorded also in Afghanistan (7200 deaths in 1976), Tajikistan (3900 deaths in 1993), India, South Africa, Jamaica, Ecuador, China, Great Britain, the United States, and in other countries.

The interval of the human toxic daily doses of PAs is 0.1–10 mg/kg body weight (bw). According to the World Health Organization (WHO), 0.015 mg/kg bw is the lowest total daily dose of PAs (on the basis of compfrey alkaloids) that has produced adverse effects. In the case of an adult individual weighing 70 kg, it corresponds to about 1 mg of PAs per day. Since the content of PAs in the leaves and roots of compfrey varies in the interval 450–8300 mg/kg, the actual degree of dose that compfrey tea drinkers get also varies quite as much.

Among the PAs, *retronecine, senecionine,* and *petasitenine* are the most toxic substances (Figure 8.1). Their toxicity is caused by highly reactive pyrrole-metabolites that are formed in the liver under catalysis of the oxidases, especially the cytochrome P450 monooxygenase complex (CYP) as well as by the products of the hydrolysis of these pyrrole derivatives. The targets for the primary attack of these reactive compounds are not only the parenchymatic liver cells or hepatocytes, but also the endothelial cells of hepatic and pulmonary blood vessels. A subsequent hepatic vein occlusion may lead to the formation of ascites, edema, to reduced urinary volume and, especially in the case of children, to death. Survivors acquire hepatic cirrhosis. Some dehydropyrrolizidine alkaloids like monocrotaline cause an analogical occlusion of pulmonary arterioles, which can result in a pulmonary overpressure, right ventricular hypertrophy, and cor pulmonale.

Many PAs and their aforementioned pyrrole-metabolites are capable of cross-linking macromolecules, including DNA. Hence, they can be mutagenic, carcinogenic, and teratogenic to the laboratory animals. Such plants as ragworts, pestilenceweed (*Petasites japonicus*), compfrey, Japanese silver leaf (*Farfugium japonicum*), and eight different PAs have a part in the formation of the following cancers: hepatic carcinoma, hepatic hepangioendothelic sarcoma, hepatocyte adenoma, cholangiosarcoma, astrocytoma, pulmonary adenoma, enteral adenocarcinoma, adenomyoma of ileum, rhabdomyosarcoma, and so forth. It has been shown that heliotrope is teratogenic to rats.

The toxicity of PAs significantly depends on the animal species. Rats, cattle, horses, and hens are very susceptible to PAs, while sheep, guinea pigs, hamsters, and Japanese quails are more resistant. This is caused by the species differences in the metabolism of PAs. Recently, a good review of various aspects of PA effects has been published (Rietjens et al., 2005).

The main route of detoxification of PAs in the organism is hydrolysis of the ester-groups, catalyzed by esterases, followed by the excretion of the water-soluble *N*-oxides formed under the catalysis of CYP (especially CYP3A) and flavine-monooxygenase (FMO). For some PAs, the formation of *N*-oxides means metabolic activation. Detoxification of the pyrrole-metabolites by conjugation with glutathione is catalyzed by glutathione-S-transferases (GST) (Rietjens et al., 2005).

### 8.3.2 Solanine-group glycoalkaloids

Steroidal alkaloids *α-solanine* (Figure 8.2), *α-saconine and tomatine,* belonging also to the group of terpenoids, are found in such plants of the genus *Solanum* as potato (*S. tuberosum*), eggplant (*S. melongena*), and tomato (*Lycopersicon esculentum*, also *S. lycopersicon*), as well as in the berries of nightshade (*S. dulcamara, S. marinum, S. nigrum*). *The most problematic has been, of course,*

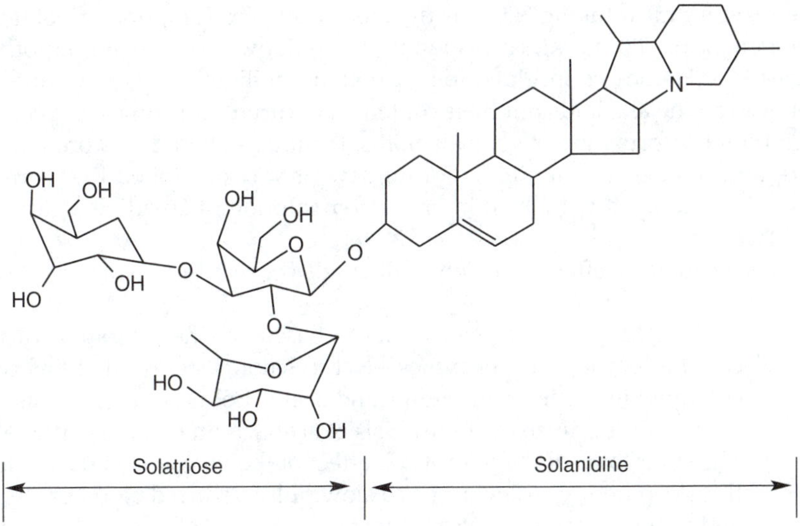

Solatriose
Solanidine

**Figure 8.2** Structure of α-solanine. Aglycone solanidine (the right part of the molecule) is linked to the trisaccharide of solatriose type.

*potato,* especially the tubers that are sprouting, green, infected by mildew, injured, or tainted. In the tubers, fruits, and green parts of potato, the content of glycoalkaloids (95% of which are solanine and chaconine—glucosides of the aglycone solanidine) can be up to 0.01%, but in the young potato sprouts it can even be up to 0.5%. The usual content of the glycoalkaloids in the tubers is 20–100 mg/kg, whereby the actual content depends on the variety, growing place, climate, light conditions, fertilization, ripeness, number of mechanical injuries, and storage conditions. During sprouting as well as when exposed to the sun, the content of glycoalkaloids can rise up to 5000 mg/kg. The content of glycoalkaloids in the winter cultivars is relatively higher. *Most of the alkaloids are located in the surface layer and in the peel;* that is why the small tubers are relatively more toxic. In potato chips, summary contents of glycoalkaloids in the interval 20–600 mg/kg have been reported (Rietjens et al., 2005). The content of the glycoalkaloids in potato tubers is usually estimated by high performance liquid chromatography (HPLC).

The total daily toxic dose of solanine and chaconine (LOEL) for humans is 2–5 mg/kg, the lethal dose being 3–6 mg/kg. These two doses are extremely close to each other, partly even overlapping. $\alpha$-saconine is more toxic than $\alpha$-solanine, both alkaloids are expected to be teratogenic. *The intoxication symptoms* (headache, vomiting, diarrhea, neurological symptoms, weakness, and even death), part of which can be related to the irritative and anticholinesteratic effect of the alkaloids appear at the total concentrations of the glycoalkaloids exceeding 200 mg/kg of tubers. This number also represents the highest legislatively allowed content of alkaloids in potatoes. This number is planned to be reduced to 100 mg/kg together with the breeding of new potato cultivars containing less glycoalkaloids. By the action of light, mechanical injuries, and fungal diseases, this number can rise several times. Taking 300 g as the mean daily consumption of potatoes, the content of 200 mg/kg corresponds to the daily dose of 1 mg/kg of the ingested solanine-group glucosides. Cooking, boiling, or microwaving do not destroy the alkaloids but their content is reduced three times by peeling. To inhibit the formation of glycoalkaloids, the tubers should be stored in the dark and treated with inhibitors of mildew or wax or dipped into corn or olive oil or sprayed with lecithin or a water solution of an edible detergent like Tween 85.

The acute toxic effect of the glycoalkaloids consists of two parts:

1. Their *effect on the nervous system* is based on the inhibition of the cholinesterase-group enzymes—butyrylcholinesterase that has concentrated in the liver and lungs and acetylcholinesterase, the task of which is to catalyze the hydrolysis of acetylcholine in the nerve cells. The symptoms of the neurotoxic effect of the glycoalkaloids are apathy, sleepiness, gasping, fast and slow pulse, reduced blood pressure, and in severe cases, coma and death.
2. Solanine and chaconine *destroy the membranes of erythrocytes and other cells,* causing development of dyspepsia (abdominal pain, nausea,

vomiting, and diarrhea) and hemorrhage. Solanine and chaconine are also suspected in teratogenicity.

A review of potato glycoalkaloids has been recently published (Friedmann, 2006).

## *8.3 3 Xanthine alkaloids*

*The xanthine-subgroup of the group of purins* includes closely related alkaloids such as *caffeine, theobromine,* and *theophyllin,* the main alkaloid components, respectively, in the seeds of the coffee tree (*Coffea arabica* or *C. canephora*), fruits (coffee beans), in the seeds of the cacao tree (*Theobroma cacao*) fruits (cacao beans), and in the leaves of the tea plant (*Camellia sinensis*). Caffeine has actually been found in more than 60 different plants.

Over the years, *controversial opinions have been expressed concerning the toxicity of coffee,* the well-known beverage, hot water extract of coffee beans. The main substance that has been studied in this respect is caffeine (Figure 8.3), whose toxic effects have been sought and sometimes even found. In recent years, several other bioactive components have been discovered in coffee. *Caffeine has been found also* in the leaves of the tea plant and in the seeds or nuts of the cocoa tree (*Cola nitida*). Caffeine is likely, at least in North America and Europe, to be the most consumed pharmacologically active substance. Caffeine has a well-known stimulating effect on the central nervous system (CNS), heart, and respiratory system, and it is also an efficient diuretic. Caffeine is used in the composition of analgetics, as an asthma drug, and in the case of apnea of early newborns. It is equally bioavailable from coffee and tea, but it is absorbed more slowly from Coca-Cola. Caffeine is absorbed fast from the stomach; the top-concentration of 100 mg caffeine administered with coffee (1.5–1.8 $\mu$g/mL) arrives in the plasma 50–75 min after drinking. The half-life of caffeine in the plasma of adult individuals is normally 3–4 h and 9 h, if overdosed. Normal coffee-drinkers have a caffeine level of 0.2–2 $\mu$g/mL in plasma. Pharmacological as well as toxicological effects start to appear when the concentration of caffeine reaches the level 10–30 $\mu$g/mL. Caffeine is distributed quickly into tissues and crosses placental and brain–blood barriers easily. Caffeine is almost entirely metabolized in the liver; the end products of the multistep bioconversion are eliminated with 98% yield with urine.

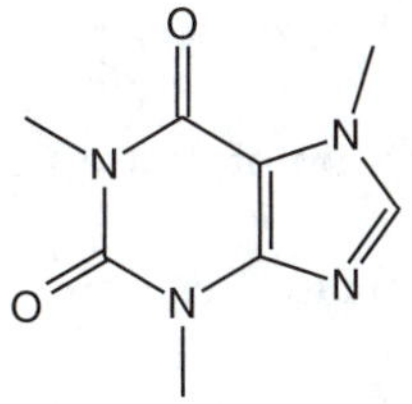

***Figure 8.3*** Structure of caffeine.

Caffeine and other methylxanthines inhibit enzyme fosfodiesterase, resulting in intracellular accumulation of cyclic adenosine monophosphate-(AMP), blocking of adenosine receptors and increase of release of $Ca^{2+}$ ions from the terminal cisterna of the sarcoplasmatic reticulum.

*Acute intoxication* with caffeine first causes headache, nausea, gastrospasm, sleepiness, excitement, and palpitation of the heart, which start at the administration of 0.5–1 g of caffeine to an adult human. Still higher doses give rise to tremor, delirium, a rise of blood pressure, hyperthermia, coronary arrhythmia, heart attack, coma, and death. The lethal dose is about 5 g for a child and 10 g for an adult. This amount is present in about 75 cups of coffee, 125 cups of tea, or in 200 cola beverages. At the same time, a recovery has been recorded even after ingestion of 30 g of caffeine. Since sensitivities to caffeine are very different, the lowest acute dose at which the adverse effects start is in the interval 5–15 mg/kg.

The most important symptoms of the *chronic toxicity* of caffeine (caffeinism or coffee syndrome) are din in the ear, arrhythmia, nausea, abasia, and breast pain. The threshold dose of chronic effect is 500–600 mg/day for an adult human weighing 70 kg. The recommended daily consumption of caffeine is below 300 mg.

*Caffeine has also been known to cause carcinogenicity and osteoporosis.* Although both, caffeine and theobromine are mutagenic to the bacteria and are capable of potentiating genotoxicity of other compounds, despite long-term studies no proof of human mutagenicity, carcinogenicity, or teratogenicity either for coffee (consumed 2–5 cups/day) or the corresponding quantity of caffeine has been obtained. Even vice versa, *the latest epidemiological studies have shown a cancer risk lowering effect of coffee.* Analogical results have been obtained by chronic animal studies with rodents (hamster, rat, and mouse) when coffee, administered to the animals in big quantities, reduced the formation of spontaneous tumors in the organs such as liver, kidneys, lungs, and enteric wall. It also has been shown that both coffee and some of its constituents reduce the carcinogenic effect of such well-known carcinogens as nitrosoamines and 1,2-dimethylhydrazine. Both green and roasted coffee inhibit the carcinogenesis induced by 7,12-dimethylbens[$\alpha$]anthracene (Cavin et al., 2002). The particular substance causing this effect and the inhibition mechanism are not completely clear yet, but most likely, this is a complex of two coffee-specific lipid diterpenes—caffestol and caffeol (C + K), anticarcinogenic effect of which can be based on mechanisms like:

- Induction of phase II enzymes, participating in detoxification of carcinogens
- Suppression of expression of phase I enzymes, participating in activation of carcinogens
- Specific inhibition of phase I enzymes, participating in activation of carcinogens
- Stimulation of the intracellular antioxidant defense mechanism

*No proof exists for a direct link between coffee consumption and the intensity of osteoporosis* (Heaney, 2002). Studies performed two decades ago showed that one cup of coffee is able to remove up to 5 mg of calcium from the organism, which is indispensable for normal physiology of bones. Further studies revealed that consumption of beverages containing caffeine could really in some, but by far not in all cases, reduce the mass of bones and increase the risk of bone fractures. Both physiological studies and human studies of the calcium balance showed very weak but still obvious effect of caffeine on the absorptivity of calcium and no effect on 24 h excretion of calcium with urine. The negative effect of caffeine can be explained simply by the inverse proportionality of administration of calcium and drinking of caffeine-containing beverages. Since administration of a small amount of calcium actually causes an increase of the risk of bone fractures then abundant consumption of coffee is often simply a marker of the small administration of calcium. The negative effect of caffeine on the absorption of calcium is so small that it can be neutralized by drinking 1–2 tablespoonfuls of milk. All studies whose results pointed to the consumption of caffeine-containing beverages as a risk factor of osteoporosis were performed with human populations consuming calcium essentially (twice) below the norm. So far, no firm data exist about the harmful effect of caffeine on the shape of bones as well as on the calcium balance in the organisms of humans, continuously consuming a daily norm, about 1300 mg of calcium.

### *8.3.4 Ephedrine alkaloids*

*Ephedrine alkaloids that belong to phenylethylamines,* are found in the species of genus *Ephedra*—*E. sinica* (Chinese ephedra), *E. intermedia,* and *E. equisatine,* growing in the steppe (prairie) and desert areas. The most important ephedrine alkaloid is *ephedrine* (Figure 8.4). The other notable ones are pseudoephedrine, norephedrine, methylephedrine, methypseudoephedrine, and norpseudoephedrine. *Ephedrine is used also as a constituent of food additives;* especially popular are those *boosting weight loss and athletic performance.* The content of ephedrine and related alkaloids in the powders of ephedra and in standardized extracts of Chinese origin called *Ma Huang* is 6–8% (Rietjens et al., 2005).

Structurally, the ephedrine molecule is very similar to the molecules of adrenaline or epinephrine; the mechanism of action of ephedrine alkaloids is

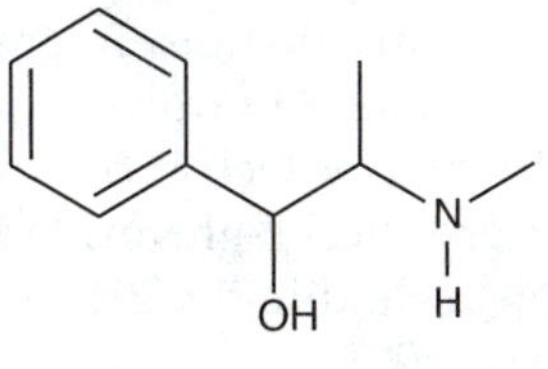

N
H
OH

***Figure 8.4*** Structure of ephedrine.

based on the same similarity. As an adrenaline agonist, ephedrine causes a sympathomimetic effect that triggers heart acceleration, a rise in blood pressure, and stimulation of the CNS. Doses exceeding 50 mg, when administered to adult humans can cause heart palpitation, nausea, dizziness, headache, sweating, neuropathy, and tremor. The effect of stimulation of the CNS can manifest in loss of appetite, sleepiness, nervousness, and euphoria. Ephedrine doses over 500–1000 mg evoke nausea, vomiting, fever, psychoses, spasms, convulsions, respiratory distress, coma, heart infarction, and death. A single dose of 2000 mg is lethal for any adult person. Chronic exposure to ephedrine may cause behavioral disorders and psychoses (Andraws et al., 2005). The ephedrine alkaloids may have an intense synergism with various drugs (see Section 17.2.2).

Since consumption of ephedrine alkaloids is connected with an unacceptably high and scientifically substantiated health risk, most countries have recently banned their use in food additives as well as foods.

Very similar to the alkaloids of ephedras is *synephrine* that is found in the fruits of various citrus plants like bitter orange (*Citrus aurantium*) and mandarin (*C. reticulata*). Adrenaline agonist synephrine is a constituent of all foods produced from the citrus plants; its average content in these products is 0.25%, but it may even be up to 2%. In Europe, synephrine is used as a sympathomimetic drug (Osedrin, Sympathol) for increase of heart capability. After prohibition of products of ephedra by the U.S. Food and Drug Administration (FDA), the companies, dealing with dietary drugs, turned to synergine in their search for less dangerous "ephedra-free" alternatives. Whether this alkaloid really has fewer adverse side effects, such as an increase in blood pressure or tachycardia, needs still to be established. So far, the consumers who are suffering from elevated blood pressure or other disorders related to heart function should be careful with the synephrine-containing products as well (Rietjens et al., 2005).

## 8.4 Cyanogenic glucosides: Toxicity mechanism of HCN

These glucosides of low toxicity, containing cyanic (–C ≡ N) group, and *releasing highly toxic hydrocyanic acid* (HCN) at their hydrolysis are not only found in many plants (over 300 species), but also in mushrooms, bacteria, and even in animals. The list of such dietary plants and foodstuffs include cassava, sweet potato, yam, corn, millet, sorgo, sugar cane, peas, beans, almonds, lemons, apples, pears, cherries, apricots, and so forth. From the 20 cyanogenic glucosides identified so far, the most interesting are four—*amygdalin, dhurrin, linamarin*, and *lotaustralin*. Hydrolysis of cyanogenic glucosides is promoted by physical destruction as well as by stress (drying, freezing, cooking) and is catalyzed by the enzymes $\beta$-glucosidase and hydroxynitrile lyase, which are found both in plants and in the digestive tract of humans as well as of animals (Wallace Hayes, 2001).

Cyanogenic glucosides are synthesized in plants mainly on the basis of four amino acids (phenylalanine, tyrosine, valine, or isoleucine) and

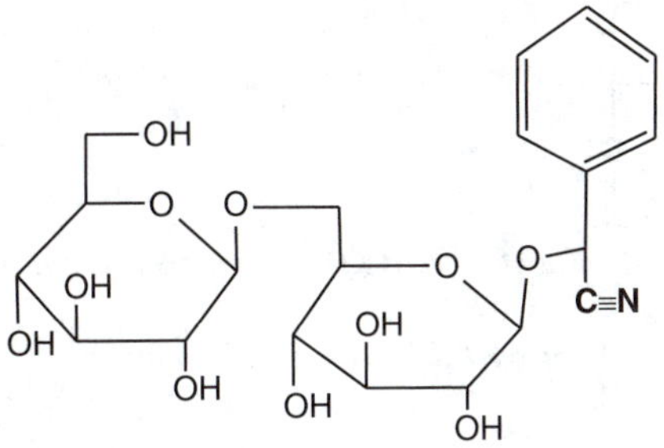

***Figure 8.5*** Structure of amygdalin.

nicotinic acid. The most well-known and widely spread cyanogenic glucoside is *amygdalin* (Figure 8.5), which is present in the bones of the fruits of drupaceous plants (*Drupaceae*). Amygdalin (*amugdalĕ* = almond in the Greek) is an odorless and a bitter substance, readily soluble in water and ethanol. The amygdalin content in various fruits is different—in bitter almonds, it is 2.5–3.5%; in peach seeds, it is 2.0–3.0%; in apricot or plum stones, it is 1.0–1.8%; and in cherry stones, it is 0.8%. Smaller concentrations of amygdalin can be traced in the seeds of apples and pears, in the berries of the bird cherry (*Prunus padus*), and in the leaves of laurel (*Laurus nobilis*). The content of amygdalin in the bark of the bird cherry is higher, that is, up to 2%, because of which it gives a specific strong smell. In the blossoms of the bird cherry *prunasin*, a metabolite of amygdalin that gives the blossoms a specific soporific smell can be found. Amygdalin is persistent in the stones of the drupaceous for years. It is not destroyed even during drying, boiling, cooking, or stewing of the stones.

Serious, even fatal, health injuries have occurred when bitter almonds were consumed. The sweet almond contains little amygdalin and is harmless. The cyanic acid content is also high in the sweets containing marchpane. So in the raw pasta of marchpane, the total content of HCN (the sum of free and bound) is 50 mg/kg. It has been estimated that the average daily dietary consumption of HCN is 46–95 $\mu$g per person.

Eating of 2–3 stones of apricot, plum, or cherry does not cause any intoxication. Humans can obtain a lethal dose from 50 to 60 or about 100 stones of bitter almond and plum or apricot, respectively.

*Question*: Is it necessary to remove the stones from the fruits of plum or cherry before preparation of jam or stewed fruit?

*Answer*: Although the cookbooks recommend it, from the toxicological viewpoint it is not obligatory. The content of cyanic acid in these products is much lower than the harmful content. The minimal lethal dose of HCN is 0.5%–3.5% and 2–10 mg/kg for humans and animals, respectively. The toxic and lethal concentrations of HCN in blood plasma are 0.2 and 1.0 $\mu$g/mL, respectively.

The linseeds and the leaves and roots of cassava (*Manihot esculenta*) contain cyanogenic glucoside *linamarin* (Figure 8.6), at enzymatic hydrolysis

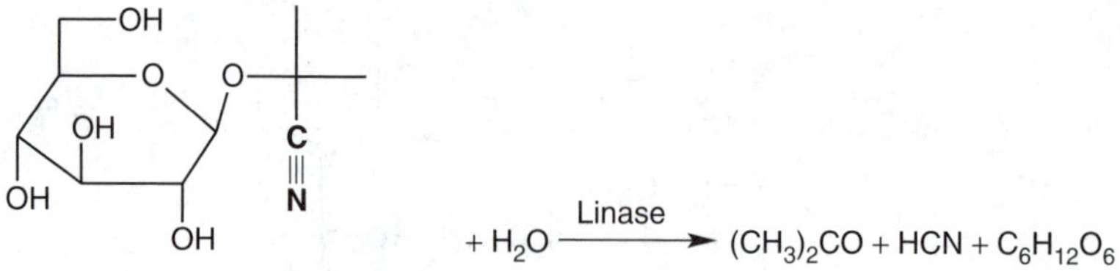

$$+ H_2O \xrightarrow{\text{Linase}} (CH_3)_2CO + HCN + C_6H_{12}O_6$$

*Figure 8.6* Structure of linamarin.

of which by linase or linamarase also cyanic acid is produced. Linamarase attacks the glucoside only after the plant cells have been destroyed. The content of linamarin in the linseeds can reach 0.7%. The death of a horse can be caused by 3 kg of linseeds. Two-month-old lambs died after eating 125 g linseed flour per day. Adult cows are less sensitive to the linamarin since cyanic acid is bound in their rumen. Boiling of linseeds for 10 min causes the loss of toxicity due to the denaturation of linase. The glucosidic bound of linamarin is relatively stable, in addition to the enzymatic hydrolysis; linamarin is disintegrated only by high pressure, temperature, and mineral acids.

The roots of cassava may contain 10–1120 mg/kg of free and bound HCN, but the seeds of lima bean (*Phaseolus lunatus*) may contain even 100–3000 mg/kg. Cassava is an important source of carbohydrates in the diet of people in Africa and South America. The toxic components linamarin and lotaustralin are rendered harmless during enzymatic hydrolysis taking place by cutting and milling of the roots. Nevertheless, cyanic acid intoxications occasionally still occur in these parts of the world.

*The acute toxicity of HCN* originates from its high affinity to the enzymes containing metalloporphyrin, especially to *cytochrome oxidase*. Already such a low concentration of HCN as 33 $\mu$M can, through a complete blocking of the mitochondrial electron transport chain (stronger than binding of molecular oxygen to $Fe^{3+}$ ion in the molecule of cytochrom $a_3$), prevent the use of oxygen for the synthesis of adenosine triphosphate (ATP). The symptoms of an acute cyanide intoxication of humans are hyperventilation, headache, nausea, and vomiting, general weakness, coma, and death through cytotoxic anoxia. *Lethal doses of the cyanide* are 200–300 mg in the form of potassium or sodium, cyanides, or 50 mg in the form of hydrogen cyanide, in the case of the adult humans. Toxic and lethal plasma concentrations of cyanide ion are 0.2 and 1 $\mu$g/mL, respectively. The actual sensibility to cyanide substantially depends on the age, body weight, and health conditions of the individual. Contrary to the widespread misconception, cyanide ion, unlike a similar molecule of carbon oxide (CO), does not bind strongly to the heme proteins like hemoglobin containing $Fe^{2+}$.

*Animals can develop acute cyanide intoxication* by eating, for example, young sorgo (*Sorghum bicolor*) in tropical Africa or arrowgrass (*Triglochin* ssp.) in the temperate zone.

Treatment of acute cyanide poisonings includes the artificial respiration and conversion of blood hemoglobin to methemoglobin by inhaled amylnitrite and intravenously administered sodium nitrite. Methemoglobin competes with cytochrome oxidase for HCN. The latter diffuses from the cells to the blood, where it forms cyanomethemoglobin. Simultaneously, intravenously administrated sodium thiosulfate converts the free cyanide in blood into harmless thiocyanate ($SCN^-$), which will be excreted by kidneys. Another safe and effective antidote of cyanide is hydroxocobalamine (vitamin $B_{12a}$), the drug of choice in the case of cyanide poisoning through smoke inhalation. The cyanide ion has a higher affinity to hydroxocobalamine than cytochrome oxidase (Geller et al., 2006).

To cope with small concentrations of cyanide that are present in various foods, an organism has enzyme rodanide synthetase capable of converting cyanide ions into a thiocyanate ion. *Chronic exposure to the cyanide ion* is relatively rare. Chronic nerve injuries (ataxic neuropathy), konzo (irreversible paralysis of the upper motor neurons), diabetes, goiter, dizziness, and vomiting have been registered as the symptoms of chronic cyanide ion poisoning in Africa and in other regions in the case of a continuous consumption of food containing cyanogenic glucosides. These intoxications are partly attributed to the malnutrition of these populations leading to a deficiency of sulfur-containing amino acids in their food and, hence, to a deficiency of cyanate ion, the antidote of toxic cyanide ion. The thyreotoxic effects of cyanide ion can be explained by the influence of thiocyanate ion (antagonist of iodine) forming during intoxication. This effect may be especially strong in the regions where the iodine content in food as well as in drinking water is low (Rietjens et al., 2005).

## 8.5 Phytoestrogens

Phytoestrogens are plant compounds that are structurally and functionally similar to 17$\beta$-estradiol and, hence, bind to estradiol receptors. Although their affinity for binding is 20–200 times less than that of estradiol, they compete with estradiol for receptor sites. It results in 500–10,000 times weaker *in vivo* estrogenic effect than the effect of an endogenous estrogen. As a result, phytoestrogens are able to act as both agonists and antagonists. The phytoestrogen content in plants is the highest during blossoming. Owing to the weak binding, the phytoestrogens are capable of hindering the effect of the endogenous steroidal estrogens and, in higher doses, of inducing antigonadotropic effects at the levels of hypothalamus, hypophysis, and gonads of the individuals of both sexes. *Phytoestrogens are*

- *Various flavonoids* (isoflavones genistein, glycetein, daidzein, formononetin, biochanin A) and their glucosides (genistin, glycetin, daidzin) from soy, fava bean, kudzu, alfalfa, peanut, and so forth
- *Coumestans* like coumestrol and 4-*O*-methylcoumestrol from legumes and clover
- *Stilbenes* like resveratrol from grapes and grape wine, peanuts, and so forth

- *Lignans* from flax, pumpkin, and sesame seeds, rye, soybeans, broccoli, beans, and some berries
- *Resorcylic acid lactones* like mycotoxin zearalenone (see Section 11.5)
- *Equol*—4′,7-isoflavandiol, an isoflavandiol, metabolized from daidzein or formononetin by bacterial flora in the intestines

So far, no clear effect on humans that certainly is weaker than the effect of the well-known endocrine disrupters like PCBs or dioxines has been shown. *Genistein from soy* has been the main suspect. Babies can be in contact with isoflavones with an intensity of 4 mg/kg through the artificial food prepared on the basis of soy. The question whether this exposure is sufficient for manifestation of long-term adverse effects requires further investigation. In addition to the aforementioned effects, genistein inhibits protein tyrosine kinase that is linked to a number of growth factors and other enzymes participating in cell proliferation and differentiation. Relative estrogenic potency (*in vitro* assay): estradiol (1186), genistein (1).

On the other hand, phytoestrogens are capable, maybe just thanks to their antiestrogenic nature, of defending the human organism from coronary heart diseases (CHD), mammary prostate, and colon cancer as well as postmenopausal osteoporosis. Phytoestrogens are neither mutagenic in the Ames test (Section 5.3) nor carcinogenic at oral administration. A comprehensive review of the health effects of various dietary phytoestrogens has been written by Cornwell et al. (2004).

## 8.6 Glucosinolates

*Glucosinolates include over 100 thioglucosides* that have been found up to the concentration of 60 mg/g from the numerous plants of the order *Brassicales* (especially in the family *Brassicaceae*) like cabbage, broccoli, Brussels sprout, cauliflower, turnip (rutabaga), radish, black radish root, horseradish, mustard, rape, and so forth. Glucosinolates are, for example, *sinigrin* (Figure 8.7), *progoitrin*, or *glycorapferine*, *epi-progoitrin*, and *glycobrassicin*. Their biological activity is caused by both the initial compounds and compounds like isothiocyanate or mustard oil, nitrile, oxazolidinthione (OZT) and thiocyanate ions, which

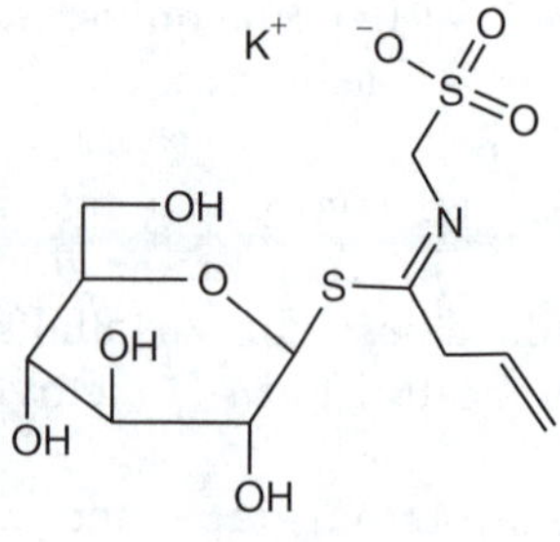

***Figure 8.7*** Structure of glucosinolate.

are formed from glucosinolates in the human digestive tract by the enzymatic catalysis with bacterial myrosinase (thioglucosidase). Plants use substances derived from glucosinolates as natural pesticides and as a defense against herbivores. Like most proteins, myrosinase is denatured during boiling.

Although there is no human data, the *thiocyanate ion inhibits assimilation of iodine by the animal thyroid gland.* As a result, iodine-reversible hyperplasia, thyroid hypertrophy (the so-called "cabbage and legume goiter") and growth retardation are developed. OZT, inhibiting iodine binding to the respective initial compounds, also inhibits the synthesis of thyroxine. The result will be the development of the so-called "brassica seed goiter," which is not curable by the administration of iodine. In addition to goiter, progoitrin, and epigoitrin, it causes dilatation of liver and kidneys and even death in the dietary concentration of 2.6%. Hyperplasia of the bile duct, hepatocyte necrosis, and megalocytosis of renal tubular epithelium have been diagnosed in the case of the animals.

Isocyanates exert an embryocidal effect and cause weight reduction of the embryos. Isothiocyanates and some glucosinolates (e.g., sinigrin), but not thiocyanates, are mutagenic in the Ames test (see Section 5.3).

At the same time, due to the formation of isothiocyanates from indols, indolyl-3-carbinol, 3-indolyl-acetonitrile, and 3,3′-diindolylmethane, eating of large quantities of the cruciferous plants can have an anticarcinogenic effect. There is an opinion that the anticarcinogenicity of glucosinolates can be caused by the induction of the phase II enzymes (quinone reductase) in the digestive tract and liver and stimulation of the apoptopsis. A recent review of glucosinolates has been published by Tripathy and Mishra (2007).

## 8.7 Coumarin

*Coumarin* (1,2-benzopyrone) is a constituent of several essential spices like cinnamon. *It is found in large concentrations in the oil of cinnamon tree* (*Cinnamomum*) bark (7,000 mg/kg) and leaves (40,000 mg/kg), in the oil of cassia (*Cassia*) leaves (83,000 mg/kg), in lower concentrations in lavender and peppermint oils (20 mg/kg), in green tea (1.1–1.7 mg/kg), in bilberries (0.0005 mg/kg), and in chicory. As a pleasant hay-smelling substance, coumarin (Figure 8.8) is contained also in plants like white and common melilots, holy grass, and sweet vernal grass.

*The use of coumarin as a food additive was stopped* after its hepatotoxicity to rats and dogs was discovered. The EU Scientific Committee on Food

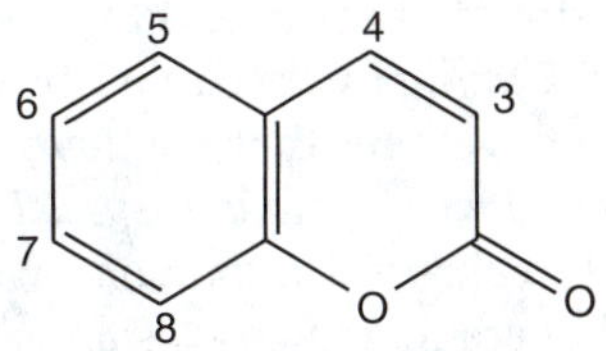

5 4 6 3 7 8 O O

***Figure 8.8*** Structure of coumarin or benzo-2-pyrone.

(EU-SCF) fixed 2 mg/kg for the limit of coumarin content in foods and alcoholic beverages in 1999. Enforcement of this regulation would have meant removal of a number of products from the food market.

Fortunately, the latest studies have revealed that the *metabolic systems of humans and rat treat coumarin in completely different ways* and that sometimes it is not correct to transfer the results of the animal tests directly to the humans. Coumarin is absorbed, on peroral administration, from the human digestive tract completely and to a great extent has the first pass metabolism in the liver; only some percentage of the initial compound reaches the systemic bloodstream. In the case of rats, most of the dose is first excreted into the bile and further with feces, and in the case of humans, with urine. In the case of rats, on peroral administration, coumarin generates hepatic and cholangial adenomas and carcinomas as well as renal adenomas. In the case of mice, pulmonary adenomas and carcinomas as well as hepatic adenomas are caused. *The main metabolite*, formed in the case of *the rodents* in participation of CYP is *3,4-epoxide of coumarin*, which is the actual culprit of the coumarin's acute as well as chronic toxicity. *In the case of humans, the main metabolite is 7-hydroxycoumarin*, formed by the action of CYP2A6, where formation of the epoxide is of minor importance. Which metabolic route is then realized in the case of humans, deficient in CYP2A6? It cannot be excluded that it is still the formation of the epoxide.

Recently, the former decisions concerning coumarin were revised by the European Food Safety Authority—EFSA, which set the tolerable daily intake (TDI) value—0.1 mg/kg bw. The estimated maximal daily dietary consumption of coumarin is 0.07 mg/kg or 0.02 mg, according to a more realistic scenario. Since both these numbers are smaller than TDI, the removal from the market of at least part of the coumarin-containing products turned out to be unnecessary (Rietjens et al., 2005).

## 8.8 Thujones

In various oils as well as parts of a number of plant monoterpenes with a menthol-like smell, $\alpha$- and $\beta$-thujones, are found. These plants, which are used also for preparation of foods and beverages, are wormwood (*Artemisia absinthum*), salvia (*Salvia officinalis*), tansy (*Tanacetum vulgaris*), junipers (*Juniperus* ssp), cedars (*Cedrus* ssp), and so forth. The name of the substances originates from thuja or arbor vitae (*Thuja* ssp.). The relative content of the two stereoisomers depends on the plant species.

Thujones are constituents of food additives that are used for food flavoring (Figure 8.9). The estimated average daily consumption of thujones is 16 μg/kg bw in France and 4 μg/kg bw in Great Britain. Most of it comes from consumption of salvei and beverages that are flavored by salvei. *Among these beverages, absinthe is the one that has aroused special toxicological interest*. The name originates from the Greek word *apsinthion* = wormwood. Absinthe is an emerald green alcoholic beverage, which was an extremely popular "source

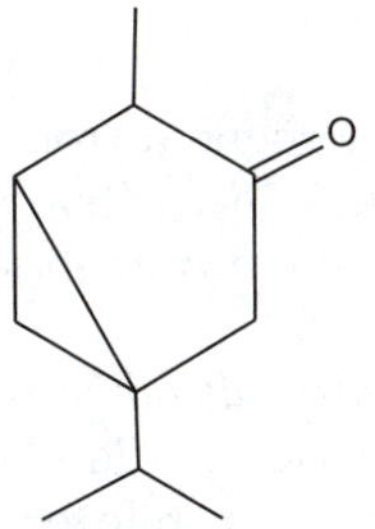

*Figure 8.9* Structure of β-thujone.

of inspiration" among actors and poets of bohemian lifestyle at the end of the nineteenth century. Wide abuse of absinthe and poisonings with thujones caused prohibition of this beverage in many countries during the 1910s. Drinking of absinthe caused convulsions, hyperactivity, excitement, hallucinations, and psychoses that could lead the drinker to suicide. In addition, consumption of wormwood causes nausea, vomiting, sleepiness, restlessness, and dizziness. High doses of thujones can elicit delirium, convulsions, paralysis, brain damage, renal insufficiency, and death.

The main toxic component of absinthe is the neurotoxic terpenoid α-thujone, the content of which is usually lower than the content of its stereoisomer. The toxic effect of α-thujone is based on its capability to block the receptors of γ-aminobutyric acid (GABA) in the brain (see Section 3.3.2). Without interaction with GABA, the normal inhibitor of nerve impulses, the neuronal signal system moves out of control.

*In the 1990s, absinthe was restored to its old popularity*. The new version of the beverage contains much less thujones (10 ppm or mg/kg) and its toxicity is already overshadowed by that of ethanol (55% v/v). In the old prohibited absinthe, the average total content of thujones had been 260 ppm. The upper allowed content of thujones in bitters is 35 ppm in the EU (Rietjens et al., 2005).

*But still, not all is clear about absinthe*. From a critical review of the literature, it has been concluded that chronic abuse of the "old" absinthe could not cause any syndrome distinct from alcoholism (see Section 16.3). It has been shown recently that thujone concentrations of about 260 ppm are actually unreachable by the nineteenth-century method of preparation of absinthe. However, much of the literature is focused on thujone as the only potentially toxic component of absinthe. It is known that besides the aforementioned herbal ingredients, different manufacturers of absinthe sometimes used strange or even toxic additives such as methanol, sweet flag (*Acorus calamus* L.), tansy (*Tanacetum vulgare* L.), nutmeg (*Myristica fragrans*), antimony, aniline green, copper sulfate, and cupric acetate indigo.

There is still a possibility that other constituents found within wormwood or other ingredients of absinthe are responsible for the earlier health problems other than the ethanol-borne ones (Padosch et al., 2006).

## 8.9 Anisatine

The case of *anisatine*, a sesquiterpenoid from the fruits of the *Japanese star anise* (*Illicium anisatum*) is a very instructive example of how toxicological problems can be provoked by either intentional or unintentional adulteration of goods anticipated for food.

The fruits of *Chinese star anise* (*Illicium verum*) have been used in various cultures for preparation of herbal teas. In Southern Europe, the tea is used also for the treatment of gastropathies in children. Over 60 persons complained about nausea and vomiting, and in 2001, 22 patients were hospitalized with the diagnosis of tonic–clonic insult after drinking of star anise tea in the Netherlands (Oudesluys-Murphy and Oudesluys, 2002). Electrocardiograms pointed to a diffuse brain damage, the complaints were similar to the symptoms of intoxication with *Japanese star anise*. The investigation revealed that it was very likely, incidentally, that instead of fruits of Chinese star anise, the fruits of Japanese star anise containing a toxic compound anisatine must have been present in the drug. Other cases have also been reported concerning the adverse neurologic reactions in infants seen with the home administration of star anise tea, once again caused by a proven contamination of Chinese star anise with Japanese star anise (Ize-Ludlow et al., 2004).

*Anisatine, acting as a noncompetitive antagonist of GABA*, also causes reduction of heartbeat frequency and hallucinations (Figure 8.10). To exclude the danger stemming from possible mistakes as well as adulterations, the drugs containing Chinese star anise are not recommended for children.

## 8.10 Toxic amino acids

Generally, our diet does not contain any common individual amino acids in too high, toxically irrelevant concentrations. But plants contain also rare nonproteinous amino acids that may have adverse effects on the organism (Wallace Hayes, 2001, p. 499). Next, we will discuss some examples of possible poisonings with such amino acids.

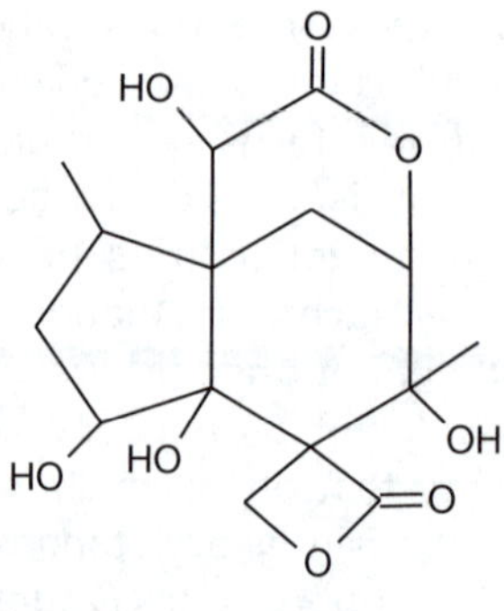

**Figure 8.10** Structure of anisatine.

*Hypoglycin A* (*β*-methylene-cyclopropylalanine) and its *γ*-glutamylconjugate *hypoglycin B* are contained in the fruits of *ackee* or *akee* (*Blighia sapida*), native to tropical West Africa. In America, it was first introduced to Jamaica and later on to the other islands of the Caribbean Sea. By now, it has been introduced also to Florida in the United States. Eating of the raw fruit induces *hypoglycemia* due to the inhibition of the gluconeogenesis cycle. It is only the fleshy arils around the seeds that are edible. The fruit must be picked after it has opened naturally, and it must be fresh and not overripe.

The symptoms of the poisoning are vomiting, convulsions, hypothermia, and coma and fatalities are also possible. Hypoglycin binds irreversibly to coenzyme A and carnitine thus reducing their bioavailability and, consequently, inhibiting beta-oxidation of fatty acids. Beta-oxidation normally provides the body with ATP, NADH, and acetyl-CoA that is used to supplement the energy produced by glycolysis. Glucose stores are consequently depleted leading to hypoglycemia (Blake et al., 2006).

*3,4-Hydroxypyridin* (3,4-DHP), a metabolite of *mimosine* [3-*N*-(3-hydroxypyridone-4)-2-aminopropionic acid], that is contained in legumes like *Leucaena leucocephala,* growing on the Hawaiian Islands, is a ruminants' goitrogen. It causes growth retardation of ruminants as well as loss of hair and decrease of bone strength of birds. The basis for the toxic effect of 3,4-DHP is its chelating ability of zinc and magnesium, resulting in a decrease of the levels of thyroid and other hormones in the plasma and inhibition of a number of the enzymes that are involved in the synthesis of DNA (Hammond, 1995).

*Djenkolic acid,* similar to cysteine in the structure of the molecule, is contained in the beans of the Jungle tree (*Pithecellobium lobatum*) growing on Sumatra and Java. Despite the similarity, djenkolic acid is not capable of replacing cysteine and does not metabolize completely. Djenkolic acid crystallizes in the kidneys, thus causing renal intoxication. The symptoms of poisoning called *djenkolism* are pain in the region of kidneys, dysuria, and later anuria (hematuria). Beans of *P. lobatum* contain a lot of vitamin B and this is the main reason why regardless of their well-known toxicity, the beans are still being eaten (Areekul et al., 1978).

*3,4-Dihydroxyphenylalanine* and pyrimidinaglycones of glucosides *vicin* (Figure 8.11) and *convicin* that are found in raw horsebean (*Vicia faba*) cause a hemolytic disease *favism* in the case of individuals genetically deficient in glucose-6-phosphate dehydrogenase (G6PD). The symptoms of the poisoning are paleness, fatigue, dyspnea, nausea, abdominal and back pain,

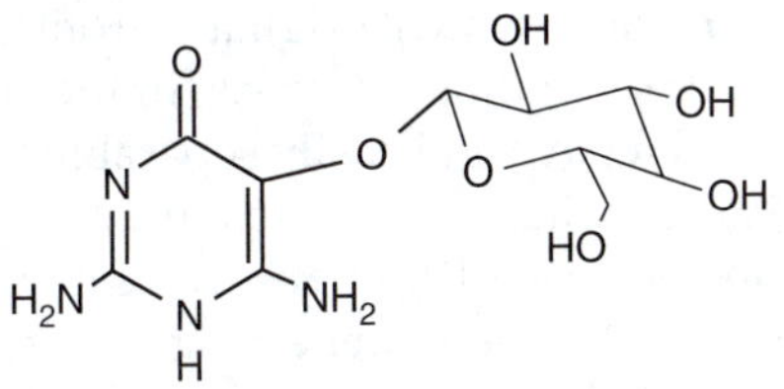

OH
O
OH
O
N
O
OH
HO
$H_2N$
N
H
$NH_2$

***Figure 8.11*** Structure of vicin.

fever, and tremors. More serious symptoms such as hemoglobinuria, jaundice, and renal insufficiency may rarely occur. Favism can be caused also by the inhalation of the pollen of horsebean. The latent period of the intoxication is 5–24 h.

*Cooking abolishes the hemolytic effect of the beans, which has the following mechanism.* The beans contain *two heterocyclic compounds—vicin* and *covicin* that are capable of oxidizing essential intracellular reducer glutathione especially in the erythrocytes. When the oxidation proceeds with an intensity, which influences the redox-situation of the cells, the oxidation of the membrane lipids of erythrocytes starts and is then followed by hemolysis. In the case of genetically normal individuals, the oxidized glutathione is reduced by glutathione reductase that in turn depends on the reducing equivalents provided by G6PD. The cells of the individuals who are genetically deficient in G6PD are not capable of a sufficient regeneration of the reduced form under the stress of vicin/covicin. Men are mainly endangered. These individuals, most abundantly found living in the Mediterranean region and in Asia, as well as in Africa, may *exert* favism with a fatal end. *The deficiency of G6PD is the most abundant enzyme-deficiency in the world,* thus affecting about 400 million people worldwide. On the island of Sardinia in the Mediterranean Sea, almost 70% of men are G6PD-deficient. Favism is widely distributed among the Jews of Israel and Greeks of Cyprus, in the Near East and in China, in tropical Africa, but is almost unknown for the nations of North Europe and the Indians of North America.

Favism can also be elicited by antimalarian drugs like primaquine, by antibacterial sulfonamides, nitrofuranes, naphthalene, and acetylsalicylic acid (aspirin). All the drugs mentioned are oxidants.

*But it is partly a blessing in disguise.* Namely, G6PD-deficiency is one of the three genetic disorders connected with erythrocytes, which guarantee tolerance to the malaria parasite (*Falciparum malaria*). The other two are thalassemia and sickle cell anemia. Experiments have shown that for survival and reproduction, the parasite needs the host cell G6PD; in the absence of the enzyme, the parasite will die. Since the mortality rate of favism is low but that of malaria is high, the protecting effect of this enzyme deficiency is very essential. It is interesting that on the world map, the region of G6PD deficiency coincides rather well with the area of distribution of the malaria parasite. *Falciparum malaria* has obviously participated in making the respective natural selection (Greene, 1993).

*Methylselenocystein, selenocystein, selenomethionine* and several other selenium-containing amino acids, which originate from plants grown in soil that has a high content of selenium, can cause, in the case of long-term consumption, formation of defective and easily separable hooves and hair of the animals. The reason is the inclusion of these amino acids into the structural proteins of animals. In the case of humans, nausea, vomiting, diarrhea as well as hair loss have been observed after eating of nuts of *Lecythis ollea,* containing selenocystathione in high concentrations, up to 5 mg/g (Ferry et al., 2004) (see also Section 9.2).

*L-2,4-Diaminobutyric acid* (DABA), *β-N-oxalyl-L-α,β-diaminopropionic acid* (*β*-ODAP), *3-cyanoalanine, 4-glutamylcyanoalanine and* their homologs, which are found in the seeds of *Vicia sativa* and several species of genus *Lathyrus,* participate in the pathogenesis of *neurolathyrism. Lathyrus sativus* (grass pea) is a high-yielding, drought-resistant legume consumed as a food in Northern India and its neighboring countries as well as in Ethiopia. Long-term ample eating of the seeds causes muscular rigidity, weakness, paralysis of foot muscles, and death. These acids are able to bind irreversibly to the glutamate receptor. The following increase of glutamine deliberation at the respective nerve ends will cause a degeneration of blood vessels and necrosis of the neurons (Yan et al., 2006).

Recently, some low-toxin lines of grass pea have been developed that may prove safe as food for both animal and human consumption.

## 8.11 Toxic lipids

The adverse effects related to dietary lipids can be divided into three main groups:

1. Withdrawal from habitual diets
2. Inclusion of new lipids into the diet
3. Hereditary errors in the lipid metabolism

### 8.11.1 Erucic acid

Rapeseed and mustard oils and other seed oils of *Brassica* genus contain *erucic* (*cis*-13-decosenoic) *acid,* which is a 22-C monounsaturated fatty acid (Figure 8.12). When in the diet of rats, when they were immediately weaned from the mother's milk, the caloric concentration of erucic acid exceeded 20%; growth retardation, fatty infiltration of cardiac muscle, infiltration of mononuclear cells, and fibrosis were observed (Kramer, 1988). Ducklings acquired hydropericard and cirrhosis, and guinea pigs splenomegalia and hemolytic anemia. Oxidation of glutamate and inhibition of synthesis of ATP in the cardiac mitochondria by dietary erucic acid have been proposed as the mechanism of these emergencies. Respective studies have indicated too high levels of erucic acid linked to the formation of fatty deposits in heart muscle in animals. Although in the case of humans, a long-term use of the Lorentzo oil (a mixture of oleinic and erucic acids) for the therapy of adrenoleucodystrophy and adrenomyeloneuropathy causes thrombocytopenia and lymphopenia,

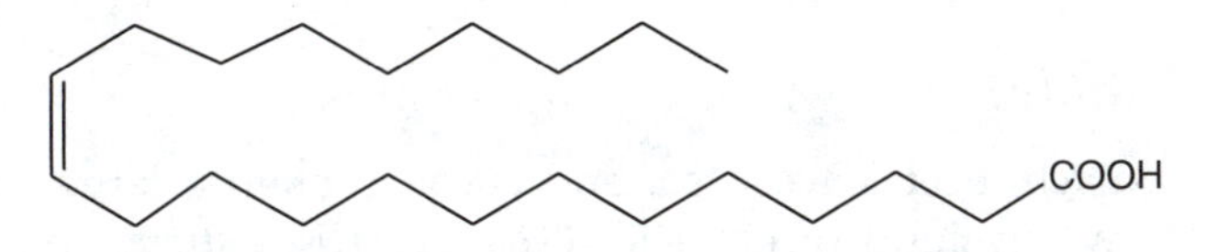


***Figure 8.12*** Structure of erucic acid.

no data about any adverse effects of dietary erucic acid are available so far. Canola is a low erucic acid species of rapeseed developed for human consumption and the oil is often used for cooking.

### 8.11.2 Sterculic and malvalic acids

*Sterculic* ($C_{19}$) and *malvalic* ($C_{18}$) acids are natural components of the oils produced from the seeds of the plants belonging to the order *Malvales* like the cotton shrub (*Gossypium* spp.) or kapok tree (*Ceiba pentandra*) in which they are found in various proportions of up to 70%. These as well as similar *cyclopropene fatty acids* (CEFA) are carcinogenic and they remarkably increase the carcinogenicity of aflatoxin in the rainbow trout. CEFAs are found in *Malvales* seed oils (*Sterculiaceae, Gnetaceae, Bombacaceae, Tiliaceae, Malvalaceae,* and *Sapindaceae*) and baobab, kapok, and mowrah seed oils that are used as human food in Madagascar. They are associated also with the reduction of egg production in hens fed with the cottonseeds, causing the pink color of the eggs, as well as growth retardation and reproductive disorders in rats. CEFAs interfere with animal fatty acid (desaturation of stearic acid), protein, and carbohydrate metabolism. Such toxic acids are either deactivated or removed by a normal industrial treatment of the oils. The biological effects of cyclopropenic acids have been reviewed by Andrianaivo-Rafehivola et al. (1994).

### 8.11.3 Polyunsaturated fatty acids

Well-known is the blood cholesterol lowering effect of *polyunsaturated fatty acids* (PUFA), leading to decrease of risk of CHD. Simultaneously, adverse effects of PUFAs, like generation of vitamin E deficiency, have been discovered. PUFAs are easily oxidized by cooking and storage, producing various mutagenic epoxides, enols and aldehydes, alkoxy- and hydroperoxy radicals. Epoxides of PUFAs can be formed by autooxidation and by CYP. Epoxides of linoleic acid are toxic, whereas the epoxides of arachidonic acid (EETs) have a wide range of biological effects. The oxidation products of PUFAs influence the signal transmission in the cell, thus increasing the cell proliferation and induce tumor formation. Inhibition of the immune response and an increase in the production of tumor promoters such as prostaglandins and bile acid by PUFAs have also been observed (Bhattacharya et al., 2006). For example, the incidence of pancreatic cancer substantially increases in the case of a diet containing 20% of corn oil, but not in the case of a diet containing 18% of hydrogenated (saturated) coconut oil and only 2% of corn oil (Guthrie and Carroll, 1999).

## 8.12 Oxalates

Rhubarb, spinach, beet leaves, tea, and cocoa contain a large amount of (0.3–2.0% of wet weight) oxalic acid. *Oxalic acid is a strong acid* that irritates tissues. *Oxalates form tiny insoluble crystals with sharp edges, which also irritate tissues*. So, high levels of oxalic acid/oxalates in the diet lead to an irritation of the digestive system, particularly of the stomach and kidneys. They may also contribute to the formation of kidney calcium oxalate stones. As a result of an accumulation of calcium oxalate, degeneration of kidneys as well as blood vessels and necrosis may occur. Oxalic acid is capable of chelating blood calcium resulting in hypocalcemia, coagulation disorders, and tetany. Oxalates also influence the absorption of calcium, iron, magnesium, and copper and inhibit the metabolism of succinate dehydrogenase and of carbohydrates. To get the lethal dose of oxalates, one should eat either 5 kg of rhubarb, 2.5 kg of tomato, or 0.5 kg of spinach leaves.

The symptoms of oxalate poisoning are weakness, flushing in the mouth and throat, respiratory distress, abdominal pain, nausea, vomiting, diarrhea, convulsions, coma, and death as a result of collapse of the coronary heart system. Oxalate intoxication is much more common in animals. $LD_{50}$ (rat) is 375 mg/kg.

Foods containing oxalic acid/oxalates should be consumed moderately. People suffering from a kidney disease and stones, rheumatoid arthritis or gout are recommended to avoid foods that are high in oxalates or oxalic acid, such as chocolate, cocoa, coffee, tea, most berries (especially strawberries and cranberries), most nuts (especially peanuts), beans, beets, bell peppers, black pepper, parsley, rhubarb, spinach, summer squash, and sweet potatoes. The toxicity of oxalic acid has been reviewed by Von Burg (1994).

## 8.13 Fluoroacetates

*Sodium monofluoroacetate is an extremely toxic substance,* interrupting the tricarboxylic acid or citrate or Krebs cycle—one of the main elements of the metabolic system of all plants and animals and most of the microbes. Fluoroacetate is included into this cycle in the shape of pseudosubstrate *fluoroacetyl-CoA*. By condensation of the latter substance with oxaloacetate, fluorocitrate that blocks the next enzyme of the Krebs cycle is formed. As a result, the cycle stops and the respective cell and organism will die from a lack of metabolic intermediates as well as energy. Actually, the action mechanisms of fluoroacetate are more versatile; not all the toxic effects can be explained by the disturbance of the Krebs cycle. Fluoroacetate toxicity to the liver, brain, and other organs has been recorded. Fluoroacetate passes the blood–brain barrier and its peroral $LD_{50}$ is 2–10 mg/kg (humans), 0.1–5 mg/kg (rats), and 0.05 mg/kg (dogs). Various animals have various symptoms of the toxic effect—in the case of rabbits, goats, horses, and sheep, no damage of the CNS has been observed, but death is caused by the cardiac toxicity of fluoroacetates. In the case of cats, pigs, rhesus monkeys, and humans, both cardiac and CNS toxicities are observed; death is usually caused by convulsions of the respiratory muscle. In the case of dogs and guinea pigs, epileptic convulsions are dominating; death is caused by stopping of respiration. In the case of rats and hamsters, depression of the respiratory center of the brain and sinusoidal bradycardia are dominating.

*Fluoroacetate has been used as an antirodent pesticide in different countries.* Spraying of fluoroacetate was prohibited in the United States in 1972. But, for example, in New Zealand and Australia, it is still in use as an antirodent agent. Fluoroacetate is a metabolite of many other fluorinated pesticides like diisopropylfluorophosphate (DFP) as well as anesthetics and anticancer drugs and a decomposition product of phosphoorganic substances. As a decomposition product of fluorinated carbohydrates, fluoroacetate can be found in air and water. Fluoroacetate is formed in cereals and in water plants, it can be found in some Australian, African, and South American plants like *Acacia georginae, Dichapetalum cymosum, D. toxicarum, Gastrilobium grandiflorum, Oxylobium parviflorum,* and *Palicourea margravii*. Eating of these plants has caused the death of domestic animals. *Especially sensitive are dogs* that have been intoxicated by eating poultry meat contaminated with fluoroacetate. Endemic Australian animals like the skink (*Tiliqua scincoides*) or ostrich emu (*Dromaius novaehollandiae*) have acquired a tolerance to this compound. There is an opinion that for plants, fluoroacetate is a toxicant preventing them from being eaten by herbivorous animals. A recent review of various aspects of fluoroacetate toxicology has been published by Goncharov et al. (2006).

## 8.14 Bracken toxins

Bracken (*Pteridium aquilinium*), a plant belonging to the family of ferns, grows in suitable places all over the world, except the Antarctic, high mountains, and deserts. Its rhizome, rich in starch, is used for food in several places like New Zealand. The rhizome of the bracken is used also in brewing beer and for preparation of an excellent paste. In Japan, the shoots are eaten and the starch of the rhizomes is used as candy. The specific smell of bracken deters various insects. Bracken is used also as an herb—the milled rhizome has been used for prevention against parasitic worms, the American Indians ate the raw rhizomes in case of bronchitis. *The whole plant is still toxic.* Eating of bracken by herbivorous animals causes a number of both acute and chronic toxicity syndromes—thiamine deficiency in the case of horses and pigs, acute hemoraghy in the case of cows, degeneration of the retina in the case of sheep, neoplasms of bladder, stomach, and esophagus in the case of ruminants (Shahin et al., 2003). *Bracken is the only known higher plant that causes animal cumulative cancers.* It has been established that the toxic compounds of the bracken can reach milk. It is supposed that the wide distribution of gastric cancer in Japan is caused by a widespread eating of bracken.

The carcinogenic constituent of bracken, which also causes degeneration of the retina and myeloid aplasia, is *ptaquiloside* (Figure 8.13), a glucoside containing a reactive cyclopropylic group (Rasmussen et al., 2003). The studies performed in Japan show that 19 out of the 31 species of fern contain various potentially carcinogenic ptaquilosides. At hydrolysis of the ptaquilosides, pterosines or, in milder conditions, dienone is formed. The last compound

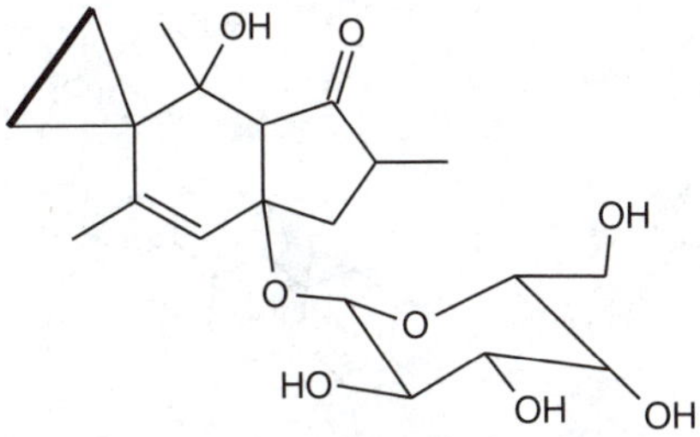

*Figure 8.13* Structure of ptaquiloside. On the left side of the molecule, the triangular cyclopropyl group is located.

is expected to be a DNA-reactive carcinogen that is capable of alkylating adenine in the position of *N*-3 in a definite sequence of the DNA molecule. Compounds containing analogical cyclopropyl groups have also been found in fern *Hypolepsis punctata* and in fungi, *Omphalotus illudens* and *Streptomyces zelenis*.

In addition to the carcinogenic ptaquiloside, raw bracken contains enzyme thiaminase that decomposes vitamin $B_1$ or thiamine. Consequently, eating of bracken in large quantities may cause beriberi. Ruminants are less sensitive to the action of thiaminase since they synthesize thiamine themselves.

## 8.15 Saponins

Saponins are glycosidic saponaceous surface-active substances (from Latin *sapo* = soap). They are found in soybeans, sugar beet, peanuts, spinach, broccoli, alfalfa, potato, apples, and in other plants and fruits. In terms of chemical structure, they are divided into two large groups—*steroidal* and *triterpenic saponins*.

- The aglyconic part of the *steroidal saponins* is based on pentano-perhydrofenanthrene, which is also the basis for the heart glucosides and sexual hormones. When the molecule includes an atom of nitrogen, we have a steroidal alkaloid (solasodine, tomatidine—see Section 8.3.2).
- The aglycones of *triterpenoid saponins* are mostly pentacyclic terpenoids, where the isoprene moiety ($-C_5H_8$) is repeated six times. This type of saponin is further divided into three subtypes: $\alpha$-amyrine type (ursane), $\beta$-amyrine type (oleanane), and lupeol type (lupan). Besides that, tetracyclic triterpenoid saponins like damarran of cycloartan are found in plants. The group of $\beta$-amyrines also contains saikosaponins, which have been extensively studied in recent years.

Saponins are very toxic substances; due to their surfactant activity they are able to decompose the cellular membranes. They can cause blood

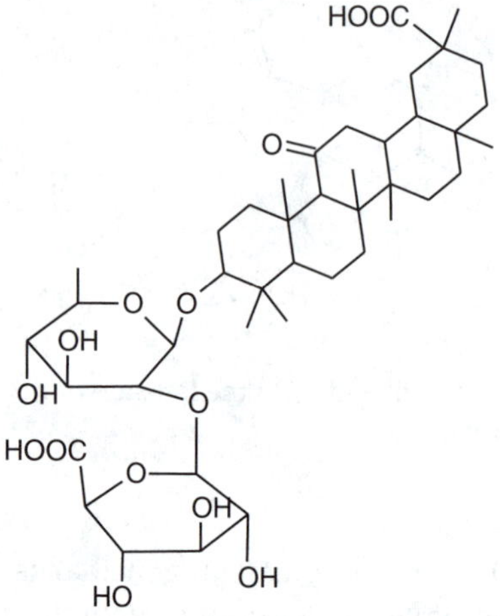

*Figure 8.14* Structure of glycyrrhicinic acid or glycyrrhizin.

hemolysis, especially in the case of cold-blooded animals. For warm-blooded animals, small perorally administered doses of saponins are generally harmless, since they are decomposed by enteric microflora. In addition, they are poorly absorbed and their effect is inhibited by blood plasma. Nevertheless, in the case of large doses of saponins nausea, vomiting, diarrhea, and dizziness can occur.

*Glycyrrhizin* or *glycyrrhicinic acid* (Figure 8.14) is an interesting triterpenoid saponin that is present in the roots and rhizomes of licorice (*Glycyrrhiza glabra, G. uralensis*) in the form of ammonium and potassium salts. Mixtures containing licorice extract are used for the preparation of drugs and candies. Standardized licorice extracts contain 20% of glycyrrhicinic acid. Enzyme glucuronidase catalyzes the hydrolytic decomposition of glycyrrhicinic acid into glycyrrhetic acid and diglucuronic acid. Glycyrrhicin has a number of useful effects on the organism—it defends the liver against the toxic effect of tetrachloromethane, has an antivirus effect (HIV, influenza A and B, herpes simplex 1 and 2, hepatitis B and C) both *in vitro* and *in vivo,* and helps to regenerate the inflammatory tissue. The medicinal usage of licorice extracts is rather broad: from cough and inflammations of the upper respiratory tract to gastritis and gastric ulcer. Licorice extract, containing glycyrrhizinic acid, which is 200 times sweeter than saccharose, is used for making licorice candies, toothpaste, chewing gum, chewing tobacco, and various alcoholic beverages.

*At the same time, glycyrrhizin or, more correctly, its enteral hydrolysis product glycyrrhetic acid has toxic effects as well.* Because of the lipophilicity of the molecule, glycyrrhetic acid is slowly excreted via the kidneys and it participates in enterohepatic recirculation. It inhibits the enzyme 11-$\beta$-hydroxysteroid dehydrogenase-2 (11-BOHD-2) in distal kidney tubules, where it converts the steroid hormone cortisol to cortisone. Cortisol binds to the mineralocorticoid receptor but cortisone does not. A decreased activity of the enzyme leads to an excess of cortisol and an overstimulation of the mineralocorticoid receptor. This causes water and sodium retention and an increased excretion of potassium. Long-term consumption of high doses of glycyrrhizin increases blood pressure, and through increasing sodium content in blood (hypernatremia), it causes water retention in the body. Hypokalemia is also formed in blood (Rietjens et al., 2005).

In many cases, intoxication is caused due to lack of knowledge. People are used to eating candies containing licorice extract, to drinking beverages like anisette, and to smoking tobacco processed with licorice extract. The toxic dose of glycyrrhizin depends on the individual. The European Commission recommends 100 mg as the upper daily limit for this compound. An incident of a woman who ingested daily 200 g licorice candies for treating constipation is documented. The patient was hospitalized with a very high blood pressure and general weakness. Studies performed in Finland showed that eating of licorice can cause early birth. Licorice cannot be consumed by persons with type 2 diabetes, who have a high arterial blood pressure and low potassium concentration in blood. Licorice is not recommended also in the case of hepatitis and during pregnancy. One should not eat licorice with drugs that increase blood pressure, with corticoids, and heart stimulators.

*The symptoms of glycyrrhizin poisoning are* elevated blood pressure, edema (mainly in the face and ankles), sensation of burning, weakness, dark urine, loss of the menses, alterations in libido, cardiac arrythmia, and so forth.

In the case of intoxication, consumption of licorice should be stopped. Medical treatment is usually unnecessary.

## 8.16 Grayanotoxin

*Grayanotoxin has been found in rhododendron* (*Rhododendron*) and other plants of the family *Ericaceae*. The other names of the substance are *andromedotoxin, acetylandromedol*, and *rhodotoxin*. From the plant nectar, grayanotoxin passes to the honey and may be the causative agent of *honey-* or *rhododendron-intoxication*. Emanating from the areas where rhododendrons are grown, the intoxications are mainly caused by highland or coastal honey.

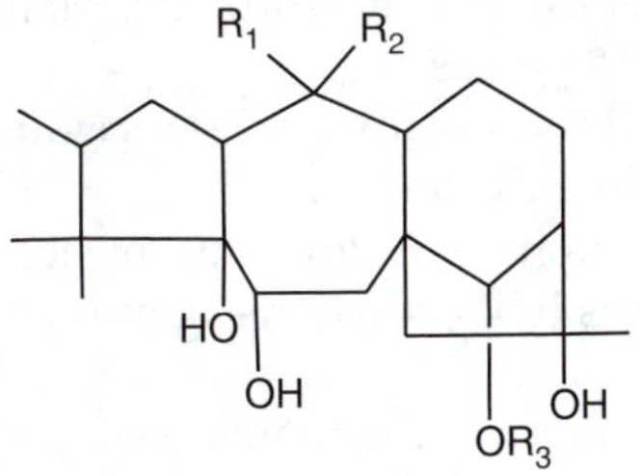

**Figure 8.15** General structure of grayanotoxins.

From a chemical point of view, grayanotoxin is a polyhydroxylated cyclic diterpene (Figure 8.15). It binds to the specific sodium channels of the cell membranes, hindering their inactivation and leaving the cell excited. *The symptoms of the poisoning*, appearing after (until 3 h) a 3-hour latent period are ample salivation, sweating, vomiting, dizziness, weakness and paraesthesia of the limbs and face, low blood pressure, and bradycardia. In the case of higher doses, coordination disorders, severe progressing muscular weakness, bradycardia, and paradoxically ventricular tachycardia as well as the Wolff–Parkinson–White syndrome can appear. Despite the potential heart problems, the intoxication is seldom fatal, lasting usually 24 h. Medical aid is mostly not needed and in some cases, atropine treatment is recommended for relieving the symptoms.

According to the Roman writer Pliny the Elder and later to the Greek geographer and historian Strabo, the nations inhabiting the coasts of the Black Sea had used the toxic rhododendron honey as an unconventional weapon against the armies of Cyrus the Great (401 BC) and Pompey the Great (69 BC).

In recent times, documented poisonings have occurred in Turkey and Austria. In 1984–1986, 16 patients were treated for honey intoxication in Turkey. The Austrian case was also caused by a Turkish honey (US FDA—http://www.cfsan.fda.gov/~mow/chap44.html).

*A number of grayanotoxic species are native to the United States.* Of particular interest are the western azalea (*R. occidentalis*), found from Oregon to Southern California, the Californian rosebay (*R. macrophyllum*) found from British Columbia to Central California, and *R. albiflorum* found from British Columbia to Oregon and in Colorado. In the eastern half of the United States, grayanotoxin-contaminated honey may be derived from other members of the family *Ericaceae* like mountain laurel (*Kalmia latifolia*) and sheep laurel (*K. angustifolia*).

## 8.17 Mushroom toxins

### 8.17.1 General principles

Eating of mushrooms may be connected with various hazards:

1. Real mushroom poisoning:
   a. With persistently toxic mushrooms. Many mushrooms lose their toxicity during processing.
   b. With old mushrooms. Toxic substances are formed during decomposition of the mushrooms.
2. Poisonings by environmental toxicants, bioaccumulated in mushrooms.
3. Allergic-supersensibility reactions against mushroom constituents.

Of the several thousands of mushrooms growing in forests and meadows, only some hundreds are intrinsically toxic. By far, not all of these hundreds are harmful for people, since nobody eats them. But quite many poisonings have occurred with mushrooms picked in forests and mistaken for edible

***Table 8.1*** Main Classes of Mushroom Toxins

| Type of toxin | Genus and species of mushroom |
|---|---|
| Cyclic peptides | *Amanita phalloides, Galerina* spp., *Lepiota* spp. |
| Cyclic peptides/orellanine | *Cortinatius* spp. |
| Muscarine | *Amanita muscaria, Inocybe* spp., *Clitocybe* spp. |
| Muscimol/muscazone/ibotenic acid | *A. muscaria, A. pantherina* |
| Gyromitrin (monomethylhydrazine) | *Gyromitra esculenta,* other *Gyromitra* |
| Coprine | *Coprinus orellanus, C. plicatilis* |
| Psilocybine | *Psilocybe, Paneolus,* magic mushrooms |
| Various irritants of the digestive tract | *Agaricus, Chlorophyllum, Marasmius, Naematoloma,* and many others |

species. For example, it is rather difficult to distinguish the extremely toxic *white death cap* (*Amanita virosa*) from the brilliant edible meadow mushroom (*A. campestris*) or from the Gypsy mushroom (*Rozites caperatus*).

*Really fatally toxic* are *two mushrooms of the genus Amanita—the green death cap* (*A. phalloides*) and *white death cap* (*A. virosa*), which contain *amatoxins* ($\alpha$-, $\beta$- and $\gamma$-amanitins) and *phallotoxins* like phalloidin, phalloin, phallacidin, and phallolysin. In addition, *A. virosa* contains virotoxins. Amatoxins are also constituents of the other species of Amanita such as *A. verna, A. ocreata, A. bisporigera, A. suballiacea, A. tenulifolia, and A. hygroscopica*. Some species of genus *Amanita* such as *A. rubescens* and *A. vaginata* are even edible after scalding. Amatoxins have also been found, although in lower concentrations, in the genera *Galderina* and *Lepiota*.

Amatoxins are thermostable and insoluble in water; they cannot be destroyed either by boiling or cooking or by drying of the mushrooms.

Phallotoxins induce severe gastroenteritis, appearing 4–8 h after ingestion of the white death cap, but due to negligible absorptivity, this group of toxins has no essential role in mushroom poisonings.

According to the molecular structure of the main toxins, mushrooms can be divided into several classes (Table 8.1).

Good reviews of toxic mushrooms have been written by Berger and Guss (2005), by far the most comprehensive review, and also by Tegzes and Puschner (2002).

### *8.17.2 Amatoxins*

*Amatoxins* or *amanitins* (Figure 8.16) are divided into five subtypes, $\alpha$ and $\beta$ subtypes being the most important ones. Amatoxins are cyclic octapeptides (MW≈900) that after absorption via amatoxin-transporting system block the synthesis of mRNA in the eukaryotic cells by inhibition of enzyme RNA polymerase II. This inhibition causes a continuous fall of the mRNA concentration and rate of protein synthesis in the cell and cell death during the first 24 h after ingestion of the mushroom. These processes take place first in the digestive tract and thereafter, in the liver and kidneys. Liver necrosis

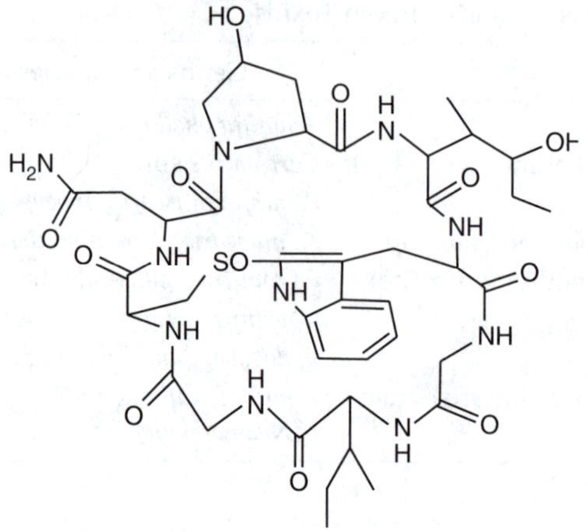

*Figure 8.16* Structure of α-amanitin.

(the most critical injury) and decomposition of the renal tubular cells may also occur. Thionic acid is used for curing the amatoxin-caused intoxication. The toxin of *Amanita pantherina* damages the nervous system.

Development of the amatoxin poisoning can be divided into three phases:

1. A characteristic latent period from eating to the appearance of the first clinical symptoms, then abdominal (ventral) spasms, vomiting, and profuse diarrhea (rice water reminiscent of cholera) will start. Water loss may be sufficient for profound dehydration and even for circular collapse. This gastrointestinal phase lasts from 6 to 12 h.
2. When the first acute phase is over, the second phase usually starts 24 h after eating. Although externally the patient looks clinically recovered, the laboratory investigation shows continuous liver damage (elevated serum aminotransferases and prothrombin time). This phase can last from 2 to 3 days.
3. The hepatic and renal damage become clinically observable and may progress into fulminated deficiency of liver. Death can follow on the 3rd to 7th day of the intoxication.

Lately alternative toxicity mechanisms of amatoxins have been presented. It has been supposed that α-amanitin acts in synergy with the endogenous cytokinins like tumor necrosis factor (TNF) that induces cell apoptosis.

*Amatoxins are potent poisons,* and an acute dose of 0.1 mg/kg bw can be lethal for an adult person. Intake of a green death cap weighing 20–25 g that contains 5–8 mg of amatoxins may end in fatality. $LD_{50}$ of amatoxins intraperitoneally administered to white mice is 0.3–0.7 mg/kg bw, and the same is 1.5–3.0 mg/kg bw in the case of phalloidin. It is interesting that no toxic response was observed at peroral administration of phalloidin to mice. The absorption rate of amatoxins from the intestines depends on the animal species, and varies in humans, dogs, and guinea pigs; amatoxins are not absorbed at all in the case of rodents. In humans, amatoxins are absorbed fast. In one case of intoxication, they were detected in urine 90 min after ingestion. Distribution to other parts of the body is also fast, the half-life in plasma is 27–50 min in the case of dogs, and 80% of the dose reaches urine during 6 h. Elimination takes place mainly through the kidneys during a period of up to 3 days. The fact that amatoxins are detectable in urine for a long time after they have disappeared from plasma indicates the binding of the toxin with the renal tissue. High toxin levels have been detected also in the gastric-duodenal liquid, but due to the small volume of the latter, this elimination route is less important. High concentrations of amatoxins were measured in the liver after 9 days and in the kidneys (especially high) 22 days after the ingestion. Although amatoxins are rather strongly linked to plasma proteins, they are toxic even in this conjugate form.

Therapy of amatoxin intoxication consists mainly in supportive care (rehydration, balance of electrolyte loss, administration of activated charcoal), and also to accelerate toxin elimination from the body, replacement of the toxin molecules from protein complexes by sulfonamides or benzylpenicilline is also recommended.

*Incidence of amatoxin intoxications largely depends on the geographical location.* Above all, it is a problem not only in Europe with 50 to 100 yearly fatal cases, but also in North America (Virginia). Single casualties have been recorded in Asia (Thailand, Korea, and Japan), Central America (Mexico), and so forth. The death rate was 22%, in the case of infants going up to even about 50%, in the 1970s in Central Europe (Karlson-Stiber et al., 2003; Mas, 2005).

### *8.17.3 Muscarine*

*Muscarine* is found in mushrooms of the genera *Inocybe* and *Clitocybe* in concentrations of up to 3 mg/g of the dry weight. Lesser muscarine (about 0.003%) can be found in fly agaric (*Amanita muscaria*), a decorative toadstool with a bright red cap, which has given the name to the toxic substance. Muscarine (Figure 8.17) is an alkaloid belonging to the group of betaines that interacts with muscarinic cholinoreceptors; the mechanism of its physiological effect

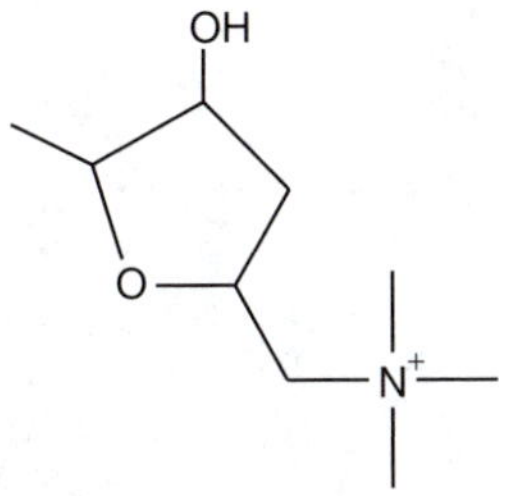

OH

O

$N^+$

***Figure 8.17*** Structure of muscarine.

resembles the effect of the other M-cholinomimetics. *The toxicity symptoms of muscarine* (salivation, epiphora, visual disorders, headache, vomiting, diarrhea, bronchospasm, bradycardia, low blood pressure, and shock) develop during 30–60 min after ingestion of the mushroom. *Atropine*, a blocker of M-cholinomimetic effect of muscarine, is used as an antidote in addition to the supportive aid (Tegzes and Puschner, 2002).

### 8.17.4 Isoxazoles

The other and very likely *even more important group of toxicants of fly agaric* are cholinomimetic and hallucinogenic compounds *ibotenic acid, muscimol* (Figure 8.18), and *muscazone*, which belong to the group of *isoxazoles*. These substances are capable of causing excitability, restlessness, hallucinations, and delirium. The latter disorder may be harmful both to the consumer of the mushroom and to others around them. Sleepiness as well as sedation can also occur. In general, the physiological effect of these compounds resembles the effect of ethanol. The first symptoms of intoxication appear during 30–90 min, the peak arrives 2–3 h after the intake of fly agaric. The very first symptom is dizziness, followed by confusion, coordination disorders, swimming, and euphoria, further hyperkinesis, muscular convulsions, spasms, and delirium. Deep sleep and coma, lasting for 4–8 h can also occur. The intoxication can turn lifethreatening when cardiac arrhythmia, convulsions, and cardiac arrest set in. The therapy is mainly supporting.

Various data exist concerning the concentration of isoxazoles in fly agaric. So a total content of 180 mg per 100 g of dried mushrooms of the three isoxazole derivatives has been found; according to another study, the content of ibotenic acid was 25 mg in 100 g of fly agaric. The high storage stability of these compounds can be proved by the fact that the last figure was measured 7 years after the picking of the mushrooms.

*Ibotenic acid resembles glutamic acid* and in an animal organism, it tends to act similarly. *Muscimol*, formed from ibotenic acid by decarboxylation is in turn similar to the GABA (Figure 8.18c). Because of this similarity, muscimol activates GABA-receptors and acts as a sedative. A number of other effects of muscimol on the CNS can be explained by its capability to act as an agonist of GABA. The toxic doses of ibotenic acid and muscimol are 30–60 mg and 6 mg, respectively. In a human organism, a small part of ibotenic acid is metabolized to muscimol, the remaining part is excreted with urine as the

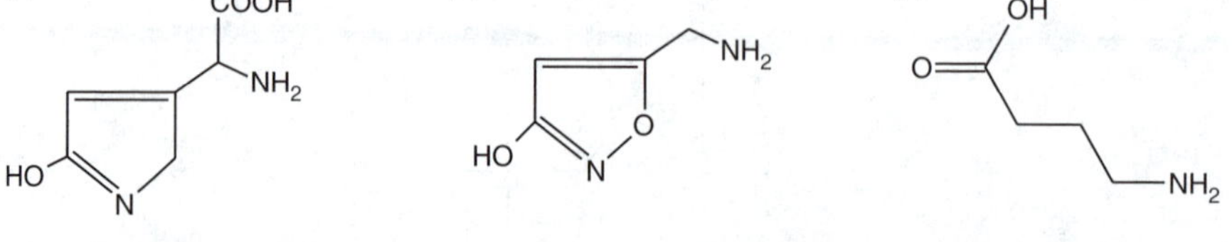

(a) COOH $NH_2$ HO N

(b) $NH_2$ O HO N

(c) OH O $NH_2$

***Figure 8.18*** Structure of (a) ibotenic acid, (b) muscimol, and (c) GABA.

initial molecule. From muscimol, one-third is excreted without any alteration, the remaining part is oxidized.

Fatalities caused by isoxazoles are still rare. Isoxazole-containing mushrooms have been used at least during the last 3000 years at ritual and religious ceremonies.

*Amanita muscaria* contains no amatoxins. A recent review of the substances identified in the genus *Amanita* has been written by Li and Oberlies (2005).

### 8.17.5 Other mushroom toxins

There are also some other toxic mushrooms such as *turbantops* or *brain mushrooms* (*Gyromitra*), *false morel* (*G. esculenta*) and *G. infula*. The first-mentioned mushroom is very toxic until uncooked, but edible after long-time scalding or drying. These mushrooms contain *gyromitrin* (*N*-methyl-*N*-formylhydrazine) that is converted into *monomethylhydrazine* (MMH) during hydrolysis. MMH elicits at first intermittent gastrospasms and weakness, and thereafter, hepatic insufficiency and hemolysis that can lead to fatalities. MMH is a highly reactive compound. It is used as a rocket fuel and industrial solvent. Since the aforementioned substances are highly volatile, their content in the mushrooms is substantially reduced during boiling (Tegzes and Puschner, 2002).

It takes 2–3 h for a specific physiological effect, similar to the effect of the antialcoholism therapy to establish, when *ink caps* (*Coprinus*) are eaten in parallel to ethanol consumption. The reaction is unpleasant and productive of mortal fear—fast pulse, gasping, lack of air, flushing of the face and neck, widening of neck veins, metal sensation of taste, tachycardia, nausea, and vomiting. The active constituent of the ink caps is the disulfiram-like compound *coprine* (Figure 8.19), that is toxic only in the presence of ethanol. Disulfiram [tetraethylthiuram disulfide (TETD)] has been used for more than 50 years as a deterrent to ethanol abuse in the management of alcoholism. Symptoms of the coprine intoxication appear typically about 30 min after the consumption of ethanol and may occur during 5 days after eating of ink caps. More about ink caps is written by Michelot (1992).

Some toxic species of the family *Cortinarius* like *C. orellanus* and *C. speciosissimus* are capable of evoking renal insufficiency through a damage of the renal tubules. The symptoms of poisoning may appear only 36 h to 14 days after eating these mushrooms. The main symptoms are nausea, paresthesia, anorexia, gastrointestinal disorders, headache, severe burning thirst, oliguria,

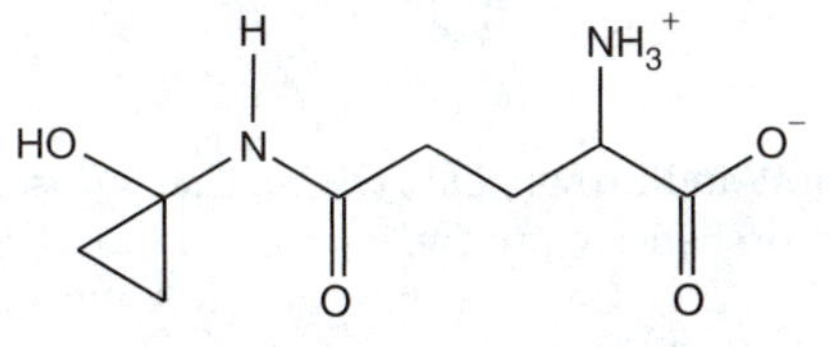

HO N H NH$_3^+$ O$^-$ O O

***Figure 8.19*** Structure of coprine.

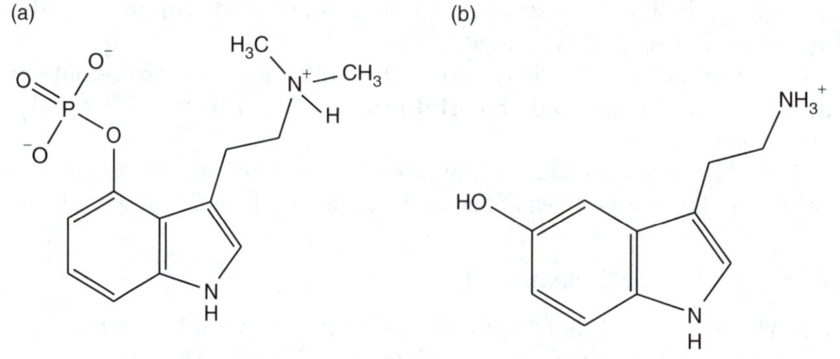

*Figure 8.20* Structure of (a) psilocybin and (b) serotonin. Molecules of both compounds contain indolyl cycle.

and renal insufficiency. The normalization of the renal function may occur either spontaneously and slowly or by the means of a prolonged alternating dialysis. Due to the introduction of dialysis, the fatalities are nowadays rare. The toxically active principles of *Cortinarius* are cyclopeptides *orellanine* and *orelline* (Karlson-Stiber and Person, 2003).

Mushrooms of the genus *Psilocybe* contain *psilocybin* (Figure 8.20) and its metabolite *psilocin* belonging to the class of indolalkaloid (subclass tryptamines) compounds. Psychoactive tryptamines are found also in the mushrooms of the genera *Paneolus, Gymnopilus,* and *Inocybe*. Since psilocybin resembles an essential neurotransmitter serotonin (Figure 8.20), these toxins are competitive agonists of a definite *serotonin* receptor. *The actual toxicant is psilocin* that is formed from psilocybine by dephosphorylation catalyzed by alkaline phosphatase immediately after ingestion. Symptoms of intoxication appear during 30–60 min after the intake of the mushrooms and are as follows: hilarity, unmotivated laugh, enforced movements, muscular weakness, sleepiness, hallucinations, and sleep. Recovery is spontaneous. Although generally there is no reason to fear death, the high possible variability of the hallucinogen content can evoke unexpected severe side effects. Quite often, the patients complain about psychical after effects (Berger and Guss, 2005).

Some species of *genus Lepiota* (parasol mushrooms) are also very toxic, but these intoxications are rare. Nausea, vomiting, and diarrhea can also be caused by some species of the genus *Russula* and white-spored agaric (*Tricholoma*).

## References

Andraws, R.J., Chawla, P. and Brown, D.L. (2005). Cardiovascular effects of ephedra alkaloids: a comprehensive review, *Prog. Cardiovasc. Dis.*, **47**, 217–225.

Andrianaivo-Rafehivola, A.A., Gaydou, E.E. and Rakotovao, L.H. (1994). Review of the biological effects of cyclopropene fatty acids, *Oloagineux,* **49**, 177–188. (in French).

Areekul, S., et al. (1978). Djenkol bean as a cause of urolithiasis, *Southeast Asian J. Trop. Med. Public. Health*, **9**, 427–432.

Berger, K.J. and Guss, D.A. (2005). Mycotoxins revisited: Part II, *J. Emerg. Med.*, **28**, 175–183.

Bhattacharya, A., et al. (2006). Biological effects of conjugated linoleic acids in health and disease, *J. Nutr. Biochem.*, **17**, 789–810.

Blake, O.A., Bennink, M.R. and Jackson, J.C. (2006). Ackee (*Blighia sapida*) hypoglycin A toxicity: dose response assessment in laboratory rats, *Food Chem. Toxicol.*, **44**, 207–213.

Cavin, C., et al. (2002). Cafestol and kahweol, two coffee specific diterpenes with anti-carcinogenic activity, *Food Chem. Technol.*, **40**, 1155–1163.

Cornwell, T., Cohick, W. and Raskin, I. (2004). Dietary phytoestrogens and health, *Phytochemistry*, **65**, 995–1016.

Dąbrowski, W. and Sikorski Z.E., Eds. (2005), *Toxins in Food*, CRC Press, Boca Raton, FL.

Ferri, T., et al. (2004). Distribution and speciation of selenium in *Lecythis ollaria* plant, *Microchem. J.*, **78**, 195–203.

Friedman, M. (2006). Potato glycoalkaloids and metabolites: roles in the plant and in the diet, *J. Agr. Food Chem.*, **54**, 8655–8681.

Geller, R.J., et al. (2006). Pediatric cyanide poisoning: causes, manifestations, management, and unmet needs, *Pediatrics*, **118**, 2146–2158.

Goncharov, N.V., Jenkins, R.O. and Radilov, A.S. (2006). Toxicology of fluoroacetate: a review, with possible directions for therapy research, *J. Appl. Toxicol.*, **26**, 148–161.

Greene, L.S. (1993). G6PD deficiency as protection against falciparum malaria: an epidemiologic critique of population and experimental studies, *Yearb. Phys. Anthropol.*, **36**, 153–178.

Gumbmann, M.R., et al. (1986). Safety of trypsin inhibitors in the diet: effects on the rat pancreas of long-term feeding of soy flour and soy protein isolate, *Adv. Exp. Med. Biol.*, **199**, 33–79.

Guthrie, N. and Carroll, K.K. (1999). Specific versus non-specific effects of dietary fat on carcinogenesis, *Prog. Lipid Res.*, **38**, 261–271.

Hammond, A.C. (1995). Leucaena toxicosis and its control in ruminants, *J. Anim. Sci.*, **73**, 1487–1492.

Heaney, R.P. (2002). Effect of caffeine on bone and calcium economy, *Food Chem. Toxicol.*, **40**, 1263–1270.

Ize-Ludlow, D., et al. (2004). Neurotoxicities in infants seen with the consumption of star anise tea, *Pediatrics*, **114**, 653–656.

Karlson-Stiber, C. and Persson, H. (2003). Cytotoxic fungi—an overview, *Toxicon*, **42**, 339–349.

Kramer, J.K., et al. (1988). Testing a short-term feeding trial to assess compositional and histopathological changes in hearts of rats fed vegetable oils, *Lipids*, **23**, 199–206.

Kumar, O., Sugendran, K. and Vijayaraghavan, R. (2003). Oxidative stress associated hepatic and renal toxicity induced by ricin in mice, *Toxicon*, **41**, 333-338.

Li, C. and Oberlies, N.H. (2005). The most widely recognized mushroom: Chemistry of the genus *Amanita, Life Sci.*, **78**, 532–538.

Mas, A. (2005). Mushroom, anatoxin and the liver, *J. Hepatol.*, **42**, 166–169.

Michelot, D. (1992). Poisoning by *Coprinus atramentarius, Nat. Toxins*, **1**, 73–80.

Oudesluys-Murphy, A.M. and Oudesluys, N. (2002). Tea: not immoral, illegal, or fattening, but is it innocuous? *Lancet*, **360**, 878.

Padosch, S.A., Lachenmeier, D.W. and Kröner, L.U. (2006). Absinthism: A fictitious 19th century syndrome with present impact, *Subst. Abuse Treat. Prev. Policy*, **1**, 14.

Rasmussen, L.H., Kroghsbo, S., Frisvad, F.C., and Hansen, H.C.B. (2003). Occurrence of the carcinogenic bracken constituent ptaquiloside in fronds, topsoils and organic soil layers in Denmark, *Chemosphere*, **51**, 117–127.

Rietjens, I.M., et al. (2005). Review—Molecular mechanisms of toxicity of important food-borne phytotoxins, *Mol. Nutr. Food Res.*, **49**, 131–158.

Shahin, M., Smith B.L. and Prakash A.S. (2003). Bracken carcinogens in human diet, *Mutat. Res.*, **443**, 69–79.

Tegzes, J.H. and Puschner, B. (2002). Toxic mushrooms, *Vet. Clin. Small Anim.*, **32**, 397–407.

Tripathi, M.K. and Mishra, A.S. (2007). Glucosinolates in animal nutrition: A review, *Anim. Feed Sci. Technol.*, **132**, 1–27.

Vasconcelos, I.M. and Oliveira, J.T.A. (2004). Antinutritional properties of plant lectins, *Toxicon*, **44**, 385–403.

Von Burg, R. (1994). Oxalic acid and sodium oxalate, *J. Appl. Toxicol.*, **14**, 233–237.

Wallace Hayes, A., Ed. (2001). *Principles and Methods of Toxicology*, 4th edn., Taylor & Francis, Philadelphia and London.

Yan, Z.-Y., et al. (2006). *Lathyrus sativus* (grass pea) and its neurotoxin ODAP, *Phytochem.*, **67**, 107–121.

# chapter nine

# Geochemical pollutants that plants absorb from soil

Plants, growing in the wild or cultivated for food or animal fodder, are capable of concentrating various compounds from the soil, some of which can prove to be toxic to the eater. Also, groundwater that is used for drinking or preparation of food may contain large amounts of compounds that are toxic. In this chapter, we will deal with three chemical elements—arsenic, selenium, and fluorine, which animals need in small quantities for their normal life. Humans also need selenium and fluorine, whereas the necessity of arsenic for humans has not been proven as yet. Arsenic compounds have been used for centuries as a means of intentional poisoning.

## 9.1 Arsenic

*Arsenic* (As) is a chemical element, a metalloid, which is absolutely necessary for the normal life of some animals such as rats, goats, and young birds. *Arsenic is widely distributed in the lithosphere.* It is linked to the ores from where zinc, copper, gold, and lead are extracted. Mining of these ores is one of the bases of human exposure to arsenic. Arsenic has found use in various pesticides. Although seafood can be contaminated by arsenic, it is usually present in seafood in the form of organic methyl-, dimethyl- or trimethyl-derivatives that are significantly less toxic than inorganic arsenic. Compounds of $As^{3+}$ are much more toxic than derivatives of $As^{5+}$; the latter form of arsenic can be reduced *in vivo* to $As^{3+}$ or methylated. Elementary arsenic is nontoxic, but its derivatives can cause disorders of both central and peripheral nervous systems and heart function, injuries of peripheral blood vessels, upper respiratory tract, liver, skin, and gastrointestinal tract, and interfere with chromosomes and the hematopoiesis system. *Intoxication with arsenic compounds lacks specific symptoms* that would help diagnosis. The most characteristic ones are typical for many toxicants—headache and abdominal pain, nausea, vomiting, and weakness. The lowest acute dose causing intoxication is 5 mg and the lethal dose is 50–500 mg arsenic oxide ($As_2O_3$) per individual. Long-term exposure to daily doses of 2–5 mg also causes emergence of symptoms like nausea, muscular weakness, and so forth. Half-life of arsenic in the human organism is 10–48 h.

*The most important source of arsenic for humans is food*—fish and meat (on average 40 $\mu$g/day). In addition, humans assimilate about 10 $\mu$g of arsenic daily from drinking water and less than 1 $\mu$g from the air. Arsenic is excreted from the organism quickly via the kidneys, but inorganic arsenic is capable of accumulating in the bones, skin, and muscles with a half-life of 2–40 days. Arsenic can cross the placental barrier.

Groundwater and water from thermal springs may also contain arsenic dissolved from rocks. *Groundwater itself can be the main source of arsenic poisonings* (arsenisms), since in different places all over the world the level of arsenic in groundwater is extremely high. According to the standard of World Health Organization (WHO) and US EPA, the *maximum allowable content* of arsenic in drinking water is 10 $\mu$g/L or 10 ppb, the ordinary number is 1–2 $\mu$g/L. Millions of people throughout the world are suffering from the toxic effects of arsenicals due to natural groundwater contamination as well as industrial effluent and drainage problems. In Taiwan, concentration of arsenic as high as 1.8 mg/L in groundwater, exceeding the WHO's standard value by 180 times, has been registered. In fact, even higher levels of arsenic have been discovered. Natural contamination of groundwater with arsenic, which can cause skin hyperpigmentation, keratosis, goiter, cancer, and a disorder of the peripheral blood circulation such as blackfoot, has been established also in places like East India, Bangladesh, Argentina, Mongolia, Mexico, southeast Hungary, in some regions of Finland (surroundings of Tampere and Hämeenlinna), and so forth. The arsenic poisoning that occurred in Bangladesh at the beginning of the 1970s through groundwater has been called the largest mass poisoning of a population in history (Scheindlin, 2005). In the United States, parts of Massachusetts, Michigan, Minnesota, California, South Dakota, Oklahoma, New Mexico, Texas, and Wisconsin have a concentration of arsenic exceeding 10 $\mu$g/mL in groundwater. In two counties of Nevada, the arsenic concentration in drinking water is about 90 $\mu$g/mL. About 3 million people in the United States drink arsenic-contaminated water. Exposure to high doses of arsenic has caused skin cancer in Taiwan as well as in Argentina. An ecological study carried out in China revealed a link between the content of arsenic in drinking water and the lethality to skin, urinary bladder, hepatic, pulmonary, renal, and prostate forms of cancer. Animal studies also have revealed the teratogenicity of arsenic.

*Acute poisonings with arsenic have also happened or been arranged.* A few examples are

- *Intoxication by beer prepared with sugar contaminated with arsenic* in the United Kingdom, in 1900–1901, when at least seventy people in Manchester, Salford, and the surrounding areas were fatally poisoned. Investigation revealed that the beer contained arsenic in two various sugars used in the brewing process. Tracing further back, it was found that the sulfuric acid used in hydrolyzing starch and cane sugar had a high arsenic content (Scheindlin, 2005).

- *By spring water, which was industrially contaminated with arsenic* in Japan and by drinking cow's milk also in Japan in 1955 that was stabilized with an industrial-grade sodium phosphate, which contained 5–8% of arsenic (expressed in arsenic acid), that was fatal for more than 100 infants (Dakeishi et al., 2006).
- *From history,* famous poisoners like Caesar, Nero, the family of Borgia, Terofania di Adamo, Katarina de Medici, and many others are known, who used arsenic oxide for their criminal acts. Arsenic intoxications have been suggested as the possible cause of death for Napoleon Bonaparte and Pyotr Tchaikovsky.

The mechanism of arsenic toxicity, including carcinogenicity is not clear yet. Most likely, here we have a complex effect where an essential role is played by oxidative stress in participation of the reactive oxygen species (ROS) (see Section 3.3.6).

A comprehensive review of all aspects of arsenic toxicology has been published by Mandal and Suzuki (2002).

## 9.2 Selenium

*Selenium* (Se) is a nonmetallic nonsubstitutable chemical element that is absolutely necessary for all living organisms. *Deficiency of selenium in food* can elicit liver necrosis, cardiomyopathia, ischemic cardiac disease, joint-muscular disease (Kaschin–Beck disease), white muscle disease, fertility reduction, several (prostatic, pulmonary, dietary tract, skin) forms of cancer, and so forth. *Selenium is an essential antioxidant* that belongs to the active center of several antioxidant enzymes like glutathione peroxidase and protects cell membranes against the toxic effect of free radicals. Lately, it has been established that selenium is also necessary for plants. Selenium raises the stress tolerance of potato cells, decelerates the aging of plants, and intensifies the use of light in photosynthesis. Selenium content in various foodstuffs is rather variable. The largest amounts of selenium can be found in fish (eel, herring, salmon, tuna) and less in meat, hen eggs, and soy beans, and is especially low in cereals.

Humans must get a daily dose of 50–200 $\mu$g of selenium with food. Overdoses of selenium, especially in the form of selenates and selenides, cause the syndrome of selenosis. *Long-term contact with selenium overdoses* has caused liver damage, enlargement of spleen and pancreas, anemia, growth retardation, and increased mortality of rats. In the case of humans, selenium overdoses cause fatigue, dullness and coolness of hands, muscular tremor, digestion disorders, nausea, vomiting, irritability, tooth damage, loss of hair and nails, skin pigmentation disorders, liver damage, and disorders of the peripheral nervous system (PNS).

Adverse effects of selenium start to appear at its chronic doses of over 1 mg/day. *A casualty was registered in 1984* when a 57-year-old woman in New York used selenium tablets for over 3 months, each of which contained, due to a mistake of the manufacturer, 27 mg of selenium (182 times) more than expected. After 11 days, a completely healthy woman started to lose her hair, to develop various injuries of nails, fatigue, nausea, and garlic-scented breath. The estimated cumulative dose of selenium she ingested over the 77 days was 2387 mg. The selenium toxicity was probably minimized by a simultaneous ingestion of large doses of vitamin C, which reduces selenite to elemental selenium, which is poorly absorbed (Anonymous, 1984).

The same health problems are being faced by people living in a region of China, where the daily dietary dose of selenium is 4 mg (Tan et al., 2002). Usually, it is difficult to acquire toxic doses of selenium, but it is most likely to happen in the case of people who have an industrial exposure to this element. Selenium is used as a constituent of fertilizers, feed compositions, plastics, paints, lacquers/varnishes, and as a catalyst in the pharmaceutical, electronic, and glass industries.

*Inorganic natural selenium* is found everywhere in small quantities, mostly as a supplement in copper, silver, lead, and nickel ores. Its content in different rocks varies in rather large limits (0.1–675 mg/kg). Selenium content is especially high in vulcanous rocks, and is low in sandstone and limestone. In soil, selenium can be found in various forms—as elementary selenium ($Se^0$), as selenide ($Se^{2-}$), selenite ($Se^{4+}$), or selenate ($Se^{6+}$) and in the form of seleneorganic compounds. Water-soluble compounds of selenium like selenates and selenides are very mobile in the soil. They can be transported into surface or groundwater and they may become hazardous to the environment. The worldwide medium content of selenium in various soils is 0.1–0.2 mg/kg, but in the arid and semiarid regions of China, Mexico, Colombia, Canada, and United States (especially in the western states), it can reach the level of 5000 mg/kg (Presser et al., 1994). The plants growing in these regions may assimilate much selenium and become toxic. In this sense, China is a very diverse country and it has soils that contain both the poorest and the richest in selenium in the world. In the thirteenth century itself, Marco Polo associated the loss of the hooves of his horses during traveling in West China with eating definite plants, mainly milk vetch or loco weed (*Astragalus*) in regions that have the highest selenium content in the world. But also China, as well as in Tibet, Siberia, and North Korea, there are regions where Kaschin–Beck disease and, only in China, endemic cardiomyopathia, which is caused by selenium deficiency, are widespread. In the Hubei province of China, selenium is concentrated in coal (up to 8%) thus causing human selenosis (of humans). Hubei is the only place in the world where human selenosis originates directly from nature (Zhu and Zheng, 2001). Nature-linked selenosis of the waterfowl has been observed in several places in the world.

*Plants are divided into three groups* according to their capability to assimilate selenium from soil:

1. Plants containing 1–10 mg/kg dw of selenium. These include the genera *Astragalus* (*Leguminosae*), *Machaeranthera* (*Compositae*), and so forth.
2. Plants containing 0.1–1 mg/kg dw of selenium. These include the genera *Aster, Atriplex, Castilleja, Menzelia,* and so forth.
3. Plants containing less than 0.03 mg/kg dw of selenium. These include most of the cultivated plants.

Selenium is assimilated from soil mostly in the form of selenates. In plants, they are converted into selenoorganic compounds like dimethylselenide, which is excreted as a gas, and amino acids like L-selenocysteine and L-selenomethionine. These selenoaminoacids reach animal and human organisms in the same form, whereby in proteins L-selenomethionine randomly replaces L-methionine. An analogical replacement of L-cysteine by L-selenocysteine does not occur. L-Selenocysteine is regarded as the 21st coded amino acid. Thirteen proteins have been found that normally contain selenium. Up to six of them, including four peroxidases, are parts of the antioxidant system of the cell. This is the reason why selenium deficiency significantly reduces the antioxidant level of the cell as well as the efficiency of different parts immune system of the organism. It has been established that optimal concentrations of selenium have an anticarcinogenic activity that can be caused by both antioxidant and immune system potentiating properties of selenium. In animal studies, selenium has demonstrated an antiangiogenic effect. Selenium is also capable, as an antagonist, of inhibiting the activity of a number of toxic heavy metals like cadmium or arsenic.

Homeostasis of selenium is regulated by excretion via kidneys. At very high doses, the volatile forms of selenium (mainly dimethylselenide) are eliminated via lungs; the exhaled air has a smell typical of garlic. Methylselenol and trimethylselonium are excreted with urine.

## 9.3 *Fluorine*

*An extremely chemically active element, fluorine* ($F_2$) is found everywhere in nature, mostly in the form of compounds. Fluorine reacts with hydrogen at room temperature in the dark with an explosion, forming an extremely toxic and corrosive hydrofluoric acid ($H_2F_2$), the salts of which are fluorides. Fluorides appear in the air as a result of volcanic activity, after which they can be transported by rain into the surface water. Fluorides leach in water also on destruction of rocks. The concentration of fluorides is 0.01–0.3 mg/L in rivers, about 1.5 mg/L in seas, and 0.02–1.0 mg/L in other water bodies. In Estonia, the maximum allowable content of fluorides in the groundwater used for drinking is 1.4 mg/L, whereas a concentration between 0.7 and 1.2 mg/L is considered to be optimal.

When the content of fluorine in drinking water is below 0.5 mg/L, formation of tooth caries is favored. Fluorides have a prophylactic anticaries effect up to their content of 1.5 mg/L; at higher concentrations, the toxicity of fluorine starts to reveal itself. The highest concentrations of fluorine in groundwater (over 10 mg/L) have been registered, once again, in China.

Of foodstuffs, the highest concentration of fluorine is in sodium chloride (8.5 mg/kg on the average). Sea fish contains normally 1.2–1.5 mg/kg of fluorine. 0.01–0.1 mg/kg is regarded as a normal content of fluorine in solid foodstuffs. In respect of the fluorine content (0.1 mg/g), notable are the leaves of the tea shrub (*Camellia sinensis*). This content allows one to ingest 0.4–0.8 mg of fluorine within 3–4 cups of tea. In the case of normal content of fluorides in drinking water, the total recommended daily dose of fluorine for humans is derived from food (0.56 mg) and drinking water (1 mg).

Fluorine is used in the chemical industry for the production of plastics and pesticides. The most famous fluorine-containing plastic is the extremely chemical resistant polytetrafluoroethylene (teflon). Fluorine is added to toothpaste as an anticaries agent.

The absorbed fluorine as well as fluorides accumulate partly in bones, and the remaining part is excreted rather quickly with urine, sweat, and feces.

The *acute toxicity* of fluorine and fluorides manifests itself locally in a severe corrosive effect; gaseous fluorine damages lungs, skin, eyes, and heart; and hydrogen fluoride irritates the eyes, skin, and lungs. The injuries can be life threatening, and as much as 5 mg of the ingested sodium fluoride is lethal. *Fluorides also have chronic toxicity*. Long-term consumption of drinking water containing 2 mg/L or more of fluorine may cause *fluorosis of the teeth*. Large populations in the developing world suffer the effects of chronic endemic fluorosis. More than 200 million people worldwide are expected to drink water with fluorine in excess of the WHO-recommended guideline value (1.5 mg/L).

Depending on body weight, daily doses of over 20–80 mg, taken over the years, may cause a chronic fluorine intoxication. Such doses can be obtained in places with a fluorine concentration in the drinking water of approximately 10 mg/L or working in an industry using fluorides. By far, the fluorine-amended toothpastes are not recommended everywhere for use. An excess of fluorides may also cause neurological diseases like Alzheimer disease and dementia.

## References

Anonymous. (1984). Epidemiologic notes and reports selenium intoxication, *MMWR Morb. Mort. Wkly Rep.* **33**, 157–158.

Dakeishi, M., Murata, K. and Grandjean, P. (2006). Long-term consequences of arsenic poisoning during infancy due to contaminated milk powder, *Environmental Health: A Global Access Science Source* 5:31; http://www.ehjournal.net/content/5/1/31

Mandal, B.K. and Suzuki, K.T. (2002). Arsenic round the world: a review, *Talanta*, **58**, 201–235.

Presser, T.S., Sylvester, M.A. and Low, W.H. (1994). Bioaccumulation of selenium from natural geologic sources and its potential consequences, *Environ. Manage.*, **18**, 423–436.

Scheindlin, S. (2005). The duplicious nature of inorganic arsenic, *Mol. Interv.*, **5**, 60–64.

Tan, J., et al. (2002). Selenium in soil and endemic diseases in China, *Sci. Total Environ.*, **284**, 227–235.

Zhu, J. and Zheng, B. (2001). Distribution of selenium in a mini-landscape of Yutangba, Enshi, Hubei Province, China, *Appl. Geochem.*, **16**, 1333–1344.

# chapter ten

# *Environmental pollutants*

*Environmental pollutants include toxic metals, radionuclides, polychlorobiphenyls (PCB), and dioxins* that can be found in nature in trace amounts as a result of human unconcerned industrial, military, and other activities. The results of carelessness are now continuously returning to us like a boomerang in the form of anthropogenous pollutants. This chapter will consider mainly such toxic effects of these substances that manifest in the case of humans through contacts with food or beverages.

## 10.1 Toxic chemical elements

### 10.1.1 Mercury

Most *mercury* (*Hg*) finds itself in the atmosphere after combustion of fossil fuels like coal or petroleum, both of which contain a remarkable amount. Mercury is used in the production of chlorine, polymers, and paints. Mercury exists in three principally different forms:

1. *Elementary mercury* ($Hg^0$) that is absorbed as vapor and that causes damage to the central nervous system (CNS).
2. *Salts of inorganic mercury* ($Hg^+$ and $Hg^{2+}$) that are poorly absorbed, but the absorbed part is capable of causing renal damage.
3. Easily absorbable *organic mercury* ($R–Hg^+$), where R is mainly a methyl, ethyl or phenyl group.

The lion's share of dietary mercury intoxications originate from the Hg-organic fungicides like dimethylmercury, chloride and phosphate of methylmercury, and chloride and acetate of phenylmercury. Since these extremely toxic compounds are lipid soluble, they absorb easily and bioaccumulate in erythrocytes and in the CNS. During the last century, methylmercury was used as a seed disinfectant. From the seed grain, methylmercury was transmitted into domestic animal feed that caused the death of a large number of animals. Thus, use of methylmercury was prohibited.

*Inorganic mercury* has been used in the industrial production of plastics, paper, and electric batteries. Waste waters contaminated with inorganic mercury can reach rivers and lakes. Methylmercury and even more volatile dimethylmercury (preferably at alkaline pH) found in fish have been synthesized by anaerobic microorganisms from the salts of inorganic mercury deposited in the bottom sediments of water bodies. The most well-known casualty of this type happened in the late 1950s at *Minamata Bay* (Japan), where a local chemical factory started to deposit its waste waters that contained inorganic, presumably low-toxic, mercury chloride into the bay. The fish caught in the bay were extremely toxic and caused local fishermen and their families to fall ill with the so-called *Minamata disease*. As a result, about 700 cases of Minamata disease were recorded, 70 of them with a fatal end. Respective studies showed that both samples of the bottom sediments and fish caught in the bay contained toxic methylmercury (Kudo et al., 1998). Analogical mass intoxications, including those with fungicides containing mercury also have been registered in other parts of the world. High contents of mercury have been discovered in meat and in some milk products. In Japan, problems have occurred with rice plants treated with compounds of mercury.

*The mechanisms of the toxic action of various forms of mercury* are similar to each other. They are based on the reaction of both elementary and ionic particles of mercury *with free sulfhydrylic (SH) groups* of the substances. These SH-groups may belong to protein molecules as functionally essential structural units. Formation of this complex can cause an inhibition of the activity (e.g., enzymatic activity) of this protein. Especially susceptible to methylmercury is the cerebral cortex, where the inhibition of the protein synthesis will be followed by the death of cells and necrosis of the nervous tissue.

*Both acute and chronic intoxications* are characteristic for mercury. Symptoms of *acute poisoning* with inorganic compounds of mercury are necrotic damage of the cells of renal tubules; in the case of mercury-organic compounds, there occur nonspecific disorders of the CNS like paresthesia, numbness around the mouth and in limbs, tremor, ataxia, difficulties in swallowing and articulation, neurasthenia, general weakness, fatigue, concentration difficulties, loss of sight and hearing, coma, and death.

*The chronic toxic effect* of inorganic mercury appears as glomerular nephropathia caused by immune complexes and organic mercury as brain edema, decomposition of brain gray cells, formation of the glial tissue, and brain atrophy. Mercury is also toxic to the embryo (teratogenicity) and to babies through the mother's milk. The half-life of methylmercury in the organism is about 70 days, being localized mainly in the liver and in the brain. The brain bears 10%–20% of the body burden of methylmercury. The toxicity of mercury has been recently reviewed by Magos and Clarkson (2006).

Bioaccumulation and biomagnification are characteristic for mercury. Its concentration in sea fish (up to 5 mg/kg) may be 1,000 times higher than in seawater.

Nevertheless, the *level of mercury in nature has stabilized* during the last 50 years and new casualties with mercury have been caused either by eating ish grown in the vicinity of industrial enterprises, or by other emergencies.

In addition to fish, some risk to human health may also be caused by vapors of mercury from *tooth amalgam* and *Hg-thimerosal,* which is used as an antiseptic in vaccines (Clarkson, 2002). The allowable tolerance dose is 0.35 mg of Hg per week in the case of an adult weighing 70 kg, of which maximally 0.2 mg can be the extremely toxic methylmercury. The estimated dietary exposure to mercury in different European countries in the 1980s and 1990s was in the range of 0.7 in the Netherlands and 13.5 $\mu$g/day in Belgium (Nassredine and Parent-Massin, 2002).

In the European Union (EU), mercury belongs with lead, cadmium, and arsenic as toxic elements that are monitored in food raw material by a special program.

### *10.1.2 Lead*

*Lead* (Pb) is a long-known toxicant. Lead poisoning was described long ago by Hippocrates in 300 BC. There is a hypothesis that the fall of the Roman Empire was at least partly caused by lead intoxications. From the skeletons of the old Romans, high concentrations of lead have actually been found. It could be caused by use of lead plumbing and utensils as well as use of lead in the glaze of pottery. Industrial lead intoxications that for centuries were connected with mining and melting of lead, became rather frequent at the end of the nineteenth century. Although the toxicity of lead is nowadays well known, the issue is still on the agenda. For example, it was found recently that in the blood of 13% of babies in Glasgow (UK), the content of lead exceeded 10 $\mu$g/L, the limit in the WHO's guideline (Watt et al., 1996). This number is sufficiently high and a matter of worry.

Lead is assimilated, above all, from food, water, and air. Although its concentration in food can be higher than in air, the pulmonary absorption of lead is considerably more efficient than that from the gastrointestinal tract. Infants are more susceptible to lead than adults since they assimilate higher amounts of lead from the digestive tract. In the case of children, lead poisoning can appear as encephalopathy together with mental fixation, brain attack, and brain paralysis. The brain is the first target organ for lead, especially in the case of infants. It is not completely clear yet, whether a single case of lead intoxication is sufficient to cause permanent brain damage.

Concentrations of lead in different food products vary greatly. *It is estimated that approximately half of human lead intake is through food*—with around a half of it originating from plants where the content of lead is higher. Still in some places, existing lead pipelines cause elevated lead concentrations in drinking water. The dietary intake of lead in the EU is in the interval of 16 $\mu$g/day in Spain and 280 $\mu$g/day in Italy (Nassredine and Parent-Massin, 2002). The last number is especially worrying.

*Absorption.* Metallic lead as well as its salts absorb slowly and incompletely from the mouth cavity. Only 2–20% of the ingested lead is absorbed and the remaining part is excreted from the organism with feces. Acidic diets as well as diets that are deficient in calcium, zinc, or protein are able to increase the absorption of lead. Ninety percent of the absorbed lead is transported as a protein-complex into erythrocytes, where its half-life is 2–3 weeks. The concentration of lead in the blood serum is very low.

*Deposition*. Part of lead is redistributed in the form of di- or tri phosphate into the liver and kidneys, and further to the bile and bones. It is possible to see in a light microscope the lead–protein complexes in the kidneys. In bones, lead is deposited in hydroxyapatite, where lead is determinable by x-rays even after centuries. The presence of lead in the bone marrow hinders hematopoiesis. Lead crosses the placenta and accumulates in the embryo. This process can result in a delay of the embryonal nervous system and in a spontaneous abortion or early birth. Owing to the immaturity of the blood–brain barrier of newborns, lead is capable of accumulating in the CNS.

*Excretion*. In the case of low concentrations in the blood, lead is excreted actively in bile and in the case of high concentrations, in urine. Some data have been obtained that show that 5% of the lead found in the blood is excreted to milk.

*Mechanisms and appearance of toxic effects*. Toxic or degenerative effects of lead manifest themselves in the nervous system, digestive tract, and in the hematopoiesis system. A normal concentration of lead in blood is in the interval of 0.15–0.7 μg/mL, the threshold concentration of toxicity is about 0.8 μg/mL; symptoms of encephalopathy (nonspecific colic and abdominal pain, fatigue, and constipation) appear at the concentration of lead 1–2 μg/mL. Anemia and CNS disorders appear after chronic contact with lead. Biochemical effects occur already at lower concentrations—lead is capable of influencing the synthesis of heme and porphyrin via the enzymes participating in these processes, and syntheses of myoglobin and cytochrome P450 monooxygenase complex (CYP) can be influenced as well. Lead binds to the SH groups of the proteins, influencing the activity of the enzymes containing these groups. Lead is also able to compete with zinc for a place in the molecules of some enzymes.

*Lead influences several phases of the heme synthesis*. Due to the effect of lead on porphyrin synthesis, the level of hemoglobin is reduced. Since the level of free erythrocytary protoporphyrin in turn rises, *aminolevulinic acid dehydrase* (ALAD) will in turn be inhibited. Enzymatic activity of ALAD, estimated by measuring the concentration of aminolevulinic acid in urine is the most sensitive measure of an organism's exposure to lead; the inhibition rate is well correlated with the content of lead in the blood. $EC_{50}$ of ALAD inhibition is 0.4 μg/mL of Pb. Most lead poisonings are subacute. *A chronic exposure to lead* causes both alterations in an infant's skeleton and nephritis; acute exposure causes renal damage.

Lipid-soluble and more easily absorbable organic lead is much more toxic than inorganic lead. For example, triethyllead that is formed from tetraethyllead during combustion of gasoline penetrates the skin easily and reaches the brain, causing encephalopathy. Intoxication develops quickly and the symptoms are the imagining of ghosts, hallucinations, and ataxia. Chronic poisoning with tri- and tetraethylleads has three main phases:

1. Disorders of the CNS as well as vegetative nervous system, impotency.
2. General weakness with tremor in the fingers, nystagmus, and equilibrium disturbance. Psychical disorders like excitedness and fear, disorders of memory and sleep, and intellectual decline start to take shape.
3. Psychosis, irreversible psychical disorders, low blood pressure, and myocardial damage.

A recent review of lead neurotoxicity has been published by Toscano and Guilarte (2005).

*Transport in the environment*. Environmental pollution by lead increased continuously due to industrialization and the use of gasoline supplemented with lead. Tetraethyllead, which is being used to increase the octane number of gasoline, is converted into lead oxide and other inorganic compounds of lead during combustion. Most of these compounds are found in the 30 m-wide strip of soil next to highways. It has been calculated that some years ago, almost all lead appeared in the environment as a result of human actions, half of which was caused by combustion of leaded gasoline. Over the last few years, the use of this gasoline has been decreased and the content of lead in gasoline has also been reduced. These measures have helped reduce lead pollution. Since lead binds tightly to soil particles, not all of toxic metal reaches the plants. Plants with a large area of leaves like spinach or cabbage may contain more lead if they are cultivated in the vicinity of lead emission areas. When the polluted plants are fed to animals, their organisms do not absorb much of the lead, most of the metal is excreted with feces. However, some lead gets accumulated in the bones and hair. An interesting fact is that the hair and bones of the people who lived in the preindustrial epoch contained a considerably larger amount of lead than those of contemporary people.

### *10.1.3 Cadmium*

*Most of the environmental cadmium* (*Cd*) *is of anthropogenic origin*, which has reached the atmosphere from the cadmium smelters and processing or burning of cadmium-containing products (plastics, paints, rubber, batteries, domestic waste, and wastewaters). Sources of cadmium pollution may also be mineral fertilizers and fungicides and both surface and tap water. Tap water can get the cadmium pollution from water pipes made from black polyethylene, copper, or galvanized iron. *Cadmium is capable of traveling* long distances with dust and rainwater, being thereafter sucked into the soil, from where it can be transported into water and plants. Since cadmium is relatively volatile, it is mainly an inhalable toxicant. At inhalation, cadmium first accumulates in the lungs, where it is absorbed to the extent of 40%. The absorbed part of cadmium is mostly bound to the blood plasma proteins. Then follows accumulation of cadmium in the liver ($\tau_{1/2}$ = 5–10 years), in the kidneys ($\tau_{1/2}$ = 20–30 years) and in the spleen. *Unlike* $Pb^{2+}$ *and* $Hg^{2+}$ *ions,* $Cd^{2+}$ *ions are successfully absorbed by plants,* especially well by root crops like turnips and leaf vegetables like spinach. In plants, cadmium is distributed quite evenly among all tissues. Removal of the outer layer of a leaf does not help reduce contamination. Cadmium is also well absorbed by some mushrooms.

Although cadmium is absorbed rather poorly from the intestines (only in range of 5–8%), *humans still receive cadmium mainly from food* (1/3 from animal and 2/3 from plant products) and cigarette smoke (good absorption), and also from drinking water and inhaled air. It is possible to get a dose of cadmium of up to 1.7 $\mu$g (of cadmium) by smoking of a single cigarette. According to the information from the United States, the highest cadmium content is in potatoes, vegetables, and cereals, and there is considerably less cadmium in fruits and beverages. In Japan, intoxications have occurred from eating rice that is contaminated with cadmium. In the raw food material of animal origin, cadmium has been found mainly in the internal organs (liver and kidneys) and in milk. Acute cadmium intoxication can be caused by acidic food and beverages (e.g., lemonade, lemon-scented ice tea, and sherbets) that are stroed in containers containing cadmium. *The highest risk groups are still smokers* who live in an industrial area or have an occupational exposure to cadmium. During long-term administration, cadmium accumulates in the human organism, mostly in the liver and kidneys. The dose causing a damage of the renal tubules is 0.2–0.3 mg of Cd per renal cortex. *About 15% of cadmium poisonings have a fatal end*. In general, the human organism is capable of metabolizing cadmium into harmless compounds, but very high doses of cadmium may evoke serious health problems. The allowable weekly tolerance dose is 0.5 mg of Cd. In Germany, the average weekly dose of cadmium assimilated by one person is 0.19 mg.

*The acute toxicity of cadmium* manifesting itself in pulmonary as well as renal or testicular damage may lead to death. Vapors of cadmium cause severe tracheobronchitis, pneumonitis, and pulmonary edema with accompanying general symptoms like sweating and shivers (chill). In the case of cadmium intoxications caused by food or drinking water, mainly gastrointestinal disorders like increased salivation, nausea, vomiting, and diarrhea appear which may cause shock and death.

Targets for the *chronic toxicity of cadmium* are the *lungs* and *kidneys*. As a result of renal damage, protein and saccharide appear in urine. Since cadmium accumulates in the kidneys in complex with protein metallothionein, the renal damage may appear as a delayed effect after single doses. *Metallothionein* is a low-molecular protein that takes part in the transportation of metals in the body. Owing to its chemical similarity with zinc, cadmium also induces the synthesis of metallothionein in complex with which cadmium is transported into the kidneys, where it will be filtrated through the glomerulus and reabsorbed by proximal tubular cells. In these cells, the complex is destroyed by proteases and cytotoxic cadmium is deliberated.

Testicular damage appears some hours after a single contact with cadmium. Results are necrosis, degeneration, and complete loss of spermatozoids. Cadmium reduces bloodflow through the testes and, due to a lack of oxygen and nutrients, ischemic necrosis will develop.

Because of a disturbance of calcium metabolism by cadmium, the bones turn fragile (osteomalacia). In a later phase, there appear hypertonia, infertility, and chronic pulmonary and renal insufficiency. Cadmium may cause also prostate and pulmonary cancer.

*Antidotes of cadmium* are the complex formers—EDTA (ethylene diamine tetraacetic acid) and its salts or DTPA (diethylenetriamine-pentaacetic acid) and compounds of calcium. $\tau_{1/2}$ of cadmium in the organism is 7–30 years and it is excreted by the kidneys after being damaged.

Cadmium toxicity has recently been reviewed by Godt et al. (2006).

## 10.1.4 Chromium

*Chromium* (Cr) is a metallic chemical element occurring in various oxidation states that are usually 0, +2, +3, and +6. Biologically, the most important forms are $Cr^{3+}$ that can be easily found in nature and are *exclusively anthropogenic* $Cr^{6+}$. Sources of environmental chromium are combustion of fossil fuels, manufacturing of several metals and plastics, galvanization, leather processing, and printing.

*In foods and food additives,* chromium is present in the $Cr^{3+}$ form. The most significant sources are processed lean meat (about 0.2 mg/kg), full-grain products (about 0.14 mg/kg), cheese, pork kidney, brewer's yeast, legumes, and spices. $Cr^{3+}$ is poorly absorbed (only 0.5–2%) and excreted mainly with feces; the small absorbed part is conjugated with blood proteins like transferrin and transported into the liver. Chromium picolinate is better absorbed. The absorbed part of Cr is excreted mainly with urine, and also with sweat and bile.

Chromium, being a constituent of some enzymes, is necessary for organisms in microamounts. It has been shown that $Cr^{3+}$ potentiates the effect of insulin, influencing in this way the metabolism of carbohydrates, lipids, and proteins. $Cr^{3+}$ hinders binding of iron to transferrin, thus influencing the metabolism and bioaccumulation of iron. A recommended daily dose of chromium is 0.05–0.2 mg for adults and 0.01–1.0 mg for infants.

*In larger doses, chromium exerts carcinogenicity and teratogenicity,* whereas $Cr^{6+}$ is more toxic than $Cr^{3+}$. The toxicity of the dietary $Cr^{3+}$ is low, partly due to its low absorptivity. For example, no adverse effects were observed during 24 weeks in case of a daily administration with food of 750 mg/kg body weight (bw) of chromic acid to the laboratory animals. Synthetic chromium picolinate, having a higher lipophilicity than other salts of $Cr^{3+}$as well as chromium chloride, did not cause any toxic effects at daily doses of 15 mg/kg bw, although an increase of the chromium content in the tissues proved its absorption. Still higher daily doses of chromium (about 100 mg/kg bw) exerted both a reproductive and developmental toxicity.

A review of chromium toxicity has recently been published by Costa and Klein (2006).

### *10.1.5 Copper*

Copper (Cu) is a metallic element that reaches the environment through the manufacture of various metals, paints, rubber, and electric accumulators, through the wastes of the printing and building material industry, and from the dumping grounds not meeting the requirements. Copper is also present in some pesticides. Copper is naturally transferred to the air from volcanoes and from forest fires, its concentration in the air is 5–20 ng/m$^3$ and in the soil about 50 mg/kg. Copper is also liberated from the ground during mineralization.

Since copper is a chemical element necessary for organisms, both its efficiency and excess are *problematic*. Copper is found in different cells and tissues, but most of all in the liver. It can occur both in an oxidated ($Cu^{2+}$) and in a reduced ($Cu^{+}$) form. Copper is a cofactor of a number of important enzymes like cytochrome-C oxidase, tyrosinase, and Cu–Zn superoxide dismutase (Cu–Zn–SOD). *In the case of mammals, most of the need for copper is covered by food*. Rich in copper are foods like liver, oysters, nuts, legumes, cereals, and dried fruits. About 6–13% of the need comes from drinking water. The absorptivity of copper depends on several factors, including its chemical form and presence of other food components. Copper is expected to be transported through biomembranes in complex with amino acids like histidine, methionine, and cysteine via the respective amino acid transport systems. Reduced glutathione and a number of organic acids also facilitate the absorption of copper; the process is hindered by zinc, iron, molybdenum, calcium, and vitamin C. In the blood, copper is transported in complex with proteins like ceruloplasmin, albumin, and transcuprein.

*The daily dose of copper* necessary for a human organism *that is equal to its safety dose*, is 2–5 mg. Copper deficiency causes anemia and loss of hair color. In case of copper overdose through drinking water, or in case of suicide attempts with copper sulfate, the *symptoms of acute intoxication* are nausea, vomiting, diarrhea and abdominal pain, hemolytic anemia, necrosis of the liver and kidneys, and death. *Symptoms of chronic intoxication with copper* are irritation of the gastrointestinal tract and liver cirrhosis accompanied by injuries of the renal tubules, brain, and other organs. In more severe cases, there occur hepatic necrosis, collapse of blood vessels, and death. Owing to its various redox-forms, copper takes part in the initiation of such neurodegenerative illnesses as Alzheimer's and Parkinson's diseases and amyotrophic lateral sclerosis. $LD_{50}$ of copper is 30 mg/kg in the case of rats. In forms of $Cu^{2+}$, $CuOH^{+}$, and $CuOH_2^{2+}$ copper is very toxic to water animals, especially to young individuals.

*The toxic effect of copper has been most often explained by the generation of the oxidative stress* (see also Section 3.3.6) that participates in the formation of reactive oxygen species (ROS) in the cell (Gaetke and Chow, 2003). In the presence of superoxide radical ions ($\cdot O_2^-$) or reducing agents like ascorbic acid (vitamin C) or glutathione (GSH), $Cu^{2+}$ is reduced to $Cu^+$. The latter is capable of catalyzing the formation of hydroxyl radicals ($\cdot OH$). *The hydroxyl radical is the most potent oxidizing radical in biological systems. It can react with almost every biomolecule.* It is able to remove a hydrogen atom both from the carbon directly linked to an amino group with the formation of a carbon-centered protein radical or from a molecule of an unsaturated fatty acid with the formation of a lipid radical. In both cases, an initiation of oxidative damage, also involving molecules of DNA, is carried out in the cell. During reactions between lipid radicals and oxygen, lipid peroxiradicals are formed, which alter the fluidity and permeability of the cell membranes and also react directly with DNA and proteins.

*The toxicity of copper is reduced by dietary antioxidants* like vitamins C and E, $\beta$-carotene, $\alpha$-lipoic acid, plant polyphenols, selenium, zinc, and so forth. A review of copper toxicity has been recently published by Gaetke and Chow (2003).

### 10.1.6 Nickel

Nickel (Ni) is a metallic element that is transported into food mainly with industrial pollution and also by equipment and utensils used for food processing. It has been found that in the course of hot processing of various foods for over 1 h, as much nickel is dissolved from stainless steel utensils as is necessary for the nickel content in food of 0.13–0.22 ppb. In the case of acidic foods, this number can be even as high as 400 ppm. A nickel alloy catalyst, used for hydrogenation of dietary oils, can also serve as a source of pollution.

*At high concentrations in food, nickel can cause some adverse effects,* such as a reduction of prolactin secretion in the pituitary gland or insulin secretion in the pancreas. At the inhalation, nickel is capable of causing cancer of the nasal cavity and lungs. The toxicity of nickel to the kidneys and liver as well as immunotoxicity and teratogenicity has been shown in animal tests. Especially toxic is colorless volatile carbonyl nickel [$Ni(CO)_4$], a dose of which of 30 ppm within 30 min is lethal for humans. A good review of nickel essentiality and toxicity has been published by Denkhaus and Salnikow (2002).

## 10.2 Radionuclides

Organisms that have been in existence on the earth for millennia have become accustomed to and are generally adapted to several types of radiation. A continuously increasing use of radiation and such casualties as the explosion of the Chernobyl atomic power plant, accompanied by an elevated exposure to radiation have turned radiation toxicology into an essential branch of toxicology. Radiation toxicology investigates the adverse effects of radiation on organisms. Cellular injuries caused by radiation may lead either to alterations in physiological activity or death of the cell, to formation of cancer, or even to the death of the organism. Radiation may originate from different sources and be either directly or indirectly ionizing. Directly ionizing radiation bears an electric charge that interacts with the atoms located in tissues either by electrostatic attraction or by repulsion. Indirectly ionizing radiation bears no charges but generates charged particles in the tissues where it is absorbed.

Radioactive irradiation can be either natural or anthropogenic. Natural irradiation is divided according to its origin, that is, into cosmic and terrestrial.

Anthropogenic radiation is divided according to the purpose of generation:

- Medical—diagnostic x-rays, radiomedicines, cancer therapy
- Production of nuclear energy, including nuclear disasters and radioactive wastes
- Nuclear weapons

*From the viewpoint of food toxicology,* chemical elements with unstable nuclei or radionuclides that decompose with a specific half-life ($\tau_{1/2}$) and that are either natural or produced by nuclear disasters are essential.

Internal radiation, which forms 11–17% of the average exposure of a population to radiation, is caused by radionuclides present in the body. Part of these sources of radiation are evenly distributed all over the body, part are concentrated in particular organs. The decomposition of the latter is locally absorbed. *The main routes of accumulation of naturally occurring radionuclides* of hydrogen, carbon, potassium, lead, polonium, radium, radon, thorium, uranium, and bismuth *are inhalation and ingestion with food or water*. In terrestrial ecosystems, radionuclides like $^{210}Ra$, $^{226}Ra$, and $^{222}Rn$ are absorbed by plants as a result of metabolic processes. In addition to the assimilation by roots, direct contamination through leaves is essential. The last process is of special importance for the entry of radionuclides into the food chain, since leaves are involved in the direct contact with animals or are eaten up by herbivores. Radionuclides are transported into other parts of the plant from the roots and leaves.

*Atmospheric radionuclides are accumulated in surface water and soil.* From there, they can be transported both into water bodies and into underground water, from where they may return to surface water and get into the biosphere. Radionuclides present in industrial and municipal wastewaters may also move first into terrestrial water bodies, further into estuaries and coastal waters. The latter are biologically very active and productive. Phytoplankton living in these water bodies converts the mineral substances including radionuclides into feed for higher organisms. This feed is first eaten by zooplankton, then by bottom fish and animals. The magnitude of the human hazard caused by radionuclides depends on the part of the plant where these particles are concentrated, whether it is edible or not. For example, such radionuclides as $^{60}Co$ or $^{65}Zn$ are concentrated mainly in edible tissues but $^{226}Ra$

and $^{90}Sr$ are usually deposited in the nonedible shells of oysters, scallops, crabs, and so forth.

Assimilation and accumulation of radionuclides depends on the entry site of the organism, on their solubility, metabolism, and particle size. The routes of entrance are inhalation, swallowing, and percutaneous absorption. Most problematic is inhalation, but many problems like solubility and efficiency of absorbance are also connected with the assimilation of radionuclides from food. The skin is usually a sufficiently good barrier that hinders the entrance of radionuclides into the organism.

The natural soil-borne radionuclides that have entered the food chain have already been in food for millennia. For example, most foods contain 1–10 Curie/g radioactive isotope of potassium with an atomic mass of 40 ($^{40}K$) and with $\tau_{1/2} = 1.32 \times 10^9$ years. From the viewpoint of general risk, it is essential to estimate the concentration of the radioactive iodine ($^{131}I$) in milk. $^{131}I$ is very toxic and its content in milk increased after the catastrophe of Chernobyl. $^{131}I$ appears in milk 3–4 days after emission of the radioactive material into the environment and milk reaches the drinker sooner than any other foodstuff contaminated by the sediments. Milk is also one of the most important sources of nutrients and calories for children, who are especially sensitive to radioactive iodine. Food can be contaminated also by radioactive strontium ($^{90}Sr$) and cesium ($^{137}Cs$), the concentration of which in soil increased after terrestrial tests of the nuclear weapon. In the EU, estimation of the content of these radioactive isotopes in foodstuffs of animal origin is included into yearly screening programs.

*Most damaging are the radionuclides that are able to penetrate into soft tissues* and hence to enter into the active metabolism. This is characteristic of $^{137}Cs$, with $\tau_{1/2} = 27$ years, that quickly absorbs into the blood circulation and thereafter into every cell of the body. Another important radionuclide $^{90}Sr$, with $\tau_{1/2} = 28$ years, is also easily absorbed from both the gastrointestinal tract and lungs and is accumulated in the bone tissue. Peroral administration of this isotope causes bone cancer and leukemia with a high incidence. This isotope is especially harmful to infants below 10 years. At high concentrations, $^{131}I$ decomposes the thyroid gland, thereby decreasing the production of the thyroid hormone. Concentrations of $^{131}I$ strongly damage the thyroid gland, still *leaving* a proliferative ability to cause formation of hyperplastic cancer. Again, children are about twice as susceptible as adults. The antidote is potassium iodine, which minimizes the reach of $^{131}I$ with $\tau_{1/2} = 8.1$ days in the thyroid gland.

## 10.3 Polychlorinated biphenyls

On chlorination of biphenyl $(C_6H_5)_2$, a mixture of chlorinated biphenyl molecules containing a various number of chlorine atoms (1–10) in different positions (positions 2-6 and 2′-6′ in Figure 10.1) is formed. Theoretically, the formation of about 210 different related compounds or congeners is possible. The toxicity of the polychlorinated biphenyls (PCB) obtained in this way

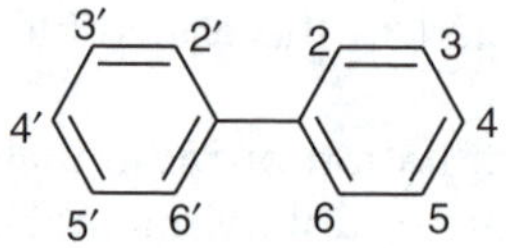

***Figure 10.1*** General formula of polychlorinated biphenyls.

depends on the number and position of chlorine atoms in the molecule as well as on the composition of the mixture. *PCBs are entirely anthropogenic substances;* no natural sources have been established. PCBs are tasteless and odorless, either colorless or slightly yellowish, water-insoluble oily substances with a low vapor pressure. They are soluble in most organic solvents, oils, and fats, are very stable and degrade slowly. Decomposition of PCBs requires a high temperature or a catalyst.

PCBs have been used since the early 1930s, widely as plastifiers of synthetic polymers, in paints, inks, thermal exchange media, hydraulic presses, vacuum pumps, and electric transformers. It has been estimated that at the end of the 1980s about 400,000 tn of PCBs could be found on the face of the Earth, two thirds of it in the seas. Although the production and external use of PCBs has been practically stopped, they can be still found in old electrical installations and in the environment (soil, atmosphere, and water), consequently also in food (eggs, game, fish, etc.). Since the expenses of decomposition of the PCB- containing material are very high, mixing of PCB oils with mineral and cooking oils was started instead. The latter items were used in the composition of animal feed until the Belgian dioxin or chicken scandal of 1999 and the analogical intoxication accidents that occurred (see Section 10.4). The study performed by US FDA established that the initial reason of several casualties connected with PCBs was an animal feed contaminated with these compounds, from which PCBs were transported to poultry and eggs.

*PCBs have a high thermal and chemical stability in water, acids, and alkalis.* They can be transported by air or water very far from the place where they entered the environment. From water, they are accumulated by fish and water animals. Thermal processing only somewhat reduces the content of PCBs in fish and other food.

Rather highly contaminated Baltic fish contains several times more PCBs than, for example, the fish from the North Atlantic. Due to some restrictions and bans of the Baltic Marine Environment Protection Commission or Helsinki Commission (HELCOM), the content of PCBs in herring has decreased 60–80% in the course of the last 25 years. The same trend with a yearly rate of 6–10% has occurred in the case of cod and perch.

*Moving higher up along the food chain,* a remarkable biomagnification of PCB residues takes place (Burreau, 2006). Since PCBs are hydrophobic and practically insoluble in water, the organisms are not capable either of eliminating them with urine or of metabolizing them into compounds of higher solubility in water. Therefore, the $\tau_{1/2}$ of PCBs in organisms is about 10 years. PCBs can migrate to food also from the packaging material. For example, a case of poisoning with a PCBs-containing repeatedly used paper package of rice oil has been established.

*The toxic effect of PCBs is generally similar to the effect of chloroorganic pesticides* (see Section 14.1). The first case of human poisoning with PCBs and dioxins happened with contaminated rice bran oil in Japan in 1968. Japanese called this intoxication "oil disease" (*yusho*). By the year 2003, high amounts of PCBs and polychlorinated dibensofurans (PCDFs) were still present in a number of patients with yusho. The patients still suffered from various mucocutaneous and subjective symptoms, and these symptoms were correlated to the blood levels of polychlorinated congeners (Furue et al., 2005).

*The symptoms of acute toxicity of PCBs are* pimples; itching; dullness of limbs; edema of eyelids; epiphora; burning of eyes, nose, and throat; skin pigmentation; and hepatic damage. PCBs are capable of crossing the placentary barrier and acting as teratogens. In case of contamination of mother's milk, defects in visual memory of babies appear.

*Chronic toxicity of PCBs.* Exposure of laboratory animals to PCBs during some weeks or months has caused anemia; alterations in leukocytes; rash; injuries of the liver, stomach, and thyroid gland; disorders of the immune system; and alterations in behavior. The chronic exposure of test animals to PCBs also increased the risk of cancer. PCBs are expected to act as endocrine disrupters, inducing estrogen-like feminization of animals also at the embryonal stage. Birds are especially susceptible to the toxic effects of PCBs. It has been shown that it is possible to reduce the adverse effects of PCBs on the test animals by administration of vitamin A.

A review of the PCB toxic effects on the nervous and immune systems has been published by Fonnum et al. (2006).

The content of PCB residues in Baltic fish is yearly determined by the Estonian national screening program of food contaminants by ECD-GC method. Eighteen fish were taken for PCB analysis in 2003; in part of the samples, the content of 12 chlororganic pesticides was simultaneously determined. The limits of detection (LOD) and quantification (LOQ) of PCBs and chlororganic pesticides were 1 $\mu$g/kg and 10 $\mu$g/kg respectively. The medium content of PCBs in Baltic fish was 36 $\mu$g/kg compared to 15 $\mu$g/kg of the Atlantic fish. Higher contents of PCBs were established in eel and flat pod (up to 36 $\mu$g/kg) in sprat and in the Baltic herring (Margna and Reinik, 2004).

## 10.4 Polychlorinated dibensodioxins and dibensofurans

These versatile toxic chemical compounds go by their general names *dioxin* and *dioxin-like compounds*.

*Dioxins are congeners or supplements* of various organic substances containing chlorine or bromine like PCBs, discussed in the previous paragraph. From altogether 75 chlorinated dibensodioxins and 153 chlorinated dibensofurans, especially toxic and environmentally stable are seven polychlorinated dibenso-*p*-dioxins (PCDD) and ten PCDFs. *Dioxins are not intentionally manufactured technical chemicals,* but are the side products of synthesis of various chlororganic compounds. They did not exist before industrialization except in very small amounts due to natural combustion and geological processes. They are formed during almost all high-temperature (200–600°C) processes taking place in the participation of substances containing either organic or inorganic chlorine. For example, dioxins are formed in the incineration process of chlorine-containing wastes if the combustion is incomplete. Dioxins can be unintentionally produced also in the chemical industry dealing with chlorine. Therefore, nowadays, dioxins can be found everywhere. Fortunately, data suggests that both the body burdens and environmental levels of dioxins are steadily declining (Charnley and Kimbrough, 2006).

Sedimentation of dioxins from the atmosphere and into dietary plants leads to an unavoidable exposure of both domestic animals and humans to these substances. Bovines accumulate dioxins from feed to their fat. The transmission rate of dioxins to milk fat and other foodstuffs of animal origin is in the interval of 0%–40%, depending on the particular compound. Dioxins appearing in the sea in composition of industrial wastewaters are accumulated by fish. For example, in the Baltic Sea, the "best" accumulators are Baltic herrings; fewer dioxins are accumulated in sprats and salmons. It has been estimated that the Europeans ingest over 90% of their dioxin from fish and meat dishes.

*Symmetric 2,3,7,8-tetrachlorodibensodioxin or dioxin* (TCDD), the most well-known isomer of dioxins, is extraordinarily toxic to rodents ($LD_{50}$ = 0.6 $\mu$g/kg) and is a very potent carcinogen (Figure 10.2). The no-observed-effect-level (NOEL) of dioxin is 1 ng/kg, if assimilated daily during 2 years. The toxicity of the other congeners and related substances is lower. For assessment of the summary toxicity of dioxin-like compounds, a value such as *toxic equivalents quotient* (TEQ) has been implemented. The TEQ-value expresses toxicity as if the mixture of dioxins were solely TCDD. The medium summary daily intake of dioxin-like compounds is 1–2 pg TEQ per kg bw. With $\tau_{1/2}$ = 8 years, it leads to contents of dioxin of about 30 ppt TEQ in the adrenal tissue. The TEQ concept was first developed by the New York State Health Department in a series of experiments in response to the need for reentry criteria of an office building contaminated by a mixture of PCBs, PCDFs, and dioxins following an electrical transformer fire (Schecter et al., 2006). In the EU, the maximum permissible level for PCDD/Fs is 4 pg WHO-TEQ per gram fresh weight.

*The symptoms of dioxin acute toxicity are* skin injuries (e.g., chloroacne) and systemic toxic effects to the liver and endocrine system. In case of a sufficiently early stopping of the exposure to dioxin, these effects are still slowly reversible and do not lead to any long-term health disorders.

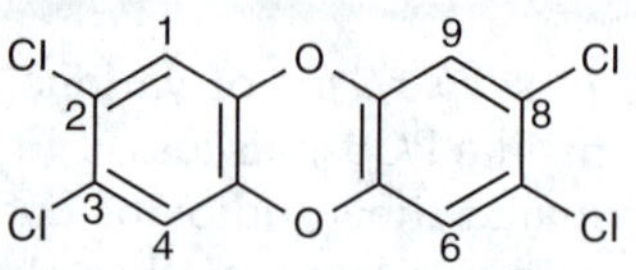

***Figure 10.2*** Structure of 2,3,7,8-tetrachlorodibensodioxin or dioxin.

The toxic effect of TCDD and many other aromatic compounds manifests itself via aryl hydrocarbon (Ah) receptors (AhR) (see also Section 3.3.2). The overall mechanism of the effect is as follows:

1. Dioxin or a dioxin-like compound binds to AhR
2. Formed complex is transported to the cellular nucleus
3. In the nucleus, the complex induces the transcription of a number of genes, taking to an increase of the synthesis of the respective proteins

Among these proteins, there are, first of all, the CYP isoenzyme CYP1A1 and also alcohol dehydrogenase, NADPH-quinone oxidoreductase, and conjugating enzymes of phase II—glutathione S-transferase and UDP-glucuronosyl transferase. Induction of CYP1A1 causes, in turn, formation of toxic (carcinogenic) metabolites of substrates of this enzyme like benzo[$\alpha$]-pyrene. AhR complex influences also the expression of other genes essential for other physiological processes like growth, differentiation, and apoptosis.

Dioxin is also a potent immunosuppressant, which via damage of the thymus, reduces the production of lymphocytes in the gland. Dioxin also exerts varied adverse effects on the mammal reproductive system—spermatogenesis, sexual behavior, and fertility. An adverse effect on the reproductive system of rats and monkeys has been proven at a daily dose of dioxin of 0.001 $\mu$g/kg. Formation of the AhR complex is also considered to have the leading role in the teratogenic effect of dioxin-like compounds, as well as in the effect on the skin, eyes, and liver.

Dioxin-like substances can be determined both directly and indirectly (Behnisch et al., 2001). Since the list is rather educating, we will present it briefly here:

1. *Method of biomarkers* for assessment of exposure to the dioxin-like compounds. Good *in vivo* biomarkers in plants, animals, and humans are, for example, enzymes CYP1A1-IA, aryl hydrocarbon hydroxylase (AHH), and ethoxyresorufine *O*-deethylase (EROD).
2. *In vivo biotests,* which are based on the AhR-depending mechanisms, induction of an enzyme like EROD/AHH, luciferase test, cell proliferation, keratinization, porphyrin accumulation, and binding of AhR or DNA. For measuring of specific biological response, either AhR-containing extracts or mammal cell cultures are used. Most of these tests are based on the induction of gene expression, the induction rate is expressed with respect to 2,3,7,8-TCDD.
3. *Enzyme-immunotests.* PCDD/PCDF is bound from solution to specific antibodies immobilized to the tubes of RIA. After removal of unbound material, the enzyme is bound to the free antibodies. The amount of the enzyme is inversely proportional to the content of PCDD/PCDF in the solution.
4. *Chromatographic methods* (GC/MS, LC-MS/MS) for identification and quantification of different dioxin-like substances.

5. *Quantitative structure–activity relationship* (*QSAR*). This is used also for calculation of TEQ of dioxin-like substances.

Dioxins have caused a number of widely known casualties like the accident of Seveso in Italy (1976), the Belgian dioxin or chicken scandal (1999), the Brazilian scandal of citric flesh bullets, and so forth (Fiedler et al., 2000).

*The disaster in Seveso* was the first major case showing the essentiality of estimation of dioxins as environmental toxicants. In the local pesticide factory, spraying of chemicals was registered after an exothermic reaction went out of control. The toxic cloud polluted large areas and about 200,000 persons inhaled toxic compounds, most important of which was 2,3,7,8-TCDD. The main symptoms of an immediate acute intoxication were skin injuries, and as a result of a delayed acute intoxication, an increased rate of cancer and teratogenic injuries were registered in the region 20 years later.

*The Belgian dioxin contamination* was discovered due to the biological effect of the toxicant. In January 1999, Belgian chicken farmers noticed that about a quarter of their chickens died before reaching maturity. In addition, disorders in the hens' nervous systems appeared and some eggs did not hatch. The laboratory studies revealed a high content of dioxin (1–2 ng/g of fat) in both the adipose tissue of hens and in their feed that exceeded the normal background numbers by about 1000 times. It was further discovered that in analogical samples of chicken feed, meat, and eggs, the concentration of dioxin-linked PCBs was at the ppm level, that is, still over 1000 times higher (Bernard et al., 2002).

Tracing revealed that the whole big scandal was caused by the mixing of 25 L of transformator oil into 107 tn of animal fat anticipated for addition to the chicken feed. From this amount, 90 tn were used for production of poultry feed, and the rest for production of milk and beef. By the beginning of October 1999, this scandal had involved 505 poultry, 1625 pig, and 411 dairy farms in Belgium. Due to the fact that birds are especially sensitive to the toxic effect of dioxins, the extent of possible human intoxication both with dioxins and PCBs, less toxic but present in poultry in 1000 times higher concentrations, was considerably reduced. All birds that had eaten the contaminated feed were destroyed. The EU then demanded from Belgium the safely certification of exported foodstuffs with a fat content exceeding 2%. The limit of the dioxin content in fat was set at 200 ppb as the sum of seven main congeners. About 50,000 chromatographic analyses of dioxin content were performed in laboratories of Belgium in 1999. The direct loss of the Belgium dioxin scandal was estimated to be $1 billion, indirect losses could have been even thrice as high. The timely utilization of contaminated transformatory oil would have cost about $1000, that is, one billion times less.

Since concentrations of PCBs in contaminated fat are sufficiently higher and the analyses are faster and cheaper than the analyses of dioxins, it is possible to use analyses of PCBs for estimation of the content of dioxins in fat in the case of the known PCB/dioxin ratio (w/w).

*Contamination of Brazilian citric flesh products with dioxins* was discovered in Europe. From July 1997 to March 1998 in Germany, in the course of local monitoring, a dioxin content that was twice as high as the usual one was measured in milk. It turned out that the same problem occurred in other EU member states as well. Investigations led to the citric flesh products imported from Brazil, the dioxin contamination of which emanated from lime erroneously used for production of complex feed. These bullets were already in use for preparation of the ruminant food in the whole of western Europe. Simultaneously, an increased content of dioxin was discovered both in beef and veal and in mother's milk. This incident demonstrates how in the case of exposure of large human populations to a toxicant, the local monitoring program, although with a definite delay, helps discover the contamination.

*A German incident of choline chloride contamination* took place in 2000. Pine sawdust that was heavily contaminated with pentachlorophenol was added to the food additive choline chloride of Belgium origin, by a Spanish premix manufacturer. The use, in food additives, of sawdust and other materials of wood origin treated with wood preservatives is prohibited in the EU.

*The Brandenburg incident* took place in Germany in 1999. During an investigation of the reasons of the high content of dioxin in hen eggs, the tracing led to the grass flour dried on wood flame. Part of the firewood was treated with preservatives and painted.

A recent review on dioxins has been published by Schecter et al. (2006).

## References

Behnisch, P.A., Hosoe, K. and Hasaka, S. (2001). Bioanalytical screening methods for dioxins and dioxin-like compounds—a review of bioassay/biomarker technology, *Environ. Int.*, **27**, 413–439.

Bernard, A., et al. (2002). The Belgian PCB/Dioxin incident: analysis of the food chain contamination and health risk evaluation, *Environ. Res.*, **88**, 1–18.

Burreau, S., et al. (2006). Biomagnification of PBDEs and PCBs in food webs from the Baltic Sea and the northern Atlantic Ocean, *Sci. Total Environ.*, **366**, 659–672.

Charnley, G. and Kimbrough, R.D. (2006). Overview of exposure, toxicity, and risks to children from current levels of 2,3,7,8-tetrachlorodibenzo-*p*-dioxin and related compounds in the USA, *Food Chem. Toxicol.*, **44**, 601–615.

Clarkson, T.W. (2002). Three modern faces of mercury, *Env. Health Perspect.*, **110**, Suppl **1**, 11–23.

Costa, M. and Klein, C.B. (2006). Toxicity and carcinogenicity of chromium compounds in humans, *Crit. Rev. Toxicol.*, **36(2)**, 155–163.

Denkhaus, E. and Salnikow, K. (2002). Nickel essentiality, toxicity and carcinogenicity, *Crit. Rev. Oncol. Hematol.*, **42**, 35–56.

Fiedler, H., et al. (2000). *Evaluation of the occurrence of PCDDE/PCDF and POPs in waste and their potential to enter the food chain*. European Commission, DG Environment.

Fonnum, F., Mariussen, E. and Reistad, T. (2006). Molecular mechanisms involved in the toxic effects of polychlorinated biphenyls (PCBs) and brominated flame

retardants (BFRs), *J. Toxicol. Environ. Health*, A, **69**, 21–35.
Furue, M., et al. (2005). Overview of Yusho, *J. Dermatol. Sci.*, Suppl. 1, S3–S10.
Gaetke, L.M. and Chow, C.K. (2003). Copper toxicity, oxidative stress and antioxidant nutrients, *Toxicology*, **189**, 147–163.
Godt, J., et al. (2006). The toxicity of cadmium and resulting hazards for human health, *J. Occup. Med. Toxicol.*, 1, 22; http://www.occup-med.com/content/1/1/22
Kudo, A., et al. (1998). Lessons from Minamata mercury pollution, Japan—After a continuous 22 years of observation, *Water Sci. Technol.*, **38**, 187–193.
Magos, L. and Clarkson, T.W. (2006). Overview of the clinical toxicity of mercury. *Ann. Clin. Biochem.*, **43** (Pt 4), 257–268.
Margna, L. and Reinik, M. (2004). Monitoring programs of food quality and safety—monitoring of contaminants, 2003. Tartu Laboratory of Health Inspection of Estonia (in Estonian).
Nassredine, L. and Parent-Massin, D. (2002). Food contamination by metals and pesticides in the European Union. Should we worry? *Toxicol. Lett.*, **127**, 29–41.
Schecter, A., Birnbaum, L., Ryan, J.J. and Constable, J.D. (2006). Dioxins: an overview, *Environ. Res.*, **101**, 419–428.
Toscano, C.D. and Guilarte, T.R. (2005). Lead neurotoxicity: from exposure to molecular effects. *Brain Res. Rev.*, **49**, 529–554.
Watt, G.C., et al. (1996). Is lead in tap water still a public health problem? An observational study in Glasgow, *BMJ*, **313**, 979–981.

# *chapter eleven*

# *Mycotoxins*

## *11.1 Overview*

Mycotoxins include over 250 detected toxins that are produced in favorable conditions by at least 120 different micro- or mold-fungi (molds). The adverse effects of many of these potentially toxic fungi are still to be studied. The most important and most studied mycotoxins, originating from the species of the genera *Aspergillus, Fusarium, Penicillium, Byssochlamis,* and *Claviceps* are briefly characterized in Table 11.1.

Reviews of toxicity, metabolism, and the impact of all the important mycotoxins have been published by Hussein and Brasel (2001) and by Yannikouris and Jouany (2002).

Several noteworthy mycotoxicological events are known from history. There are some references to ergotism in the Old Testament of the Bible and fusariotoxins are believed, for example, to be responsible for the downfall of the Etruscan civilization. The mysterious death of several archeologists investigating the Egyptian pyramids has been attributed to ochratoxin A (OTA) contained in the mummies (Yannikouris and Jouany, 2002).

## *11.2 Aflatoxins*

*Aflatoxins* (*AT*) are structurally related coumarin derivatives, altogether at least 13 types, the molecules of which contain a domain of condensed dihydrofuran (Figure 11.1). They are produced by microfungi, mostly belonging to the species *Aspergillus flavus,* which has given the name to the whole group of toxins. The four main aflatoxins, namely, $B_1$, $B_2$, $G_1$, and $G_2$ are produced by the microfungi characterized in Table 11.1 in plant feed or food raw material that has not been sufficiently dried after harvesting but stored as half-dry at relatively high temperatures. Letters B (blue) and G (green) indicate the color of the respective aflatoxin band at the thin layer chromatography (TLC) plate irradiated by ultraviolet (UV).

*Aflatoxins* $B_1$ ($ATB_1$) and $B_2$ ($ATB_2$) are produced by both *A. flavus* and *A. parasiticus*, aflatoxins $G_1$ ($ATG_1$) and $G_2$ ($ATG_2$) only by *A. parasiticus*. AT of B- and G-groups often occur together in various relative amounts, the dominating type is mostly aflatoxin $B_1$. The food materials most often

***Table 11.1*** Characterization of the Most Important Mycotoxins

| Microfungus/mold | Toxin | Acute toxicity ($LD_{50}$ mg/kg) perorally | Toxic effect | Distribution |
|---|---|---|---|---|
| *Aspergillus flavus, A. parasiticus, A. nomius* | Aflatoxins | 7.2 (rat) | Hepatic cirrhosis and cancer | Peanut, other nuts (almond, Brazilian nut), corn, and other cereals, milk |
| *Aspergillus ochraceus melleus* | Ochratoxin A | 20 (rat) | Fatty liver and hepatic damage | Barley, corn, wheat, rye |
| *Aspergillus versicolor, A. nidulans* | Sterigmatocystin | 120 (rat) | Hepatic cancer | Corn, wheat, feed |
| *Fusarium graminearum* | Zearalenone (Fusariotoxin F2) | 0.1; during 5 days (pig), (estrogenic activity) | Estrogen, infertility | Corn and other cereals, feed |
| *Fusarium oxysporum F. trincinctum* | Fusariotoxin T2 | 3.8 (rat) | Toxic aleukia, bleeding syndrome | Cereals, feed |
| *Fusarium roseum F. graminearum* | Vomitoxin | 70 (mouse) | Vomiting | Cereals, feed |
| *Fusarium moniliforme F. proliferatum* | Fumonisins | — | Leukoenceophalo-malacia, hepatic cancer | Corn |
| *Fusarium poae F. sporotrichioides F. graminareum F. culmorum F. roseum* | Trichothecenes (T-2 toxin and others) | 0.85 (rat); 1.1 (rabbit) | Atrophy of bone marrow and thymus, necrotic angina, sepsis | Corn, barley, rye, wheat, milk |
| *Penicillium expansum, P. urticae, Byssochlamis nivea, B. fulva* | Patulin | 35 (mouse) | Cellular toxin | Spoilt fruits, fruit juices |
| *Penicillium citrinum, P. viridicatum* | Citrinin | 50 (rat) | Convulsions, paralysis,coma, gasp, arrest of respiration | Wheat, barley, rye, oats |
| *Claviceps purpurea* | Ergot alkaloids | | Ergotism (gangreous or convulsive) | Mainly rye, less wheat, oats, barley |

*Source:* Adapted from Belitz, H.-D. and Grosch, W. (1999) *Food chemistry*, 2nd edn., Springer; Chan, P.K. et al., (1984) *Toxicol. Appl. Pharmacol.*, **73**, 402–410; Fuchs, R. and Peraica, M., (2005) *Food Addit. Contam.*, **22**, Suppl 1, 53–57; McKean, C., et al., *J. Appl. Toxicol.*, **26**, 139–147, 2006.

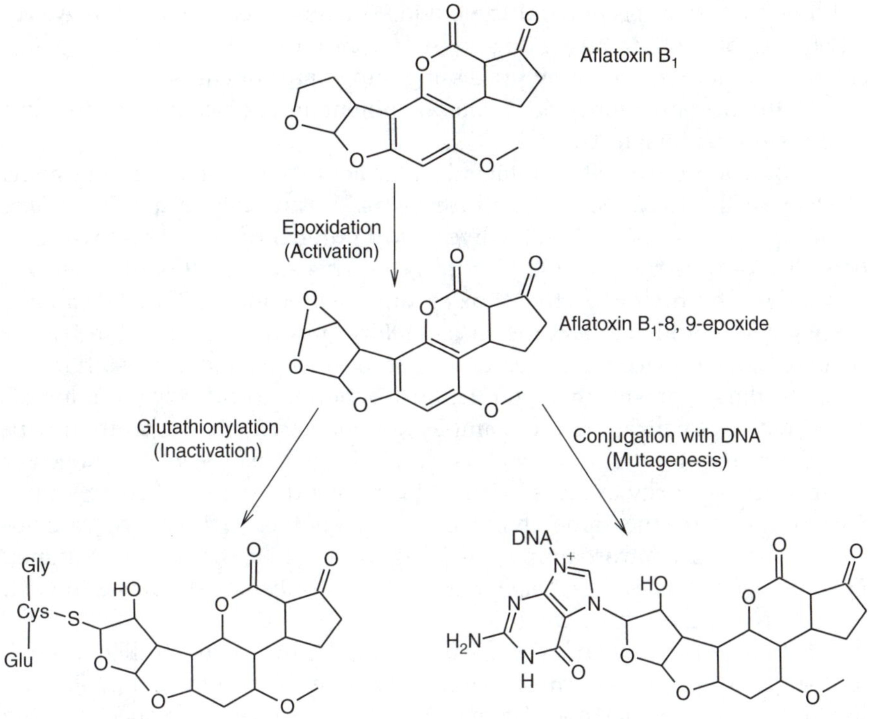

*Figure 11.1* Scheme of metabolic activation, conjugation, and inactivation of $ATB_1$.

contaminated with aflatoxins are peanuts, followed by various nuts, cereals, cottonseeds, dried fig, soybeans, almonds, paprika, and spices.

Synthesis of aflatoxins by molds growing in the tropics and subtropics is a complex process where a number of enzymes, and hence genes, are involved. Although formation of the highest concentrations of aflatoxins is undoubtedly caused by the postharvest storage of foodstuffs in unsuitable conditions (above all, at a high temperature and water activity), an earlier contamination of the plant material with aflatoxins is also possible. The spores of A. *flavus* and A. *parasiticus* may germinate, for example, in the style of peanut or corn followed by penetration of the spore to the embryo imitating the pollen tube. Formation of a fungal spawn (mycelium) may not injure the normal plant, although the synthesis of aflatoxin may already start in its tissue. Although the concentration of the aflatoxin formed in this way is never as high as in the case of a postharvest formation, about two-thirds of corn in North Carolina in 1980 already contained 20 $\mu$g/kg of aflatoxin at harvest time.

Human contact with aflatoxins takes place mainly by eating of plant material that is somehow contaminated with these toxins. When a domestic animal eats contaminated feed, aflatoxins move into its tissues and milk. In a mammal organism, the aflatoxins of B-group are converted by hydroxylation to *aflatoxin* $M_1$ ($ATM_1$), belonging to the M-group of aflatoxins. For example,

1–3% of $ATB_1$ that has entered the bovine organism reaches milk as $ATM_1$. $ATM_1$ is stable both in raw milk and in milk products. It is not decomposed either by pasteurization or by processing into yogurt or cheese.

All the aforementioned aflatoxins are potent mutagens, carcinogens, and teratogens in animal tests.

$ATB_1$ has proved to be mutagenic after activation in many test systems like HeLa cells, *Bacillus subtilis, Neurospora crassa,* and *Salmonella typhimurium.* $ATB_1$ is partly metabolized in the liver into a number of highly reactive compounds like $ATB_1$-8,9-epoxide (Figure 11.1). This step renders the aflatoxin molecule with a highly reactive epoxy group (see Section 2.4.2) that is capable of conjugating with a guanidine base of DNA. It leads to the breaking of the chain of DNA, replacement of base pairs, transversion, or to frameshift mutations. In the organism, the conjugation reaction is fortunately counterbalanced with detoxification, for example, via nonenzymatic conjugation with glutathion. The most widespread aflatoxin $B_1$ is the most potent carcinogen of all. Its carcinogenicity appears in animal tests at a daily dose of 10 $\mu$g/kg bw. For comparison—the same characteristic of another well-known carcinogen *N,N*-dimethylnitrosoamine is 7.5 mg/kg bw, that is, 750 times higher. *Epidemiological studies have shown* that hepatitis B substantially raises the risk of primary hepatocellular carcinoma.

*$ATB_1$ is also acutely toxic* ($LD_{50}$ = 0.3–18 mg/kg bw) to all studied animal, bird, and fish species. The most indifferent species is the female rat, followed by the hamster, mouse, and the male rat. The most sensitive species is the rabbit, followed by the cat, dog, and pig. Humans lie, with $LD_{50}$ = 5 mg/kg bw, in the center of the scale of the sensitivities (Moss, 2002). *Acute toxicity of aflatoxins* may reveal itself either in the death of the animal without any warning symptoms or in symptoms of anorexia, depression, ataxia, gasping, and bleeding from the body openings.

Typical symptoms of *subchronic intoxications* are jaundice, hypoprotrombinemia, hematomes, and gastroenteritis. *Chronic aflatoxicosis* that happens primarily in the case of domestic animals and also humans is characterized by cholangial proliferation, periportal fibrosis, jaundice, liver cirrhosis, weight loss, and an elevated sensibility to illnesses. Long-term exposure of animals to low doses of $ATB_1$ may lead to the formation of hepatome, cholangial, or hepatocellular carcinoma, and other malignant tumors.

In addition, $ATB_1$ inhibits the synthesis of DNA, activity of the DNA-dependent RNA polymerase, and hence, synthesis of mRNA and protein molecules. Results can be

- Formation of the so-called fatty liver, connected with a loss of ability to remove fats from the liver
- Coagulopathy caused by inhibition of prothrombin synthesis
- Reduced immunofunction

A less widespread clinical syndrome of humans exposed to aflatoxins is childhood cirrhosis in India, Taiwan, and in some African countries.

A good review of all aspects of aflatoxin toxicology has been published by Williams et al. (2004).

*Recognition of serious risks* connected with the consumption of food contaminated with aflatoxins has been followed by an acceptance of various legislative acts, regulating international trade of contaminated food and feed items. *The action limit set by the US FDA* is 100–300 ppb for corn and other feed items and 0.5 ppb ($\mu$g/kg) for milk. For other items of food, the action limit has been set to 20 ppb (FDA Compliance Policy Guides 7106.10, 7120.26, 7126.23). *EU Commission Directive* 98/53 of July 16, 1998 fixed the maximum aflatoxin content in commodities, anticipated for human food, to 2 $\mu$g/kg. For famine mitigation programs, this limit has been raised to 30 $\mu$g/kg by Food and Agriculture Organization (FAO) and World Health Organization (WHO).

*The high toxicity of aflatoxins has stimulated studies* of methods of inhibition of their formation as well as decomposition of the aflatoxins that are already formed in food. For that purpose, UV-radiation, heat, oxidizing agents such as hydrogen peroxide or sodium hypochlorite or alkaline substances such as ammonia, sodium bisulfite, or gaseous sulfur dioxide have been tested. Owing to the danger of formation of new toxic products and deterioration of the sensoric and other properties of food, the progress is connected with additional toxicological and other studies and is therefore very slow.

*Analysis methods*. For the estimation of the concentration of aflatoxins in foods and feeds, various methods have been elaborated. For the extraction of raw material, typical solvents such as acetone, chloroform, or methanol and their mixtures are used. Purification and concentration of the extract is carried out by solid-phase extraction (SPE) on microcolumns filled with $C_{18}$ or immunoaffinity sorbent. For identification and quantitation of aflatoxins, both chromatographic (HPLC, TLC) or immunochemical (ELISA) methods are used. In the case of HPLC, both normal and reversed phase columns are used; the detection is carried out either by a UV or fluorescence or MS/MS-detector; the limits of detection (LOD) and quantification (LOQ) are at the level of ng/g (Sforza et al., 2006).

## 11.3 Ochratoxins

*Ochratoxins* (OT) are a group of seven stable derivatives of isocoumarin linked via amide bond to amino acid phenylalanine, that are produced by the microfungi *A. ochraceus* and *Penicillium verrucosum*. These fungi contaminate barley, corn, wheat, oats, rye, green coffee beans, peanuts, grape juice and wine, cocoa, dried fruits, and spices. The concentrations determined are in the interval of 0.3–1.6 $\mu$g/kg of ppb in cereals; 0.8 $\mu$g/kg in coffee; and 0.01–0.1 $\mu$g/kg in wine. *OTA* (Figure 11.2) has caused proximal tubular lesion of kidneys and hepatic degeneration in the animal tests. The toxicity of OTA substantially depends on the animal species—its acute peroral $LD_{50}$ is 0.2 mg/kg in the case of dogs and 59 mg/kg in the case of mice.

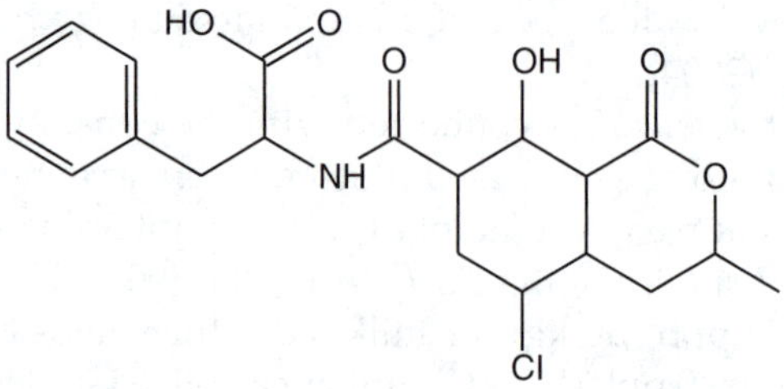

*Figure 11.2* Structure of ochratoxin A.

An existence of a direct link between ingestion of high doses of OTA and *endemic nephropathy* has been proven for humans and pigs in the *Balkan countries* (Bosnia and Herzegovina, Bulgaria, Croatia, Romania, Serbia, and Montenegro) (Fuchs and Peraica, 2005) and for pigs in the United States and in Denmark. Nephrotoxic OTA moves from fodder grain mainly to the swine blood and kidneys, but it can be found also in muscles, liver, and adipose tissue. The primary symptoms of poisoning, such as fatigue, tiredness, anorexia, diffuse abdominal pain, and severe anemia are followed by symptoms of renal damage such as a sequentially reduced concentration ability, reduced intrarenal bloodstream, reduced glomerular filtration accompanied by general and microscopic alterations of the kidneys, including necrosis, fibrosis, glomerular hyalinization, and interstitial sclerosis, followed by death via uremia. OTAs are also teratogens and genotoxic carcinogens, causing hepatoma and renal adenoma in mice.

Specific for OTA, the toxic effect on the cellular level is connected with the enzymes of glucose metabolism (reduced gluconeogenesis) and of anion transport, leading to intercellular alkalinization.

A recent review of ochratoxin toxicology focused on OT in coffee, a future key commodity in ochratoxin research and regulation has been published by Bayman and Baker (2006).

*Methods of analysis.* Mixtures of water with various organic solvents are used for sample extraction; extract prepurification is carried out on immunoaffinity columns and analysis made by HPLC-FlD, TLC, or ELISA method. In European Union (EU), the limit (MRL) concentration of OTA in food cereals and grain products is 0.005 mg/kg.

## 11.4 Sterigmatocystin

*Sterigmatocystin,* structurally related to aflatoxins and a precursor of aflatoxin $B_1$, is produced by microfungi like *A. flavus, A. versicolor; A. sydowi, A. nidulans, Bipolaris* spp., *Chaetomium* spp., and *Emericella* spp. The toxin was first detected in brown rice that was stored at normal conditions; it has also been found in wheat, green coffee beans, corn, and cheese. Like aflatoxin, sterigmatocystin (Figure 11.3) is a mutagen, teratogen, and hepatic carcinogen. $LD_{50}$ is 800 mg/kg for mice, the 10-day $LD_{50}$ is 166 mg/kg and 120 mg/

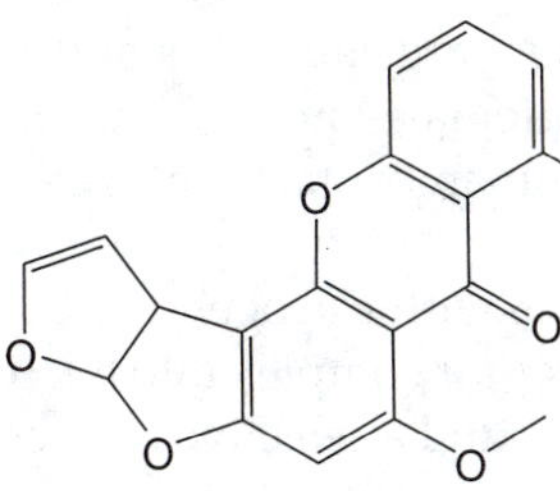

*Figure 11.3* Structure of sterigmatocystin.

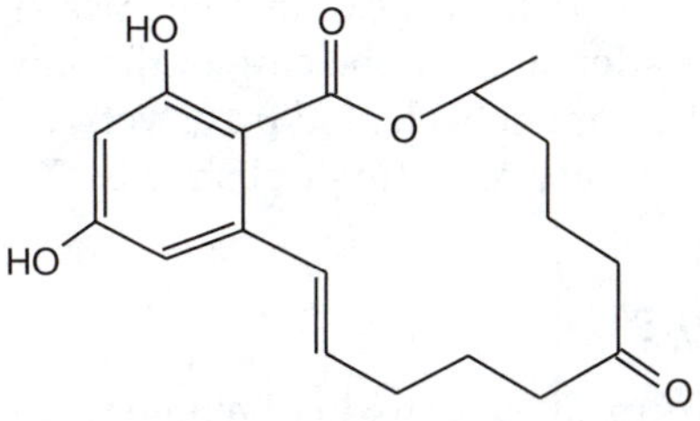

*Figure 11.4* Structure of zearalenone.

kg, respectively, in the case of male and female Wistar rats. The symptoms of chronic intoxication with sterigmatocystin are hepatomas of rats, pulmonary cancer of mice, and renal and hepatic cancer in the case of cynomolgus, rhesus, and African green monkeys (Thorgeirsson et al., 1994). It has been estimated that the carcinogenicity of sterigmatocystin is ten times lower than the same characteristic of aflatoxin $B_1$.

The bovines who had eaten feed containing about 8 mg/kg of sterigmatocystin had acquired diarrhea and exhibited reduction of milk productivity. Lethal cases were also registered.

*Analysis*. The methods elaborated for sterigmatocystin are of relatively low sensitivity in comparison with the analysis methods of most other mycotoxins. For a preliminary screening, usually TLC is used for a quantification of sterigmatocystin HPLC, most sensitive is a LC-MS method anticipated for sterigmatocystin analysis in cheese, bread, and corn products.

## 11.5 Zearalenone

*Zearalenone and zearalenol are endocrine disrupters.* They are synthesized by the microfungi *Fusarium* (mainly by *F. roseum* or *graminearum* and *F. culmorum*), contaminating primarily corn, but also wheat, barley, oats, cassava, soy, sorgo, bananas, and other fruits. Zearalenone (Figure 11.4), known also as *F-2 toxin*, has also been found in beer. Formation of the toxin on corn and other cereal grains is favored by long-period low positive temperatures and oscillations between low and medium temperatures. Zearalenone causes porcine vulvovaginitis. Its metabolites are estrogens, giving rise to hyperestrogenicity syndrome. Large amounts of the metabolites are capable of causing infertility of both female and male animals. Although the hog is the most sensitive animal, human contact with zearalenone and metabolites is also harmful.

Interaction of zearalenone with estrogen receptors is followed by transportation of the estrogen–receptor complex into the cellular nucleus, conjugation with chromatin receptors, and a selective transcription of RNA. It results in a number of biochemical effects like an increase of the muscular content of water and decrease of the lipid content; enhancing of uterus permeability in relation to glucose, RNA, and preproteins. Zearalenone is genotoxic to bacteria. It forms adducts in the cells of female mice and induces the formation of hepatocellular adenomas in the case of the same animals. Its human carcinogenicity is not completely cleared up as yet. A comprehensive review of zearalenone has been very recently published by Zinedine et al. (2007).

## 11.6 Fumonisins

Fumonisins are a group of mycotoxins belonging to the genus *Fusarium,* the most important of which are *F. monoliforme* and *F. proliferatum.* They are produced by some molds infecting especially corn. Three fumonisins—$B_1$, $B_2$, and $B_3$ are the best known, less known is $B_4$. About 70% of the total fumonisins in contaminated food or feed constitutes fumonisin $B_1$ ($FB_1$—Figure 11.5). Fumonisins cause a number of adverse biological effects such as equine leukoencephalomalacia, hepatic and renal toxicoses of horses, hogs, and rats; porcine pulmonary edema and hepatic cancer of rats. It is connected with a high rate of esophageal cancer in the Transkei region of South Africa as well as with endemic hepatic cancer in some regions of China. The mechanisms behind both the acute toxicity and carcinogenicity of fumonisins are not clear yet (Marasas, 2001).

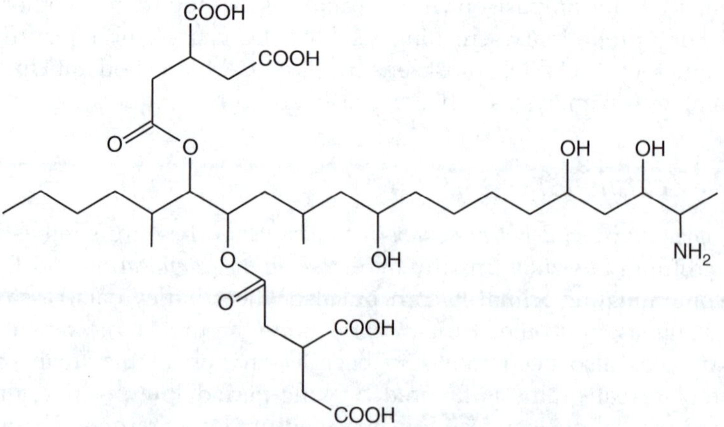

***Figure 11.5*** Structure of fumonisin $B_1$.

*Bioassay*. Chicken embryos and brine shrimp (*Artemia salina*) naulpii have been utilized in short-term toxicity bioassays. However, the chicken embryo bioassay is limited by the relatively high dose of $FB_1$ required per egg. It is anticipated that the design and simplicity of the brine shrimp bioassay will accommodate screening for $FB_1$ toxicity in contaminated samples to assess their sensitivity to $FB_1$. $LD_{50}$ is 52 $\mu$g $FB_1$/egg (Hlywka et al., 1997).

## 11.7 Trichothecenes

Trichothecenes are a group of 12,13-epoxytrichothecenes, produced on various cereal grains mainly by microfungi of the genus *Fusarium* such as *F. poeae, F. tricinctum, F. graminearum, F. nivale, F. culmorum, F. solani,* and so forth. These include T-2 toxin, neosolaniole, diacetylnivalenol, deoxyvalenol (alias DON or vomitoxin), HT-2 toxin, fusarenone and others, altogether over 20 compounds.

*T-2 toxin,* produced mainly by *F. sporotrichioides* and *F. poae,* is toxic to rats, rainbow trout, and calves. It is believed that this toxin and similar compounds had their role in *human alimentary toxic aleukia,* an illness that spread during and after World War II in several regions of the former Soviet Union, where people ate the wheat grain left over after winter or bread that was made from this grain. The illness was characterized by a complete atrophy of the bone marrow and thymus, agranulocytosis, necrotic angina, sepsis, and hemorrhagic diathesis. Up to 80% of the cases had a fatal end. T-2 toxin may also cause death of embryos, abortions, and have teratogenic effects. This toxin has been seldom found in food but cases of feed contamination have been much more frequent (Altuğ, 2003).

*F. nivale* produces trichothecenes like nivalenol and fusarenone-X in blooming heads of wheat grain, barley, rice, corn, and other cereals as well as forage crops. Thus, sheep, horses, and people are suffering in Japan from poisoning from these substances. Symptoms of human intoxication are headache, vomiting, and diarrhea, although lethal cases are unknown.

*Deoxynivalenol* (*vomitoxin*) (Figure 11.6) that contaminates wheat and corn has caused vomiting, feed refusal, and immunosuppression in the case of domestic animals, particularly swine in the midwestern United States.

Macrocyclic trichothecenes such as *satratoxins, verrucarins,* and *roridins* are produced mainly by fungi of the genus *Stachybotrys,* colonizing hay (Wallace Hayes, 2001, p. 511). Although intrinsically more toxic than other

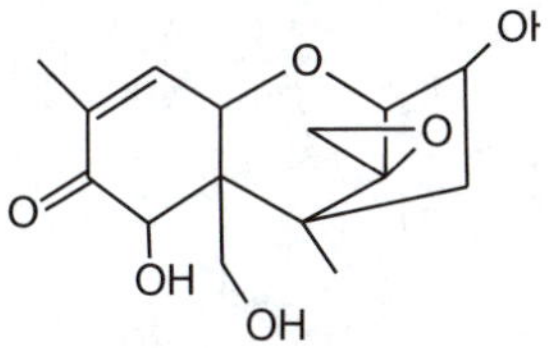

***Figure 11.6*** Structure of deoxynivalenol *alias* vomitoxin.

trichothecenes, due to their narrow expansion they are not dangerous for human health.

*Trichothecenes inhibit protein synthesis* in the cell and induce apoptosis, which explains their toxicity. In addition, at least a part of them is thought to influence the serotonergic pathways of the brain as well as induction of expression of numerous cytokines. The role of the last-mentioned effects in the intoxications is not clear yet. Although several trichothecenes are genotoxic in bacterial, yeast, and eukaryotic cell cultures, neither an initiation nor a promoter effect has been observed in the case of test animals (Rocha et al., 2005).

Since trichothecenes are quickly metabolized via deacetylation or hydroxylation, and following glucuronidation in the liver and kidneys, their residues in meat are not a cause for special human toxicological concern.

## 11.8 Patulin

*Patulin* (PAT; 4-hydroxy-4H-furo[3,2c]pyrane-2[6H]-one) is produced as a secondary metabolite by various species of the genera *Penicillium, Aspergillus*, and *Byssochlamys*. *P. expansum* is the main PAT-producing fungus causing rotting of apples and pathogenesis of many fruits and vegetables. PAT (Figure 11.7) can be found in mold apples, plums, pears, apricots, cherries, and grapes, especially when the surface of the fruit is injured. PAT also contaminates fruits, particularly apple juices. It has also been found in feeds and stored cheese. PAT is unstable in the presence of sulfhydrylic compounds and sulfur dioxide, but resistant to pasteurization. Up to 99% of PAT is destroyed during fermentation, for example, during the making of apple cider. PAT is toxic to various biological systems including bacteria, mammal cell cultures, higher plants, and animals. It has been thought to be mutagenic, carcinogenic to rats, and teratogenic to hens. PAT causes adverse pulmonary effects to dogs and is deadly for mice. It has been shown that PAT causes inflammations, ulcers, and bleeding in the gastrointestinal tract.

In *PAT the mechanism of the adverse effect* is covalent binding to cellular nucleophiles, particularly proteins and SH-groups of glutathione. As a result, covalently cross-linked over thiol and aminogroups, essentially denatured proteins such as inhibited protein tyrosin phosphatase are formed (Mahfoud et al., 2002).

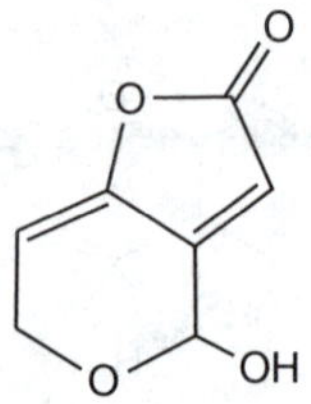

O
O
O
OH

***Figure 11.7*** Structure of patulin.

## 11.9 Citrinin and citreoviridin

*Citrinin* is a yellowish substance produced by several species of genera *Penicillium* and *Aspergillus. P. citrinum* is widespread in all regions of rice cultivation, growing particularly well on polished rice and synthesizing pigment that colors the rice grains yellow. It was established already in the seventeenth century that mycotoxin produced in large amounts by *P. citrinum* growing on rice has a substantial role in the formation of an acute form of *cardiac beriberi*.

*Citrinin* (Figure 11.8) that has been determined also in yellow peanut kernels, as well as in wheat, barley, oats, corn, rye, and fruits, *is a potent renal and hepatic poison*. The toxicity of citrinin is still lower than, for example, that of OTA, which often accompanies citrinin in cereals. Citrinin is also capable of causing vasodilation, bronchoconstriction, and elevated muscular tonus. $LD_{50}$ of citrinin is 50 mg/kg orally in rat and 35–58 mg/kg and 19 mg/kg intraperitoneally, respectively, in mouse and rabbit. It has been proposed that citrinin participates together with OTA and other so far unknown toxicants in the formation of *human fatal Balkan endemic nephropathy* (Stefanovič et al., 2006) (see also Section 11.3). At the same time, due to its instability during the processing of grain, citrinin is not considered to be capable of presenting a very significant hazard for human health in normal conditions. *Endangered are domestic animals*—especially swine that may eat grain products contaminated with citrinin. Citrinin, and again together with OTA, has caused porcine mycotoxic nephropathy in Denmark, Norway, Sweden, and Ireland, and also nephropathy of birds. Acute lethal doses of citrinin administered to rabbits, guinea pigs, mice, and swine have caused renal edema leading to necrosis. Citrinin exerts an embryocidal effect on mice and is teratogenic in high doses. There is little data concerning the carcinogenicity of citrinin in the test animals.

*Neurotoxic citreoviridin* is another toxin produced in yellow rice by *Penicillium citreoviride*. Ingestion of this substance causes vomiting, convulsions, and paralysis of hind legs and sides of animals followed by disorders in the functioning of the respiratory and cardiovascular systems, general paralysis, fever, gasping, coma, and arrest of respiration.

*Stability*. Citrinin is decomposed at dry processing at 175°C; in semi dry conditions, decomposition takes place at about 140°C itself. In the presence of water, citrinin may provide toxic products of decomposition, such as *citrinin* $H_1$, which is formed by cyclization of two citrinin molecules. Citrinin decomposes most likely also during fermentation to the extent of 90% during germination of barley and completely during the making of wort from malt.

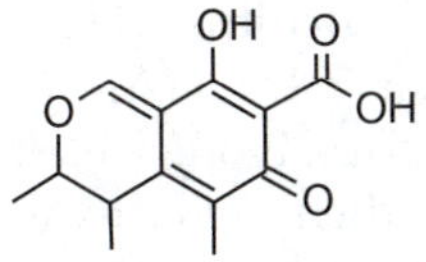

OH O OH O

***Figure 11.8*** Structure of citrinin.

Citrinin is decomposed also by propionic acid added for protection against microfungi to barley anticipated for feed.

*Analysis*. Grain samples can be analyzed first with TLC (reversed phase plates) or ELISA and, in case of a positive result, quantitatively with HPLC. Citrinin (up to 1 ppb) is visible on TLC plates as a yellow florescent band under the longwave-UV radiation, which turns blue after spraying the plate with a boron trifluoride reagent (Scott et al., 1972).

## 11.10 Ergot toxins

From history, serious acute ergot intoxications or *ergotisms* are known, appearing after eating of grain (particularly rye) contaminated with parasitic microfungus *Claviceps purpurea* (more rarely *C. paspali*) that forms nodules in the spikelets of the gramineous plants. The respective toxic substances are *ergot alkaloids,* derivatives of lysergic acid—*ergotoxin, ergotamine, ergomethrine,* and so forth, altogether about 50 different substances (Bürk et al., 2006).

Depending on the proportion of various alkaloids in the food either a *gangrenous* or *convulsive* (*CNS*) *form of the ergotism* will develop. The *gangrenous form* is characterized by prickly heat and cold sensation in limbs, and swollen, inflamed, necrotic, and gangrenous extremities that eventually slough off. The reason is a narrowing of blood vessels, cutting off the blood supply of the peripheral regions of limbs. *Convulsive ergotism* is characterized by numbness, cramps, and death.

*In medieval times,* ergotism occurred on a large scale in some regions of Europe. In 857 AD, there was an epidemic in Germany. In 944 AD, as many as 40,000 people in the south of France died from ergot poisoning. It was then a common belief that the symptoms were caused by witchcraft. In the Salem witchcraft trial of 1692 in Massachusetts, entire communities in and around Salem Village had symptoms of convulsive ergotism. This sad event is depicted in the well-known play by Arthur Miller. The last reported European outbreak with more than 200 cases and 4 deaths occurred in 1951 in the French village of Pont-St.-Esprit in Provence.

*Nowadays ergotism is rare,* but it still occurs and, if at all, then in large numbers. For example, in several regions of Ethiopia, an outbreak of gangrenous ergotism was registered in 1977–1978, when 47 victims out of 93 died. An ergotism outbreak reoccurred in the Arsi region of the same country in 2001. In this case, the concentrations of ergotamine and ergometrine (Figure 11.9a) were, respectively, up to 2.7 and 1.2 mg/100 g of barley, containing wild oats contaminated with ergot (Urga et al., 2002). An outbreak of convulsive ergotism with 78 victims took place in India in 1975.

The U.S. Department of Agriculture (USDA) has set a tolerance limit of 0.3% (by weight) of contaminated grain (Altuğ, 2003). The respective number is 0.05% (w/v) in the EU countries. If one were to assume an average alkaloid content of 0.2% w/v in grain, then the last value corresponds to 1000 $\mu$g/kg ergot alkaloids in cereal (Bürk et al., 2006).

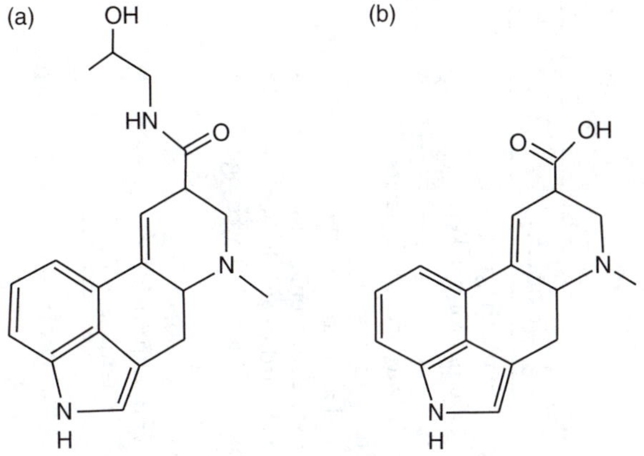

*Figure 11.9* Structures of (a) ergometrine and (b) lysergic acid.

Related to the ergot alkaloids is a semisynthetic narcotic diethylamide of lysergic acid (LSD) (Figure 11.9b) that in addition to its psychogenic properties (hallucinations, delusions, mood swings) is also mutagenic and embryotoxic.

The ergot preparations (*secale cornutum*) are medically used for the widening of the blood vessels and lowering of blood pressure.

## 11.11 Other mycotoxins

In addition to the most widespread mycotoxins, there are less widespread mold toxins that are briefly overviewed in Table 11.2.

## 11.12 Combined toxicity of mycotoxins

Foodstuffs of plant origin may simultaneously contain different mycotoxins originating either from the same or different molds. Widespread are, for example, combinations such as ochratoxin + citrinin, ochratoxin + zearalenone, ochratoxin + penicillic acid, ochratoxin + aflatoxin $B_1$, patulin + citrinin, or even more complicated like aflatoxin $B_1$ + fumonisin $B_1$ + zearalenone + deoxynivalenol + nivalenol, and so forth.

During the consumption of such food, the organism assimilates several toxins that may have a combined effect on the organism. The intensity and character of the summary effect depends on the relative absorptivities as well as on individual toxic properties. When the toxins have a similar molecular structure and action mechanism, the effects may be additive. But if the effects are sufficiently different, the summary effect may be either synergistic or antagonistic (see also Section 1.4). Actually, there is little data about the combined toxicity of mycotoxins and most of the data concerns ochratoxin (Speijers and Speijers, 2004).

***Table 11.2*** Overview of Relatively Less Important Mycotoxins

| Mycotoxin | Main producers | Typical substrate | Toxic effect |
|---|---|---|---|
| Alternaria toxins | *Alternaria alternata* | Cereal grains, tomatoes, feed | Mutagenic, hemorrhagic |
| Austamide, austadiol, austine | *Aspergillus ustus* | Stored foodstuffs | Toxic up to lethal for chicken |
| Cyclopiazonic acid | *A. flavus, Penicillium cyclopium* | Peanuts, corn, cheese | Neurotoxic, cardiovascular lesions |
| Cyclochlorotine | *P. islandicum* | Rice | Hepatotoxic, mutagenic |
| Diplodiatoxin | *Diplodia maydis* | Corn | Entero- and nefrotoxic for bovines and sheep |
| Luteoskyrin | *P. islandicum, P. rugulosum* | Rice, sorghum | Hepatotoxic, mutagenic, carcinogenic |
| Moniliformin | *Fusarium monoliforme* | Corn | Neurotoxic, cardiovascular lesions |
| Penicillic acid | *P. puberulum, A. ochraeus,* | Barley, corn, beans | Neurotoxic, mutagenic, carcinogenic |
| Penitrem A | *P. palitans* | Foodstuffs, corn | Neurotoxic |
| Roquefortine | *P. roqueforti* | Cheese | Neurotoxic |
| PR-toxin | *P. roqueforti* | Cereal grains | Hepato- and nefrotoxic |
| Rubratoxin | *P. rubrum, P. purpurogenum* | Corn, soybeans | Hepatotoxic, teratogenic |

*Source:* Adapted from Altuğ, T. (2003). *Introduction to Toxicology and Food*, CRC Press, Boca Raton, FL; Wallace Hayes, A., (2001). *Principles and Methods of Toxicology*, 4th edn., Taylor & Francis.

*Example*: Aflatoxin $B_1$ ($AFB_1$) and T-2 toxin (T-2) as important food-borne mycotoxins that have been implicated in human health and as potential biochemical weapons, interact to produce alterations in the toxic responses generally classifiable as additive. However, a synergistic interaction was noted in the case of human bronchial epithelial cells (McKean et al., 2006).

Nowadays, the content of essential mycotoxins in food is increasingly and strictly watched, and both in a legislative and analytical sense, the probability of food-borne mycotoxicoses is continuously decreasing.

## References

Altuğ, T. (2003). *Introduction to Toxicology and Food*, CRC Press, Boca Raton, FL.

Bayman, P. and Baker, J.L. (2006). Ochratoxins: a global perspective, *Mycopathologia*, **162**, 215–223.

Belitz, H.-D. and Grosch, W. (1999). *Food Chemistry*, 2nd edn., Springer.

Bürk, G., Höbel, W. and Richt, A. (2006). Ergot alkaloids in cereal products. Results from the Bavarian Health and Food Safety Authority, *Mol. Nutr. Food Res.*, **50**, 437–442.

Chan, P.K. and Gentry, P.A. (1984). $LD_{50}$ values and serum biochemical changes induced by T-2 toxin in rats and rabbits, *Toxicol. Appl. Pharmacol.*, **73**, 402–410.

Fuchs, R. and Peraica, M. (2005). Ochratoxin A in human kidney diseases, *Food Addit. Contam.*, **22**, Suppl 1, 53–57.

Hlywka, J.J., Beck, M.M. and Bullerman, L.B. (1997). The use of chicken embryo screening test and brine shrimp (*Artemia salina*) bioassay to assess the toxicity of fumonisin $B_1$ mycotoxin, *Food Chem. Toxicol.*, **35**, 991–999.

Hussein, H.S. and Brasel, J.M. (2001). Toxicity, metabolism and impact of mycotoxins on humans and animals, *Toxicology*, **167**, 101–134.

Mahfoud, R., et al. (2002). The mycotoxin patulin alters the barrier function of the intestinal epithelium: mechanism of action of the toxin and protective effects of glutathione, *Toxicol. Appl. Pharmacol.*, **181**, 209–218.

Marasas, W.F. (2001). Discovery and occurrence of the fumonisins: a historical perspective, *Environ. Health Persp.*, **109**, Suppl. 2, 239–243.

McKean, C., et al. (2006). Comparative acute and combinative toxicity of aflatoxin $B_1$ and T-2 toxin in animals and immortalized human cell lines, *J. Appl. Toxicol.*, **26**, 139–147.

Moss, M.M. (2002). Risk assessment for aflatoxins in foodstuffs, *Int. Biodeterior. Biodegradation*, **50**, 137–142.

Rocha, O., Ansari, K. and Doohan, F.M. (2005). Effects of trichothecene mycotoxins on eukaryotic cells: a review, *Food Addit. Contam.*, **22**, 369–378.

Scott, P.M., et al. (1972). Mycotoxins (ochratoxin A, citrinin, and sterigmatocystin) and toxigenic fungi in grains and other agricultural products, *J. Agric. Food Chem.*, **20**, 1103.

Sforza, S., Dall'asta, C. and Marchelli, R. (2006). Recent advances in mycotoxin determination in food and feed by hyphenated chromatographic techniques/mass spectrometry, *Mass Spectrom. Rev.*, **21**, 54–76.

Speijers, G.J.A. and Speijers M.H.M. (2004). Combined toxic effects of mycotoxins, *Toxicol. Lett.*, **153**, 91–98.

Stefanovič V., et al. (2006). Etiology of Balkan endemic nephropathy and associated urothelial cancer, *Am. J. Nephrol.*, **26**, 1–11.

Thorgeirsson, U.P., et al. (1994). Tumor incidence in a chemical carcinogenesis study of nonhuman primates, *Regul. Toxicol. Pharmacol.*, **19**, 130–151.

Urga, K., et al. (2002). Laboratory studies on the outbreak of gangrenous ergotism associated with consumption of contaminated barley in Arsi, Ethiopia, *Ethiop. J. Health Dev.*, **16**, 317–323.

Wallace Hayes, A., Ed. (2001). *Principles and Methods of Toxicology*, 4th edn., Taylor & Francis, Philadelphia.

Williams, J.H., et al. (2004). Human aflatoxicosis in developing countries: a review of toxicology, exposure, potential health consequences, and interventions, *Am. J. Clin. Nutr.*, **80**, 1106–1122.

Yannikouris, A. and Jouany, J.-P. (2002). Mycotoxins in feeds and their fate in animals: a review, *Anim. Res.*, **51**, 81–99.

Zinedine, A., et al. (2007). Review of the toxicity, occurrence, metabolism, detoxification, regulations and intake of zearalenone: an oestrogenic mycotoxin, *Food Chem. Toxicol.*, **45**, 1–18.

# chapter twelve

# Animal endogenous poisons

Upon consumption of food (meat, milk, eggs, etc.) of animal origin, it is possible to come into a contact with various toxicants originating from feed of plant origin like mycotoxins, veterinary drug and pesticide residues, endogenous toxicants of plants, and so forth. The list of endogenous animal food toxicants is not as long. Over the last years, much has been spoken and written, for example, about prions.

## 12.1 Prions

The term "prion" is associated with serious central nervous system (CNS) illnesses such as bovine *transmissible spongiform encephalopathy* (TSE), commonly referred to as "mad cow disease," human Creutzfeldt–Jacob disease, "kuru" in cannibalistic tribe, Foré in Papua New Guinea, scrapie in sheep and goats, and so forth. All these interrelated diseases cause essential alterations in the brain tissue and lead to a fatal end. Currently, no effective treatment for prion diseases is available. Several strategies have been studied for an effective antiprion medication including vaccination and screening for potent chemical compounds. Identification of the laminin receptor (LRP/LR) and heparin sulfate as cell surface receptors for prions will hopefully open new possibilities for the development of successful TSE therapies (Vana et al., 2006).

The word "prion" is an abbreviation of the phrase "proteinaceous infective particle." *Prions* were discovered and the term was adopted in 1982 by the American scientist Stanley B. Prusiner who was awarded the Nobel Prize in 1997 (Prusiner, 1982). All the pathogens known by this time (bacteria, fungi, viruses, etc.) needed nucleic acid for multiplication. It was discovered that the particle causing the Creutzfeldt–Jakob disease is stable to UV-radiation denaturating DNA, but sensitive to agents decomposing proteins.

*The prion is an otherwise normal animal membrane protein,* the primary structure of which is incorrectly folded to give anomalous infectious secondary and tertiary structures. The function of the prion protein in a normal cell is unknown. In their normal state (denoted as $PrP^{C}$ = prion protein cellular), these molecules can participate in normal cell–cell communication; in the abnormal state ($PrP^{SC}$ = prion protein scrapie), they are capable of contacting normal molecules of the prion protein to make $PrP^C$s to change into new abnormal prion molecules $PrP^{SC}$. A rise in the number of $PrP^{SC}$ molecules above a definite threshold limit causes the formation of pathological conditions in the cell. On eating food containing prion molecules, they are absorbed through the intestinal wall and start to transform the normal protein molecules into new prions. It was considered earlier that the prions belonging to one animal species are not able to transform the protein molecules of another species into prions. This statement is not actually completely true. Cases are known when people who had eaten tissues of a diseased animal fell ill with the mad cow disease. In 2004, a mouse was infected with an artificial prion, which consisted entirely of $PrP^{SC}$ molecules.

Although the nature and general properties of the prion have been elucidated by now, the mechanism of its toxic effect is to a great degree still unknown. According to one hypothesis, the conversion $PrP^C \rightarrow PrP^{SC}$ is catalyzed by a still unknown enzyme. According to another hypothesis, $PrP^{SC}$ influences in some way the synthesis scheme of $PrP^C$ in ribosome, leading to the formation of the abnormal infective molecule.

*No acute clinical symptoms* have been described for prion-caused diseases—an irreversible functional degeneration of the CNS, finally leading to the death of the patient appears only after a year-long latency period. During this rather long period, sponge-like regions are formed in the brain. The prion-related diseases can be conclusively diagnosed only after a *postmortem* microscopic investigation of the brain. An earlier diagnosis can be based on the results of studies of the patient's behavior, including his or her food, clinical symptoms, electroencephalograms, and brain magnetic resonance images.

*Prionic diseases are transmitted from organism to organism by consuming contaminated animal tissues*. The highest risk is connected with eating of bovine skull, brain, nerves directly linked to the brain, eyes, tonsils, bone marrow, and part of the small intestine. So far, no infectivity of either beef or milk has been established. Gelatin, made from the skin and bones of ill bovines is, for example, of very low risk.

It is generally accepted that the most likely route of infection of cattle with bovine spongiform encephalopathy (BSE) is by consumption of feeds containing low levels of processed animal proteins (PAPs). This likely way of infection has resulted in feed bans, which were primarily aimed at ruminant feeds, and later extended to all feeds for farmed animals. The feed bans were expected to develop into a future enforcement of the "species-to-species" ban, which prohibits only the feeding of animal-specific proteins to the same species.

Especially many cases (143 suspicious or confirmed cases during 1993–2003) of mad cow disease in humans have been registered in Great Britain. All the cases, registered so far, have been connected with humans belonging to one genotype, homozygous in relation to methionine in the codon 129 of the prion protein. But about 40% of the world human population belongs to this genotype.

To learn more about TSE, the reviews by Collins et al. (2004), van Raamsdonk et al. (2007), and Vana et al. (2006) can be recommended.

## 12.2 Lactose

*Disaccharide lactose,* a condensation product of two monosaccharides, glucose and galactose, is the main saccharide of milk. Since lactose is not absorbed from the gastrointestinal tract, it is necessary to hydrolyse the dimer into monomers in the intestines before entering the enterocytes. The intestinal hydrolysis of lactose is catalyzed by *lactase* (lactase-phlorizin hydrolase—EC 3.2.1.108), a special enzyme that belongs to the $\beta$-galactosidases. Intestinal lactase is bound via its C-end to the bristly membrane of enterocytes, where most of the glycoproteid molecule is located in the intestinal lumen. Lactase is expressed in mammal organisms solely in the enterocytes of the small intestine.

*Lactase activity is vitally high in early childhood* when milk is the main source of nutrients. All healthy newborn children express high levels of lactase and are able to digest large quantities of lactose. In the case of most mammals, including a majority of the worldwide human population, the *lactase activity starts to decrease after weaning.* This decline (*hypolactasia*), which is regulated mainly on the transcriptional level, begins usually during the second or third year of life and ceases at the age of 5–10 years; in rare cases, all the reduction takes place during adolescence. In contrast, lactase-persistent individuals have a lifelong lactase expression and are able to digest lactose. These people can drink a lot of sweet milk without any health problems. It is established that the adult-type hypolactasia is inherited in an autosomal recessive manner, whereas persistence is dominant (Troelsen, 2005).

Humans with the *adult-type hypolactasia* have the symptoms of *milk intolerance* like nausea, abdominal pain, flatulence, ample gas excretion, and sometimes diarrhea appearing after consumption of sweet milk. These symptoms are caused by moving of unhydrolyzed and unabsorbed lactose further to the colon, where it gets fermented into organic acids, primarily butyric acid and gaseous hydrogen by anaerobic bacteria. The organic acids formed disturb the absorption of water, evoke intestinal contractions, and cause diarrhea. The amount of milk, necessary for eliciting the aforementioned disorders is variable, but usually more than a glass of milk is needed to form any serious symptoms. Sour milk, buttermilk, kephir, yogurt, curds, and cheese, where the lactose content is lower, do not elicit the symptoms of milk intolerance.

*The milk (lactose) tolerability index* (0–1.0) varies among the human populations, being the highest (0.8–0.9) in the case of Swedes and Danes living in the northwest of Europe. Moving to both the south and east in Europe, the tolerability index decreases—for example, in the case of Estonians, the index is about 0.5, and in the case of southern Italians or Turks, 0.2–0.3. The same tendency emerges moving from north to south in India. Throughout other parts of the world, this index is low. Interesting exceptions are the Bedouins, pursuing a nomadic lifestyle in the desert zone of Africa and Arabia, as well as Mongols in Central Asia; Hereros and Nueres (Dinkas) in Ethiopia and Sudan; Tutsis in Rwanda and Burundi, who are distinguished from their neighboring nations by a low incidence of hypolactasia. All the aforementioned nations are cattle breeders. In the regions where the use of milk for adult nutrition has a short history (the Indians of Alaska, and Inuits of Greenland and Japan), lactose intolerance is widespread. In the United States, milk intolerance increases in the sequence—descendants of North-Europeans, Mexicans, Afro-Americans, American Indians (Swallow, 2003).

## 12.3 Phytanic acid

*Refsum disease* is a genetic illness, caused by the deficiency in peroxisomal fatty acid oxidase as well as catalase, responsible for $\alpha$-oxidation of fatty acids in peroxisomes. The organism of such an individual is not capable of metabolizing branched *phytanic (3,7,11,15-tetramethylhexadecanoic) acid,* that is contained in milk products and in the fat of ruminants. Phytanic acid is the product of chlorophyll metabolism in the rumen of the ruminats. The methyl group in $\beta$-position prevents degradation of phytanic acid by the normal $\beta$-oxidation pathway. Owing to the accumulation in many tissues of lipids containing phytanic acid, neurological symptoms like slow physical and intellectual development, blindness, and deafness appear. Withdrawal of milk products and fats of ruminants from the diet guarantees a partial regression of these symptoms (Reiser et al., 2006).

## 12.4 Avidin

*Avidin,* a protein of egg white, binds noncovalently but unexpectedly tightly to *biotin,* inhibiting the absorption of this compound from the intestines. *The avidin–biotin complex* is resistant to the action of digestive enzymes. In normal life, it does not cause any problems since it is necessary to ingest 15–20 uncooked hen eggs to elicit a biotin deficiency. Chronic ingestion of raw egg will produce such symptoms as scaly dermatitis, muscle pains, alopecia (hair loss), glossitis (red lips), mental depression, and general fatigue. Boiling, frying, or cooking of eggs denaturates the avidin molecule, which will lose the biotin-binding ability. Absorption of biotin is inhibited also by excessive alcohol, tannin-rich foodstuffs, and antibiotics (Basu and Donaldson, 2003).

## 12.5 Vitamins of animal origin

Here we will mention only vitamins, giant doses of which can be obtained from foods of animal origin. A longer overview of all toxicologically important vitamins can be found in Chapter 18.

*Vitamin A.* Large doses of vitamin A or retinol are acquired by eating definite foods in larger amounts. A toxicologically relevant amount of retinol is available in 30 g of bovine liver, 500 g of hen eggs, or 2500 g of mackerel. Especially high is the vitamin A content in the liver of polar bears and other arctic animals.

*Vitamins D* or calciferols are fat-soluble antirachitic compounds that have a hormonal effect on the human organism. Only historically, they were allocated to vitamins. Larger amounts of vitamin $D_3$ can be found in fish such as herring and particularly in the cod liver, from which vitamin-rich cod-liver oil is prepared. Owing to the hormonal effect on humans, one should be cautious using these compounds. The lethal dose of 13 mg/kg bw of vitamin $D_3$ has been established for dogs.

## References

Basu, T.K. and Donaldson, D. (2003). Intestinal absorption in health and disease: micronutrients, *Best Pract. Res. Cl. Ga.*, **17**, 957–979.

Collins, S.J., Lawson, V.A. and Masters, C.L. (2004). Transmissible spongiform encephalopathies, *Lancet*, **363**, 51–61.

Prusiner, S.B. (1982). Novel proteinaceous infectious particles cause scrapie, *Science*, **216**, 136–144.

Reiser, G., Schönfeld, P. and Kahlert, S. (2006). Mechanism of toxicity of the branched-chain fatty acid phytanic acid, a marker of Refsum disease, in astrocytes involves mitochondrial impairment, *Int. J. Dev. Neurosci.*, **24**, 113–122.

Swallow, D.M. (2003). Genetics of lactase persistence and lactose intolerance, *Annu. Rev. Genet.*, **37**, 197–219.

Troelsen, J.T. (2005). Adult-type hypolactasia and regulation of lactase expression, *Biochim. Biophys. Acta*, **1723**, 19–32.

van Raamsdonk, L.W.D., et al. (2007). New developments in the detection and identification of processed animal proteins in feeds, *Anim. Feed Sci. Tech.*, **133**, 63–83.

Vana, K., Zuber, C., Nikles, D. and Weiss, S. (2006). Novel aspects of prions, their receptor molecules and innovative approaches for TSE therapy, *Cell. Mol. Neurobiol.*, **27**, 107–128.

# chapter thirteen

# Food toxins from sea

Only a few species out of the many toxin-containing organisms (about 1200 known species) living in the sea have something to do with food toxicology. Toxic compounds are produced either by

- The edible organisms (fish, shellfish) themselves or
- The sea plankton or algae that are ingested by fish or shellfish. This group of toxic substances is called *phycotoxins*.

It is only the microorganisms inhabiting water that are the main culprits of food intoxications caused by eating most crustaceans (shellfish) and fish. To estimate the potential toxicity of seafood, various functional assays have been developed (Rossini, 2005).

## 13.1 Shellfish toxicants

Various organisms belonging to the shellfish (crustaceans) family like *clams, lobsters, mussels, oysters, scallops,* and so forth that have ingested toxic algae, particularly *dinoflagellates,* can be very toxic. Shellfish are especially toxic during an intensive blooming of algae, when seawater contains 200 or more microorganisms per milliliter. The toxicity of shellfish is proportional to the concentration of algae in water and disappears over 2 weeks after the disappearance of the toxic phytoplankton. Shellfish poisonings are divided into four groups—*paralytic, diarrhetic, neurotoxic,* and *amnesic.*

### 13.1.1 Paralytic shellfish poisoning

The toxins that cause paralytic shellfish poisoning (PSP) are more than 20 structurally related imidazoline-guanidinium alkaloids that are produced by a number of sea dinoflagellates, and linked to them are bacteria as well as fresh-water cyanobacteria swallowed and accumulated by filter-feeding bivalve shellfish and specific herbivorous fish and crabs. Nowadays, PSP has turned into a serious global toxicological issue.

The most well-known causative agents of PSP are *saxitoxins* produced by the cyanobacterium *Aphanizomenon flos-aquae,* dinoflagellates *Alexandium*

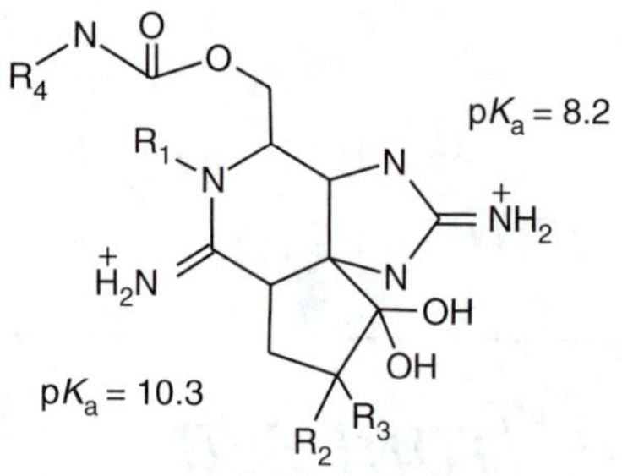

*Figure 13.1* General structure of saxitoxins. p$K_a$ values characterize the acidic dissociation of the respective ammonium groups indicating that at physiological pH values saxitoxin is mostly in the form of a stable ion bearing a double positive charge. The nonionic form, prevailing in the alkaline conditions, is unstable.

*tamarense, A. catenella, A. minutum, A. ostenfeldii, Gymnodinium catenatum* and others such as *neosaxitoxin, anatoxin* (*Anabaena flos-aquae*), and *gonyautoxins*.

*The saxitoxins* (Figure 13.1) *are a group of alkaloids* that are nonsulfated (saxitoxins), singly-sulfated (gonyautoxins), or doubly-sulfated (C-toxins). The various types of toxins vary in potency, with saxitoxin having the highest toxicity. Saxitoxin is a member of the Centers for Disease Control and Prevention (CDC) Select Agent List for its potential use as a bioweapon.

Owing to the high polarity of the molecule, saxitoxin dissolves well in water and it is stable at acidic and neutral pH values even at high temperatures. Boiling in slightly alkaline water inactivates the toxin and discarding the broth helps *avoid PSP* with saxitoxins.

*The human nervous system is supersensitive to the action of these toxins.* Saxitoxins reversibly block the inflow of $Na^+$ ions into a nerve cell via Na-channels, not having the blocking ability of $K^+$ or $Cl^-$ ions. Hence, by blocking nerve conduction they cause death by respiratory arrest. Ingestion of 1 mg of saxitoxin within 1–5 blue mussels (*Mytilus edulis*) weighing 150 g may have a weak toxicological effect but ingestion of 4 mg of saxitoxin can, in the absence of immediate medical aid, end fatally. The *action level* of saxitoxin is fixed at 80 μg per 100 g tissue (Poletti, 2003).

*The first symptoms of poisoning* that appear within a few minutes after eating are numbness of the lips, tongue, and fingertips that progresses to the legs, hands, and neck. Soon general disorders of muscular coordination will appear, followed by breath paralysis and death. Effects like cardiac deceleration, headache, increased sweating, and feeling of thirst have also been observed. $LD_{50}$ of saxitoxin is, in the case of both guinea pigs (intramuscularly) and mice (intraperitoneally), about 5 μg/kg bw.

*The name "saxitoxin" originates* from the name of the mollusk *Saxidomus giganteus* in which the toxin was first identified. In fresh water, the same toxin is produced by blue alga *Anabaena circinalis* that can forward it to freshwater shellfish like *Alathyria condola*. Nevertheless, no information concerning saxitoxin poisonings by freshwater organisms is available.

*Saxitoxins have been recorded* in only a few locations throughout the United States (New Hampshire, Alabama, and New Mexico). No occurrences have yet been reported in the Great Lakes. A few animal deaths have been linked to saxitoxins in U.S. freshwaters but most poisonings are from exposures through marine waters as the causative agent of PSPs. Saxitoxin can also be found in puffer fish notorious for its tetrodotoxin (TTX) (see Section 13.2.1). Cases of neurological symptoms, including numbness and tingling of the lips and mouth, have been reported to arise rapidly after the consumption of puffer fish caught in the area of Titusville, Florida. As a result of such cases, Florida has banned the harvesting of puffer fish from certain water bodies (CDC, 2002).

*In temperate parts of Australia,* blooms of saxitoxin-producing cyanobacteria are very prevalent. The world's largest recorded and most publicized bloom occurred in late 1991 and extended over 1000 km of the Darling-Barwon river system. This bloom focused the attention of both the government and community on the issue of blue-green algae. A state of emergency was declared with a focus on providing safe drinking water to towns, communities, and landholders. Thousands of stock deaths were associated with the occurrence of the bloom but there was little evidence of human health impacts (Baker et al., 2002).

*Anatoxins* are a group of low-molecular neurotoxic alkaloids that for the first time were found in freshwater blue alga *Anabaena flos-aquae* in Canada. Here belong secondary amines *anatoxin-a* and *homoanatoxin-a* as well as *anatoxin-a(S),* that is, the phosphoric acid ester of *N*-hydroxyguanine (Figure 13.2). Anatoxin-a has been found in blue algae such as *Anabaena planktonia, Oscillatoria* spp., *Aphanizomenon* spp., *Microcystis* spp., and homoanatoxin-a in alga *Oscillatoria formosa,* and anatoxin-a(S) in alga *A. lemmermannii.*

*Anatoxin-a* and *homoanatoxin-a,* are like acetylcholine, postsynaptic depolarizing neuromuscular blocking substances that tightly bind to the nicotinic acetylcholine receptor. Acetylcholinesterase is unable to inactivate anatoxins unlike acetylcholine by hydrolysis and the muscle will stay in the contracted state. Anatoxins cause a fast death of mammals via breath paralysis; their $LD_{50}$ for mice is about 250 $\mu$g/kg bw.

Anatoxin is the second most common cyanotoxin in the U.S. waters and has been identified in a few water samples of the Great Lakes. It has been responsible for massive die-offs of migrating birds in the midwest and in

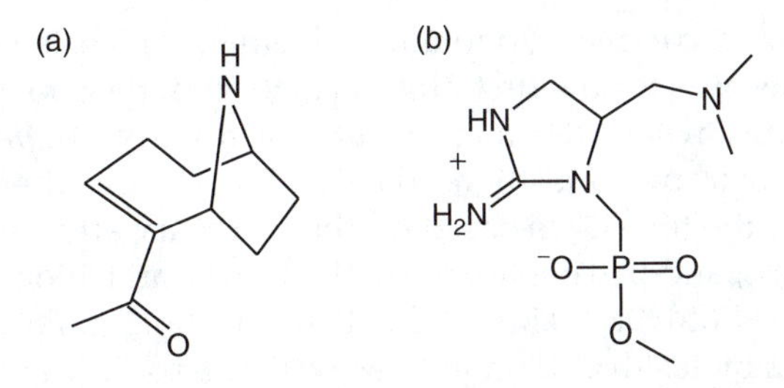


***Figure 13.2*** Structures of (a) anatoxin-a and (b) anatoxin-a(S).

intermittent but repeated poisonings of wild and domestic animals in several U.S. states, especially in the West.

*Anatoxin-a(S)* is an inhibitor of cholinesterase with $LD_{50}$ for mice 25–50 μg/kg bw. Since anatoxin-a(S) *is an organophosphorous compound* (*the only natural one*) its toxic effect is similar to the effect of synthetic organophosphorus pesticides like parathion or malathion (Matsunaga et al., 1989) (see Section 14.1). Unlike anatoxin-a, anatoxin-a(S) causes salivation, diarrhea, tremors, and mucus flow from the nose. This compound is not as common as anatoxin-a, but still has been responsible for a few animal (especially domestic dogs) and bird poisonings in the United States. It has not been identified to date in the Great Lakes.

### 13.1.2 Diarrhetic shellfish poisoning

Eating of contaminated shellfish may also cause diarrhetic shellfish poisoning (DSP). DSP has been diagnosed in the case of eating of bivalve shellfish like mussels, scallops, and oysters, which have swallowed dinoflagellates of the genus *Dinophysis* (*D. fortii, D. acuminata,* and others) and maybe also *Prorocentrum* spp. DSP is especially frequent in Japan, but is turning into a problem in Europe as well. The first DSP was diagnosed in Japan, in the province of Tohoku. Later on, information concerning the occurrence of DSP arrived from all continents, except Africa and Australia. Symptoms of the poisoning are nausea, vomiting, diarrhea, and abdominal pain. The disease starts from 30 min to some hours after ingestion of shellfish and lasts for some days in more severe cases. Fortunately, DSP is not life threatening. No special therapy exists, but a supporting treatment, such as liquid replacement, is commonly used. To mitigate the damage of epithelial tissue, administration of the usual drugs for diarrhea is recommended.

A number of different toxins causing DSP have been isolated from dinoflagellates and shellfish (Pavela-Vrančič et al., 2002). They were first divided into three groups:

1. Acidic toxins *ocadaic acid* (Figure 13.3a) and its derivative *dinophysistoxins*
2. Neutral *pectenotoxins* (Figure 13.3b)
3. Sulfated *yessotoxins* (Figure 13.3c)

*Ocadaic acid,* a complex lipophilic polyether, which is decomposed in acids and alkalis, was for the first time isolated from the sponge *Halichrondia okadai*. In Europe, where the main source of toxicity is *Dinophysis acuminata,* the main toxicant is mostly ocadaic acid. In Japan, where the source is mainly *D. fortii,* the toxicity is based on the summary effect of ocadaic acid, dinophysistoxins, and pectenotoxins. In the European Union, the maximum allowable content of DSP toxins in shellfish is 0.16 μg/g. For the estimation of DSP-toxins, a biotest based on mice as well as a specific and sensitive LC-MS/MS method has been developed (Puente et al., 2004).

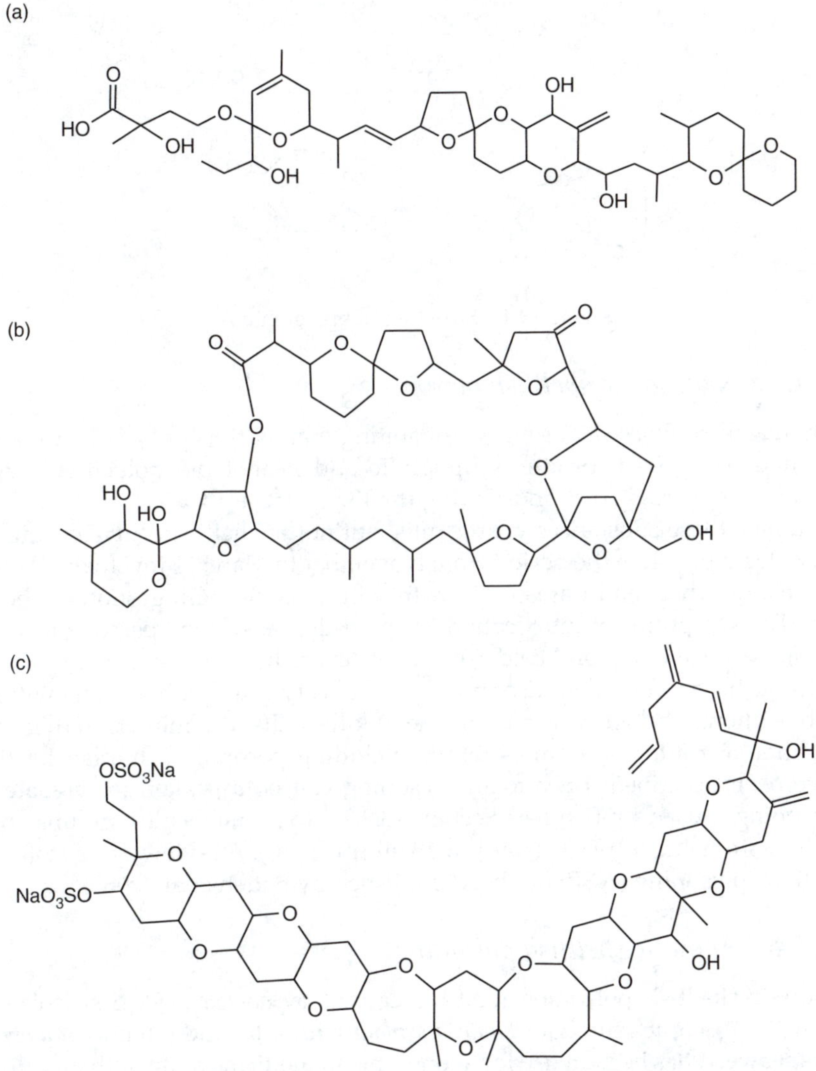

(a)

(b)

(c)

*Figure 13.3* Structures of (a) ocadaic acid, (b) pectenotoxin, and (c) yessotoxin.

*Yessotoxin* was first isolated in Japan from the digestive organs of scallops *Patinopecten yessoensis*. This toxin is also believed to originate from a microalga. Since yessotoxins are not actually diarrhetic, they are not regarded as DSP toxins any more. Yessotoxin attacks the cardiac muscle of a mouse in case of an intraperitoneal injection, and desulfated yessotoxin injures the liver. Yessotoxin is nontoxic if ingested orally. Analogs of yessotoxin have been found in the digestive gland of the mussel *Mytilus galloprovinicialis* inhabiting the Adriatic Sea (part of the Mediterranean Sea) during massive blooming of the alga *Gonyaulax polyhedra*.

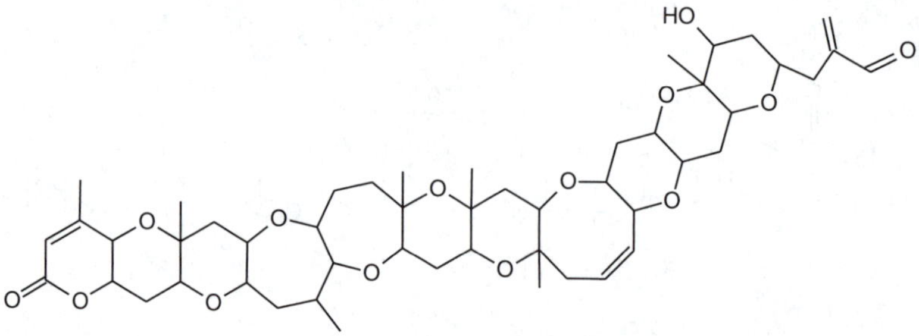

*Figure 13.4* Structure of brevetoxin A.

### 13.1.3 Neurotoxic shellfish poisoning

Dinoflagellate *Ptychodiscus brevis*, inhabiting mainly the Gulf of Mexico and southern U.S. coast, produces lipophilic and heat-stable polyether nerve poisons *brevetoxins A, B* and *C*, (Figure 13.4) that are toxic to fish, which assimilate these substances via the gills, but not to shellfish. *P. brevis* causes mass death of fish in the period from November to March. Brevetoxins cause nausea, diarrhea, and paresthesia within minutes of the ingestion of shellfish. The symptoms of intoxication are paresthesias of the lips, tongue, and throat, reversal of hot and cold sensations, fever, dizziness, supraventricular tachycardia, and broadening of pupils. A recovery takes place usually within 24 h. Although brevitoxins are capable of killing the test animals in the case of different routes of administration, including peroral, no human deaths have been described. Brevetoxin poisoning can be mistaken for ciguatera poisoning caused by fish (see Section 13.2.2). Toxin molecules are unstable both in strong acid ($pH < 2$) and in alkali ($pH > 10$). A review of neurotoxic shellfish poisoning (NSP) has been published by Baden et al. (2005).

### 13.1.4 Amnesic shellfish poisoning

Amnesic shellfish poisoning (ASP) is caused by *domoic acid*, first isolated from macroscopic red alga *Chondria armata* with a Japanese name of *domoi*. This seaweed has been in use for a long time in medicine as an anthelminthic remedy. Only lately, it was established that the same substance was the causative agent of ASP, diagnosed for the first time on Prince Edward Island in Canada. The initial sources of the toxin are diatomic algae of the genus *Pseufonitzschia* such as *P. pungens, P. multiseries, P. australis, P. pseudodelicatissima, P. delicatissima,* and *P. turgidula*. The algae are eaten by krill, which in turn is consumed by whales and other sea animals as well as birds. It is interesting that when the content of domoic acid in the algae rises very high, the krill gives up eating the algae, which in turn contributes to the flourishing of colonies of the toxic alga. A recent review of ASP has been written by Jeffery et al. (2004).

*Domoic acid*, which by its molecular structure, containing three carboxylic groups, resembles kainic acid (Figure 13.5), a glutamate receptor agonist,

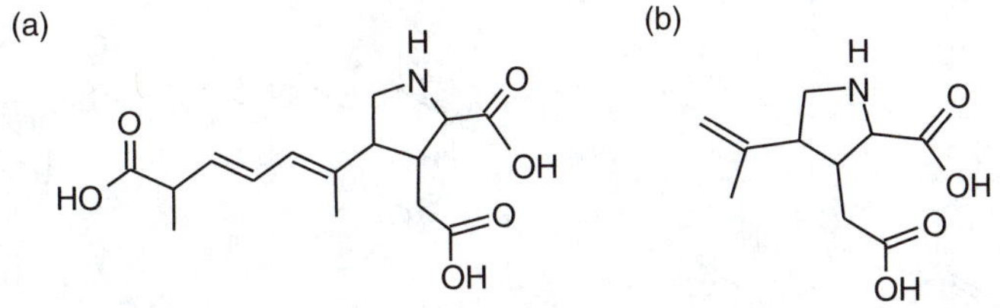

***Figure 13.5*** Structures of (a) domoic acid, and (b) glutamate receptor agonist kainic acid.

is a strong amino acid neuroexcitant, which influences the ionotrophic glutamate receptors of nervous cells, particularly in the brain. These receptors are actually ion channels that open and close the passage of cations through the nervous cell membrane by the action of L-glutamate as a ligand. These activities start several intracellular cascades either directly by letting in cations or indirectly by mediation of secondary messengers like cAMP or reactive organic species (ROS). Induction of such a neuronal misbalance by domoic acid as an agonist may cause disorders in the brain functioning or even formation of permanent brain injuries.

Domoic acid is well soluble in water and poorly soluble in organic solvents. It is toxic to mammals with $LD_{50}$ 3.6 mg/kg bw (mice, intraperitoneally). For example, in Canada, ASP caused the death of four humans in 1987, of 200 cormorants in 1991, and of 50 sea lions in 1998. In Estonia, the allowable limit of the content of domoic acid in mollusks is 20 mg/kg.

Outbreaks of blooming algae have become more frequent in not only the places where they have been described earlier but also in totally new places. Simultaneously, newer algal toxins are isolated and identified. *These trends clearly point to an increasing hazard and risk connected with eating of shellfish.* Over the last years, a number of chemical, mainly chromatographic, methods have been elaborated for analysis of phycotoxins in seafood. Since the multiplicity of phycotoxins makes a chemical determination of all of them impossible, *functional tests have been initiated* to work out a test that would *include all* (or at least most of) the phycotoxins, according to their mechanisms of action (Rossini, 2005).

## 13.1.5 Microcystins and nodularins

These toxins are produced by *freshwater blue algae* (cyanobacteria) of the genera *Anabaena, Nodularia, Nostoc, Oscillatoria,* and *Microcystis*. More than 50 various microcystin congeners are known. Nodularins have been found in the brackish water alga *Nodularia spumigena* as well as in the common mussel *Mytilus edulis*.

*Microcystins are cyclic heptapeptides* (Figure 13.6) that are both specific hepatic toxicants and hepatic carcinogens. *Nodularins are cyclic pentapeptides.* Like ocadaic acid that causes DSP (see Section 13.1.2), both nodularins and microcystins are potent inhibitors of serine/threonine protein phosphatases,

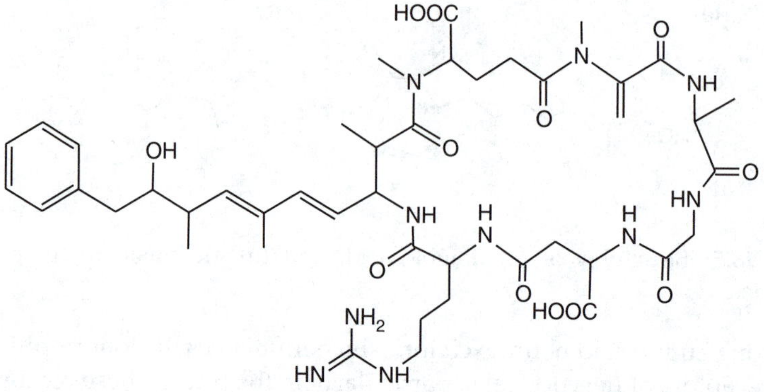

Figure 13.6 Structure of microcystin LR.

that, binding to the same sites of these enzymes as ocadaic acid, act as tumor promoters. Inhibition of the phosphatases spoils the normal balance of phosphate groups in the cellular cytoskeleton. This action is followed by a collapse of the cytoskeleton and the whole hepatocyte. Since, unlike from many other cell types, these compounds are able to penetrate the hepatocyte membranes, the liver is especially sensitive to the toxic effect of microcystins and nodularins. The process is terminated by the death of the organism through serious acute hepatic damage (Msagati et al., 2006).

In principle, human poisoning by microcystins and nodularins can be caused by eating of fish or shellfish contaminated with these toxins. But actually, *contamination of drinking water* with these substances is an even more serious toxicological issue that concerns humans as well as wild and domestic animals. There is a well-known case of poisoning with cyanobacterial toxins and deaths in a renal dialysis hospital in Brazil (Carmichael et al., 2001). Microcystins were found in the salmon and striped perch (*Embiotoca lateralis*) that died from an acute hepatic damage, and from shrimps. Shrimps as a shellfish are considered to be the vector of toxins here. Microcystins are the cause of the drinking water quality issue, related to the cyanobacterial blooms in the United States including the Great Lakes.

## 13.2 Fish toxins

Several toxic fishes that are still being using for food by gourmets are well known.

### 13.2.1 Tetrodotoxin

*This extremely poisonous substance* (acutely 10,000 times more toxic than cyanide ion—see Table 1.2) is found in various terrestrial and marine animals like porcupine fish (*Diodon hystrix*), ocean sunfish (*Mola mola*), parrot fish (*Scaridae*), in the skin secretion of the Californian newt (*Taricha torosa*), in Brazilian aposemantic diurnal yellow frog (*Bracycephalus ephippium*), in the blue-ringed octopus (*Hapalochlaena*), starfish, angel shark (*Squatina squatina*), xanthid crab (*Atergatis floridus*), and many other organisms.

But *TTX became widely famous after being discovered in the delicious puffer fish or fugu* (*Takifugu niphobles*), inhabiting mainly Chinese and Japanese waters. This fish and the first two of the above-mentioned species are the *most toxic sea organisms for humans*. Puffer fish is consumed mainly in Japan, and also in the United States as a delicacy, provided that the dish has been prepared by a specialist and does not contain a toxic dose of TTX. For that, the toxin-rich liver that may contain up to 10 mg/g TTX, intestines, skin, and sexual glands must be extremely skillfully and completely removed. Due to incompetent preparation of puffer fish, there are yearly about 200 deaths throughout the world, about 50 of them in Japan. According to other data, 646 cases of TTX intoxication were registered in Japan in the years 1974–1983, 179 of which had a fatal end. Only a few cases have been reported in the United States, and outbreaks in the countries outside the Indo-Pacific area are rare (Anonymous, US FDA).

TTX (Figure 13.7) is an extremely potent heat-stable nerve toxicant, a specific blocker of sodium channels of the skeletal muscles. The human lethal dose of TTX is about 10 $\mu$g/kg bw, that is, about 1 mg per adult individual. The complex between TTX and Na channel is extremely stable ($K_d = 10^{-10}$ nM). Except for some bacterial toxins like the botulinum toxin, only palytoxin and maitotoxin are more toxic than TTX and saxitoxin. Palytoxin can be found in some corals (*Anthozoa*) that resemble the sea anemone (*Actinia*). TTX is accompanied in nature by its analogs 6-epiTTX, 11-deoxyTTX, 4-epiTTX, 4,9-anhydroTTX, and others. The hypothesis that TTX and its analogs are actually products of bacteria, living in host organisms is widespread. This opinion is supported by the fact that TTX/anhydroTTX are produced by several bacteria of the family *Vibronaceae,* as well as bacteria *Pseudoalteromonas tetraodonis, Photobacterium phosphoreum,* and *Pseudomonas* spp. (Noguchi et al., 2006).

*Symptoms of TTX-intoxication*. TTX is well absorbed from the gastrointestinal tract; peak plasma concentration (see Figure 2.2) is achieved 20 min after the ingestion of a toxic material. TTX is characterized by a wide volume of distribution—the highest concentrations are in the kidneys, the half-life of elimination is 2–4 h. At first (10–45 min after eating), a slight numbness of the lips and tongue appears, followed by paresthesia in the mouth, dizziness,

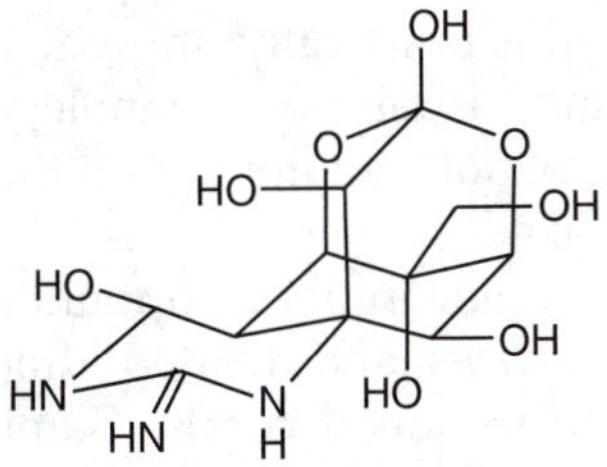

OH
O O
HO OH
HO
OH
HN N HO
HN H

*Figure 13.7* Structure of tetrodotoxin (anhydrotetrodotoxin 4-epitetrodotoxin or tetrodonic acid).

tiredness, headache, pressure sensation in the throat and breast, nausea, vomiting, partial muscular paralysis, and collapse; in the case of high doses, death caused by the paralysis of respiratory muscles arrives during 4–24 (usually 4–6) h, the shortest known period being 20 min. Salivation, muscular twitch, sweating, pains in the pleura, disorders of swallowing, afonia, and convulsions may also occur. In case of a severe intoxication, hypotony, bradycardia, depressed retinal reflexes, and constantly widened pupils may appear. Thereby the victim, being simultaneously completely paralyzed, can be conscious until death. The diagnosis is based on the clinical symptoms and on the eating card (menu). The latter is quite expensive because a dish of puffer fish costs about $400 in a Japanese restaurant in the United States.

*Analysis.* The mouse biotest elaborated for PSP is the best method for the control of the fish. A high-performance liquid chromatographic (HPLC) method with a postcolumn alkaline treatment and fluorescence detection has also been worked out for estimation of TTX. Products of alkaline degradation of TTX can also be proved as trimethylsilyl derivatives by GC-MS.

*Therapy* is only supporting, and if timely, then gastric lavage, charcoal, and maintenance of water level are recommended. No specific antidote exists.

*Legislation.* Very strict FDA guidelines have been enforced on the import of puffer fish to the United States.

### 13.2.2 *Ciguatoxin*

Poisonings caused by *the Ciguatera fish toxin or ciguatoxin* (CTX) are, for example, in the United States one of the four most important seafood intoxications caused by consumption of fish inhabiting the tropical coral reefs between 35° of north and south latitudes, especially in the Caribbean Sea, but also in the Atlantic and Pacific Oceans. With approximately 50,000 cases reported yearly, ciguatera poisoning is the most common nonbacterial, fish-borne poisoning in the United States. *Representatives of over 400 fish species can be* ciguatoxin vectors. Only some of them, first of all, the amberjack (*Seriola* spp.), moray (*Gymnothorax javanicus*), barracuda (*Sphyraena* spp.), coral trout (*Plectropomus leopardus*), and also hogfish (*Bodianus*), scorpionfish (*Scorpaenidae*), some triggerfish (*Balistes*), snappers (*Lutjanus*), and groupers (*Epinephelus*), and so forth are harmful for humans. The presence of a harmful amount of CTX in a fish is practically unpredictable. It is not possible to recognize a toxic fish either visually or organoleptically. Fortunately, most of the tropical coral reefs are not ciguatoxic, and the outbreaks of ciguatoxin poisoning are mostly localized.

*Gambiertoxins,* the less polar initial compounds of ciguatoxins are produced by *Gambierdiscus toxicus* and related dinoflagellates that inhabit on macroalgae adhered to the dead corals. Gambiertoxins are converted into ciguatoxins in the muscles of herbivorous fish and of carnivorous fish that have eaten the herbivores. Altogether more than 20 different gambiertoxins and ciguatoxins have been discovered in algae and in fish.

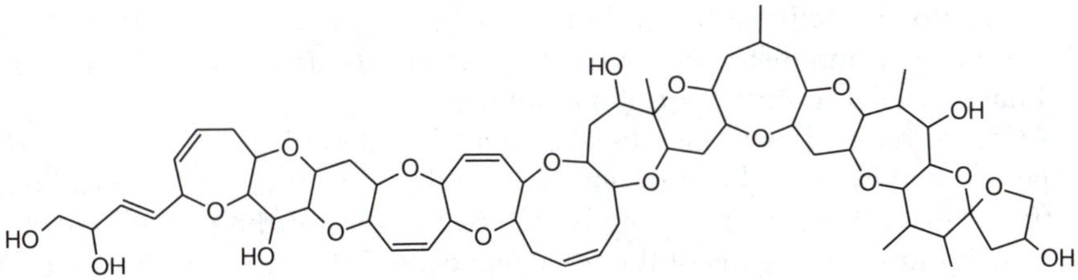

***Figure 13.8*** Structure of the Pacific ciguatoxin-1 (P-CTX-1).

These toxins are complexes of heat and acid-resistant lipophilic cyclic polyethers with molecular weight in the interval of 1038–1140 Da that are biomagnified and rendered more and more hydrophilic along the food chain. As it is fish can be intoxicated by high concentrations of these toxins that cause behavioral and morphological disorders of the fish. But these substances are seriously toxic to humans. *The main Pacific ciguatoxin* (P-CTX-1, Figure 13.8), the most toxic ciguatoxin, has intraperitoneal $LD_{50} = 0.25$ $\mu$g/kg bw in the case of mice and is able to cause human intoxication at its concentration of 0.1 $\mu$g/kg of fish meat. *The main Caribbean ciguatoxin* (C-CTX-1) that is less polar and tenfold less toxic ($LD_{50} = 3.6$ $\mu$g/kg bw) needs also a ten times higher content (minimally 1.0 $\mu$g/kg) in fish meat to elicit a toxic effect in humans.

Water-soluble *maitotoxins* as well as ciguatoxin have been found in the liver, but not in the meat of some fish like the herbivorous surgeonfish (*Acanthuridae, Ctenochaetus striatus,* or *maito*). Despite their wide benthic distribution, maitotoxins are, due to their low accumulation ability, not likely to be the causative agents of human intoxications. $LD_{50}$ of maitotoxins is 0.13 $\mu$g/kg bw (mouse, intraperitoneally), but with peroral administration they are 100 times less toxic than ciguatoxins. These differences are caused by the lower absorptivity (lower internal dose) of maitotoxins than of more polar substances.

*Symptoms of ciguatoxin intoxication* may appear less than 6 h after eating the toxic fish. The first symptoms—nausea, abdominal cramps, vomiting emanate from the digestive tract. *Further, neurological disorders will appear, namely,* headache; flushing; muscular pain and weakness; arthralgia; paresthesia; sensitivity of the lips, tongue, and mouth; and dizziness. In case of a still severe poisoning, reversal of hot and cold sensations as well as cardiac disorders may occur. The victims will usually recover after some days but serious disorders of the nervous system may remain for months or even years, recurring from time to time. *Symptoms may return,* for example, *after consumption of ethanol or eating of toxic fish.* The victims are recommended not to eat fish and drink alcoholic beverages at least for 6 months after the poisoning. In case of extremely severe intoxication, paralysis, coma, and death may occur. It is not possible to acquire immunity against ciguatoxins and the toxins tend to accumulate in the organism. A small number of the victims may acquire a reaction similar to allergy in relation to a number of foods like peanuts, poultry, and pork.

Ciguatoxins activate the sodium ($Na^+$) channels of mammals, causing the excitability and instability of the cellular membrane, and swelling of the cellular as well as mitochondrial membranes.

*CTX-poisoning is not a new disease.* It was first described in China as early as 600 BC and in the eighteenth century, and the crew of Captain James Cook suffered from it on their voyage to Tahiti. Feeding of their hogs with the heads and internal organs of the toxic fish caused the death of the animals. An analogical intoxication happened with four team members of the famous Captain Cousteau, who had eaten barracuda on Haiti. Fortunately, all four recovered completely after 2 months.

*Ciguatoxin intoxication* has, due to the growing export of fish inhabiting the coral reefs, turned into *a global medicinal issue*. According to a very approximate estimate, yearly worldwide 10,000–50,000 humans are suffering from it. Occurrence of this poisoning is actually clearly underrecognized and underdiagnosed. Indications are that an increasing bleaching of corals caused by global warming, environmental factors like earthquakes and hurricanes or anthropogenic factors like tourism, construction of docks, waste waters, and eutrophication give rise to more and more favorable conditions for the dinoflagellate *Gambierdiscus toxicus*.

Very profound and systematic overviews of all aspects of ciguatoxin poisoning from hazard identification to risk characterization have been published by Lehane and Lewis (2000) and by Lewis (2006).

## References

Anonymous, US FDA, Bad Bug Book, Tetrodotoxin (http://www.cfsan.fda.gov/~mow/chap39.html)

Baden, D.G., et al. (2005). Natural and derivative brevetoxins: historical background, multiplicity, and effects, *Environ. Health Persp.*, **113**, 621–625.

Baker, J.A., et al. (2002). Monitoring changing toxigenicity of a cyanobacterial bloom by molecular methods, *Appl. Environ. Microbiol.*, **68**, 6070–6076.

Bogialli, S., et al. (2006). Simple and rapid determination of anatoxin-a in lake water and fish muscle tissue by liquid-chromatography–tandem mass spectrometry, *J. Chromatogr. A*, **1122**, 180–185.

Carmichael, W.W., et al. (2001). Human fatalities from Cyanobacteria: Chemical and biological evidence for cyanotoxins, *Environ. Health Persp.*, **109**, 663–668.

CDC (2002). Centers for Disease Control and Prevention. Update: Neurologic illness associated with eating Florida pufferfish, *MMWR Morb. Mortal Wkly Rep.*, **51**, 414–416.

Dittman, E. and Wiegand, S. (2006). Cyanobacterial toxins-occurrence, biosynthesis and impact on human affairs, *Mol. Nutr. Food Res.*, **50**, 7–17.

Jeffery, B., et al. (2004). Amnesic shellfish poisoning, *Food Chem. Toxicol.*, **42**, 545–557.

Lehane, R. and Lewis, R.J. (2000). Ciguatera: recent advances but the risk remains, *Int. J. Food Microbiol.*, **61**, 91–25.

Lewis, R.J. (2006). Ciguatera: Australian perspectives on a global problem, *Toxicon*, **48**, 799–809.

Matsunaga, S., et al. (1989). Anatoxin-a(S), a potent anticholinesterase from *Anabaena flos-aquae*, *J. Am. Chem. Soc.*, **111**, 8021–8080.

Meier, J. and White, J, Eds. (1995). *Handbook of Clinical Toxicology of Animal Venoms and Poison*, CRC Press, Boca Raton, FL, pp. 75–83.

Msagati, T.A.M., Siame, B.A. and Shushu, D.D. (2006). Evaluation of methods for the isolation, detection and quantification of cyanobacterial hepatotoxins, *Aquat. Toxicol.*, **78**, 382–397.

Noguchi, T., Arakawa, O. and Takatani, T. (2006). TTX accumulation in pufferfish, *Comp. Biochem. and Physiol. Part D Genomics and Proteomics*, **1**, 145–152.

Pavela-Vrančič, M., et al. (2002). DSP toxin profile in the coastal waters of the central Adriatic Sea, *Toxicon*, **40**, 1601–1607.

Poletti, R., Milandri, A. and Pompei, M. (2003). Algal biotoxins of marine origin: new indications from the European Union, *Vet. Res. Commun.*, **27**, Suppl. 1, 173–182.

Puente, P.F., et al. (2004). Rapid determination of polyether marine toxins using liquid chromatography–multiple tandem mass spectrometry, *J. Chromatogr. A*, **1056**, 77–82.

Rossini, G.P. (2005). Functional assays in marine biotoxin detection, *Toxicology*, **207**, 451–462.

Wu, Z., et al. (2005). Toxicity and distribution of tetrodotoxin-producing bacteria in puffer fish *Fugu rubripes* collected from the Bohai Sea of China, *Toxicon*, **46**, 471–476.

*chapter fourteen*

# *Pesticide residues*

## 14.1 General principles

*Pesticides* (from Latin *pestis* = plague, monster + *caedere* = to kill) are toxic chemical compounds used for protection of agriculturally valuable plants against the attack of versatile growth inhibitors (weeds, parasites, insects, microorganisms). The first pesticide known from history was *sulfur*, the smoke of which was used in ancient China 3000 years ago. In the sixteenth century, Chinese repelled insects with *arsenic*. In the seventeenth century, tobacco leaves (agent nicotine) and the seeds of the *strychnine tree* (*Strychnos nux-vomica*) were used to protect agricultural plants against rodents. In the middle of the nineteenth century, *rothenon* from the roots of *Derris eliptica* and *pyrethrum* from the blossoms of chrysanthemum were used as insecticides. In the 1880s, *Bordeaux mixture* (copper sulfate, lime, and water) was used for defense against mildew; two decades later the list of pesticides was supplemented with *Paris green* (copper arsenite, aimed to kill the Colorado beetle), and with *calcium arsenite*.

*The epoch of synthetic pesticides* started in 1939 with the marketing of [(*1,1,1-trichloro-2,2-bis(p-chlorophenyl)ethane*)] *or dichlorodiphenyltrichloroethane (DDT)* (Figure 14.1). The great success of DDT stimulated the chemical industry to bring new synthetic pesticides to the market, the use of which increased until 1980. By that time, it had become clear that many substances that should have killed only pests selectively were toxic to higher animals, including humans. WHO has estimated that yearly, more than 200,000 people are killed by pesticide poisonings worldwide. On the other hand, it is rather unexpected but the truth is that DDT, which is toxic also to some warm-blooded organisms, has saved more human lives than any other

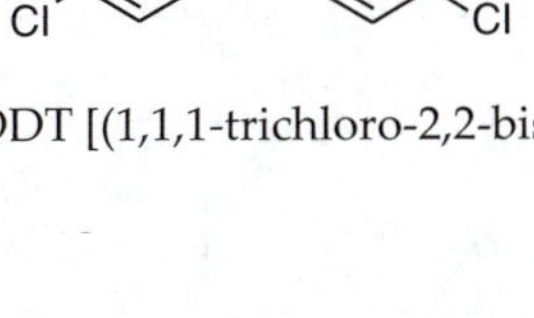

***Figure 14.1*** Structure of DDT [(1,1,1-trichloro-2,2-bis(*p*-chlorophenyl)ethane)].

synthetic chemical. The reason is that DDT, at first considered completely harmless for both humans and animals, was successfully used for the prevention of the distribution of malaria and other dangerous infective diseases like typhus, plague, and yellow fever, transmitted by insects.

By now, methods have been elaborated for the risk assessment of pesticides, taking into account the toxicity of every agent to various organisms and the probability of both occupational exposure and exposure via environment, food, and drinking water. In the case of every new causative agent, elucidation of an exact risk–benefit relationship for marketing is mandatory. The Organization for Economic Cooperation and Development (OECD) has a central role in the worldwide harmonization of the respective regulations (Solecki et al., 2005).

At present, the most important groups of pesticides are

- *Herbicides* (for defense against weeds): chlorophenoxy compounds, dinitrophenols, dipyridyls, carbamates, triazines, urea derivatives, aromatic amides
- *Fungicides* (for defense against unwanted microfungi or molds): alkyl-mercury, chlorinated hydrocarbons, dialkyldithiocarbamates, inorganic substances
- *Insecticides* (for defense against pestiferous insects): organophosphorus and organochlorine compounds, pyrethrins

In addition, *acaricides* to kill mites, *nematocides* to kill mematodes, *rodenticides* to kill rodents, and so forth are used in agricultural practice.

Although pesticides are strongly bioactive compounds and hence of elevated toxicity, their reasonable use for solving the food problems of mankind is still substantiated to reduce the harvest losses caused by pests. The overwhelming soft or organic agriculture is unfortunately inconceivable.

*Pesticides are used in various forms and by various methods*—by dusting as powder, by spraying of solution, as a smoke, or by furrow irrigation. To keep the content of pesticide residues in food and feed plants as low as possible, it is very important to strictly follow the instructions of pesticide use, including the waiting period between the last usage and harvesting, use of minimal possible doses, and so forth.

According to their chemical structure, pesticides can be divided into groups such as

1. *Inorganic compounds*. These compounds like sodium chloride, copper sulfate, sulfur, lead, and arsenite have been in use for centuries. Their toxicity to the mammals is variable, for example, the toxicity of copper is low, but that of lead is high. Mechanisms of toxicity as well as of biomagnification are also variable, the last property being still generally lower than in the case of organic pesticides; the organic compounds are degradable, the toxic elements naturally not.

2. *Organochlorine compounds*. Chlorine derivatives of the hydrocarbons where halogen atoms are linked to the aromatic carbons as in polychlorinated biphenyls (PCBs) are used as pesticides. The first such compound, DDT, was synthesized in 1874 and put into use as a pesticide in 1939. Paul Müller was awarded the Nobel Prize in medicine in 1948 for the discovery of the insecticidal effect of DDT. The worldwide yearly production of DDT grew until 1970, being then 175,000 ton.

Organochlorine pesticides are divided into the following subtypes:

a. Subtype of DDT—DDT, TDE, rotane, methyloxychlor, dicofol
b. Chlorinated cyclodins like aldrine, dieldrine, endrine, heptachlor, and chlordane
c. Others, for example, lindane, toxafene, mirex, chlorodecon

These very stable compounds are not easily degraded either by physical (heat, sun radiation) or by biological factors like bacteria. On the average, 95% of DDT or dieldrine is degraded in the environment, respectively, during 8 to 10 years. They are liable to accumulate in the edible parts of plants due to a very slow metabolism and high fat solubility in the lipid-rich adipose, nervous, hepatic, renal, and other tissues of various animals and birds. *The world record of DDT content in living material* is 36,000 ppm or 36 g/kg in the muscle of a white-tailed eagle (*Haliaeetus albicilla*) of the Baltic Sea (Koivusaari et al., 1976).

*The toxicity of organochlorine compounds to mammals is relatively low*, their biodegradation slow ($\tau_{½DDT} > 10$ years), and biomagnification high. They have both acute and chronic neurotoxicities to insects as well as mammals. Since the nervous systems of insects and animals are, to a certain extent, similar to each other, the chlororganic pesticides designed for the killing of insects are able to cause acute injuries in the nervous system of animals as well. The lethal dose depends on the species and the individual, but the nervous cells will perish when the level of DDT rises to 30 ppm.

*DDT* (Figure 14.1) *and its main metabolite dichlorodiphenylethane* (DDE) *are both very toxic to fish and birds*, shifting their hormonal balance by the *endocrine disruption* mechanisms (see Section 3.2.9). A relationship exists between the thickness (mechanical strength) of the eggshell and the content of DDE in the meat of predatory birds like the kestrel (*Falco tinnunculus*) or peregrine (*Falco peregrinus*) and that of other birds. A decrease in eggshell thickness caused by disorders in calcium metabolism increases the likelihood of fragmentation of eggs during incubation (Lundholm, 1997). Studies carried out in Great Britain in 1955–1962 stimulated by a catastrophic fall in the number of peregrines showed again an essential decrease of the thickness of the eggshells (Cooke, 1973). As the primary reason, the effect of DDE and also the direct toxicity of the chlororganic pesticide dieldrine was considered.

DDT that is deposited in the adipose tissue has no biologic effect as long as the volume of the tissue starts to decrease, for example, because of starvation or intensive physical activities. The respective increase of the DDT content in the blood leads to the strengthening of the toxic effect of DDT and reduction of the birds' populations. In addition to decreased shell thickness, DDT has been blamed for other effects of endocrine disruption, such as decreased brood, increased mortality of embryos and young birds, and abnormal aggressiveness of old birds toward young birds, abandonment of nests, and so forth.

DDT and DDE have also been accused for a substantial decline of the alligator (*Alligator mississippiensis*) population at Lake Apopka in central Florida between 1980 and 1987. Concrete symptoms of endocrine disruption were the developmental abnormalities of the gonad and abnormal sex hormone concentrations in juvenile alligators. During a heavy rain in 1980, the percolation pond of the Tower Chemical Co. manufacturing both organochlorine and organophosphorus insecticides overflowed and the wastewater discharged into the lake. It has been stated that, most probably, DDT was not the only causative agent of the reduction of the crocodilian population (Semenza et al., 1997).

DDT has been discovered to be carcinogenic to rodents but no firm proof of toxicity of DDT to humans has been obtained so far. Tumorigenic potential of DDT in nonhuman primates was also found to be negligible even after animal studies for 15–22 years (Thorgeirsson et al., 1994). Although DDT has been used wrongly and irresponsibly in agriculture, and direct human contact with DDT has been sometimes very intensive, there is no proof that DDT has caused the death of at least one human being. Nevertheless, it has been proved that as a result of spraying of fields with DDT, birds and domestic animals have died, and fish have died in the rivers as a result of similar actions against forest pests.

Other organochlorine insecticides such as *dieldrine, endrine,* and *aldrine* possess, due to participation in the enterohepatic cycle, a higher acute toxicity than DDT or lindane. *Chlordane, heptachlor, mirex, 2,3,7,8-tetrachlorodibenso-dioxin or dioxin* (TCDD) are not completely safe either. For example, chlordane acts (in case of inhalation) adversely on the nervous system, digestive tract, and liver of both humans and animals. Large oral doses of lindane may cause convulsions and death. Use of *chlordane* (Figure 14.2) is completely prohibited since 1988; during the preceding 5 years, use of lindane was allowed only in antithermitic actions.

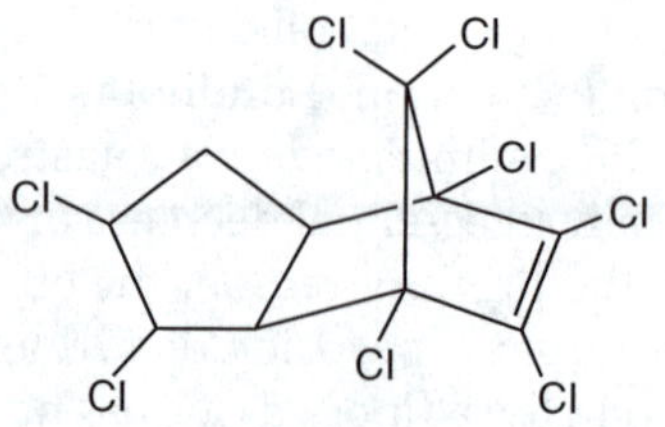

*Figure 14.2* Structure of chlordane.

3. *Organophosphorous compounds*. These include a number of phosphoric acid esters that in turn are divided into four groups according to their molecular structure. Organophosphate pesticides *parathion* (Figure 14.3) and *malathion* were first synthesized in the 1940s and their prototypes were sarin and other compounds used earlier as chemical weapons. *Organophosphorus pesticides are essentially more toxic than organochlorine pesticides,* although less toxic than their prototypes. Some of them, which are relatively safe to humans, are used as *herbicides*. Phosphoric acid esters are acutely neurotoxic to mammals, inhibiting the acetylcholinesterase-catalyzed hydrolysis of the neurotransmitter acetylcholine in synapsis. Acetylcholine, accumulated due to this inhibition, causes permanent nerve excitation. *Biomagnification of organophosphorus pesticides is low* and their *biodegradation fast* and chronic intoxications are rare. Phosphoric acid esters cause yearly about 3000 agricultural intoxications. The most toxic of them is *parathion, although malathion* is considerably less poisonous. Parathion is hydrolyzed. Very often, these compounds have a smell of garlic, which simplifies diagnosis of the poisonings. A recent review of organophosphorus pesticides has been published by Costa (2006).

4. *Carbamates*. Derivatives of carbamic acid $NH_2COOH$ such as aldicarb or carbofuran, an arylic-leaving group, appeared in the market in 1956. Carbamates are used against insects having a resistance against organophosphorus compounds. The toxicity of carbamates is mostly high in mammals and low in fish, but *carbamates are very toxic to birds and honeybees*. The mechanism of a toxic effect is similar to the effect of organophosphorus compounds, and biodegradation is fast. The main symptoms of poisoning are narrowing of the pupils, muscular weakness, spasms, hypotensia, breathing disorders, convulsions, and cardiac arrest. One of the most toxic carbamates is *aldicarb* (Figure 14.4) with peroral $LD_{50} < 1$ mg/kg bw. Aldicarb, carbofuran, methomyl, and propoxur are classified as probable human carcinogens.

5. *Pyrethroids* are esters of chrysanthemic [2,2-dimethyl-3-(2-methyl-1-propenyl)-cyclopropanoic] acid synthetic formulations, similar to

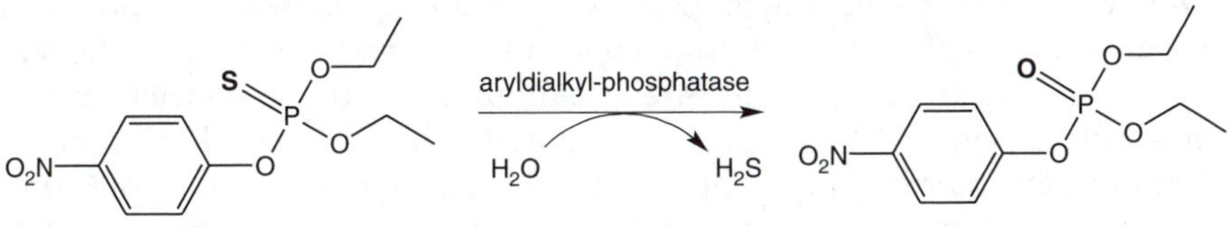

***Figure 14.3*** Enzymatic hydrolysis of parathion to more toxic paraoxon.

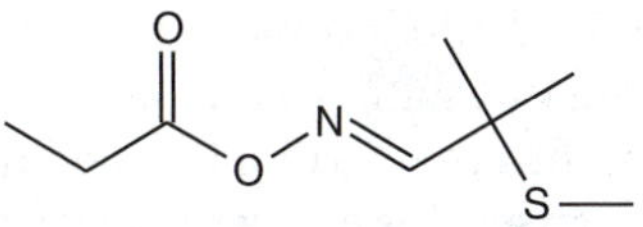

***Figure 14.4*** Structure of aldicarb.

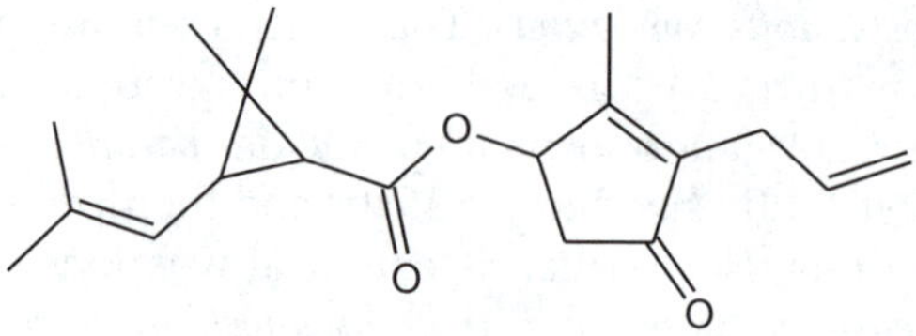

*Figure 14.5* Structure of allethrine.

*pyrethrum*—an efficient contact poison of insects from the blossoms of pyrethrum daisy (*Chrysanthemum cinerarifolium*). Pyrethroids were first produced in 1949; examples are allethrine (Figure 14.5) and dimethrine. *The toxicity of pyrethroids to warm-blooded organisms is relatively low,* their action mechanism is similar to DDT, biodegradation is fast, biomagnification is relatively fast, and some fish, shellfish, and useful insects are very sensitive to the adverse effect of pyrethroids. Contemporary knowledge of the toxicity of pyrethroids has been recently reviewed by Ray and Fry (2006). Pyrethroids are generally regarded as safe to man, and few human fatalities have been reported. Their acute toxicity is dominated by pharmacological actions on the central nervous system (CNS), predominantly mediated by prolongation of the kinetics of voltage-gated sodium channels. One significant problem is the very clear heterogeneity of pyrethroid sensitivity that is seen across sodium channel subtypes. Recent studies still suggest additional effects of pyrethroids such as developmental neurotoxicity, production of neuronal death, and action mediated via pyrethroid metabolites. The evidence for these is at present equivocal, but all the three carry important implications for human health.

6. *Biological insecticides*. These compounds are toxic to definite *Arthropoda,* for example, diflubenzurone is an inhibitor of chitin synthesis. Their toxicity to mammals is relatively low and biodegradation and biomagnification variable.

According to the recommendation 19/5 of Helsinki Commission (HELCOM) and *Convention of Persistent Organic Pollutants* (POP), the use of a number of the so-called aged pesticides (including all organochlorine) as biocides and plant protective products will be stopped and all the leftovers destroyed. These compounds are 1,2-dibromoethane, 2,4,5-trichlorophenoxyacetic acid (2,4,5-T), aldrine, aramite, HCH, beta-HCH, chlordane, chlorodecon, chlordimeform, DDT, dieldrine, endrine, isodrine, heptachlor, hexachlorobenzene, isobenzane, kelevan, lindane, mirex, nitrofene, PCP, quintosene, and toxafene. Maybe from this list, at least DDT should have been left out, which obviously is not toxic to humans and, in case of reasonable use, also to fish and birds. Regrettable are the cases of public anti-DDT hysteria, at times breaking out without any assessment of the correct scientific data. For example, opinion exists that the decline of the bird populations could be at least partly caused by a disappearance of the insects eaten by birds under the influence of DDT. In any case, this situation demonstrates how easy it is to misbalance natural systems with toxic substances.

The mechanisms of production of *collective poisoning* by pesticides have been:

- Occupational exposure
- Cutaneous contact
- Contamination of foods

Occupational exposure threatens people working in pesticide manufacture and application as well as crop management.

It is possible to distinguish four groups of *alimentary epidemics* caused by pesticides (Ferrer and Cabral, 1995):

1. Contamination of food during transport or storage
2. Ingestion of seed prepared for sowing
3. Accidental use of pesticides in food preparation because of their organoleptic similarity to foodstuffs
4. Presence of pesticides in water or food due to unsafe use of pesticides

*Mass human poisoning* with flour contaminated with parathion happened, for example, in Colombia when 88 patients out of 600 died. Analogical accidents have been reported in connection with endrine, hexachlorobenzene, pentachlorophenol, and Hg-organic pesticides (Ferrer and Cabral, 1995).

*In comparison, for example, with diseases caused by food microorganisms,* human intoxication with pesticide residues is a relatively small but still a considerable issue. Maximum residue limits (MRL) have been set up in different parts of the world to assess the safety of various food commodities. *In the European Union, MRLs are established* by the European Agency for the Evaluation of Medicinal Products (veterinary drugs) and by the European Commission (agricultural chemicals). For those veterinary drug or chemical/commodity combinations where no Community MRL exists, the situation is not harmonized and every Member State may set MRLs at the national level to protect the health of consumers. In this case, it is important to check with the country of interest for their legislation covering veterinary drugs and agricultural chemicals. *In United States, the pesticide tolerance levels* set by the EPA have been enforced by the Food and Drug Administration (FDA).

FAO/WHO is setting up a step-wise process, *acute reference doses* (ARfDs), for risk assessment of agricultural pesticides and guidance regarding selected toxicological end points. Hematotoxicity, immunotoxicity, neurotoxicity, liver and kidney toxicity, endocrine effects, as well as developmental effects are taken into account as acute toxic alerts, relevant for the consideration of ARfDs for pesticides. The general biological background and the data available through standard toxicological testing for regulatory purposes, interpretation of the data, and conclusions and recommendations for future improvements are described for each relevant endpoint (Solecki et al., 2005).

A good review of safety, toxicology, and detection of pesticide residues in food is a handbook edited by D.H. Watson (2004).

## 14.2 Insecticides

*The most important classes of insecticides* are chlorinated hydrocarbons, including PCBs, organic phosphates, and carbamates. From the toxicological point of view, the most important are compounds belonging to the first class that, being very stable and fat-soluble, are accumulated in the adipose tissue and milk fat. The analysis of mammals, including the adipose tissue of humans, indicates an exposure to this type of compound. Worldwide, due to an increasing use of tiophosphoric acid esters and pyrethroids, food contamination with chlorinated hydrocarbons is continuously decreasing. Owing to relatively fast degradation, there are no residue problems with novel classes of insecticides. But the concentration of chlorinated hydrocarbons in mother's milk is often still higher than the respective allowed daily intake (ADI). Nevertheless, feeding of suckling babies with mother's milk for 3 months overweighs the potentially harmful effect of the pesticide residues.

## 14.3 Herbicides

*Herbicides are divided into two groups*—compounds with a broad effect and selective substances. The first group consists of chlorates, iron sulfate, calcium cyanamide, and chlorinated derivatives of fatty acids. The second group is made up of growth regulators like aryloxy-fatty acids (e.g., 2,4,5-T), carbamic acid, urea derivatives (monurone), triazines (simazine), and pyridines [paraquat(PQ) and diquat(DQ)] (Figure 14.6). Herbicides are used mainly in grain farming to kill the weeds at the start of the vegetation period. Their toxic action is directed to disruption of metabolism, for example, through blocking of photosynthesis.

The problem of herbicide residues in food is happily negligible and the *toxicity of most herbicides to warm-blooded organisms is very low*. No case of dietary intoxication by herbicides is known at a normal use of these compounds. On the other hand, the negative effect of herbicides to soil microflora and arthropoda cannot be ignored. Herbicides are rather immobile in the soil; they occur in water bodies only as a result of direct spraying. Most herbicides do not accumulate in the environment and most of them contain no highly toxic additives.

*Exceptions are toxic nonselective pyridines PQ and DQ*, which are used for liquidation of all vegetation on some surface. These herbicides are toxic to humans as well, their $LD_{50}$ in humans is extremely low, approximately 3–5 mg/kg. Large doses of PQ have life-threatening effects on all human organs.

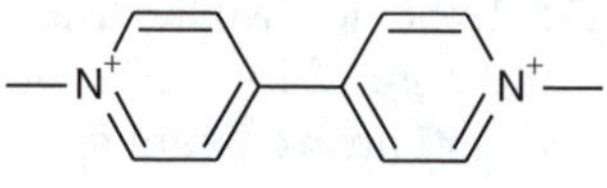

$$-N^{+}\langle\text{pyridine}\rangle-\langle\text{pyridine}\rangle N^{+}-$$

***Figure 14.6*** Structure of paraquat, used as a dichloride or a dimethylsulfate.

PQ is a well-known pneumotoxicant and provides an established model of oxidative stress connected with radical peroxidation of fatty acids (Suntres, 2002). Smaller doses of PQ have the greatest impact on the lungs because lungs are the organs where PQ is quickly transferred primarily through the generation of free radicals with oxidative damage to the lung tissue. The toxic effect of PQ can be reduced with antioxidants (Yeh et al., 2006). Poisonings from inhalation are rare. Some of the patients, who have ingested very large amounts of concentrated PQ, have died within 48 h because of circulatory failure. Animal studies conclude DQ to be moderately toxic via ingestion and dermal absorption. However, in humans it may be fatal if absorbed through the skin and harmful if swallowed or inhaled.

Residues of PQ are generally not detectable in food, except when PQ is used as a preharvest desiccant in food crops (e.g., cereals) where levels of up to 0.2 mg/kg of plant matter have been reported. The recommended ADI is 0.004 mg/kg bw day.

## 14.4 Fungicides

Fungicides have been designed to ward off fungal and mold diseases like blight of potato and tomatoes, mildew and scab of fruits. Important fungicides are inorganic compounds such as copper oxychloride and sulfur, organic compounds such as dithiocarbamates and organometallic substances. Traces of dithiocarbamates have been found in vegetables, especially in lettuce. From the toxicological point of view, the matter is not very serious, since the exposure to these compounds is low. Most fungicides do not accumulate in the environment and most of them are nontoxic to humans. Some fungicides like pentachlorophenol and organic compounds of mercury are still persistent in the environment.

*Unlike from the other fungicides, the Hg-organic compounds are toxic to a careless consumer*. It is because these substances are often used for seed disinfection to prevent the growth of molds during storage. Grains that are treated with a disinfectant and anticipated only for sowing are colored pink for caution. Despite this precaution, cases of intoxications have happened with food mistakenly prepared from disinfected seed grain.

## References

Cooke, A.S. (1973). Shell thinning in avian eggs by environmental pollutants, *Environ. Pollut.*, **4**, 85–152.

Costa, L.G. (2006). Current issues in organophosphate toxicology, *Clin. Chim. Acta*, **366**, 1–13.

Ferrer, A. and Cabral, R., (1995). Recent epidemics of poisoning by pesticides, *Toxicol. Lett.*, **82–83**, 55–63.

Koivusaari, J., et al. (1976). Chlorinated hydrocarbons and total mercury in the prey of the white-tailed eagle (*Haliaeetus albicilla L.*) in the quarken straits of the Gulf of Bothnia, Finland, *Bull. Environ. Contam. Toxicol.*, **15**, 235–241.

Lundholm, C.E. (1997). DDE-induced eggshell thinning in birds: Effects of p,p′-DDE on the calcium and prostaglandin metabolism of the eggshell gland, *Comp. Biochem. Physiol. C: Pharmacol. Toxicol.*, **118**, 113–128.

Ray, D.E. and Fry, J.R. (2006). A reassessment of the neurotoxicity of pyrethroid insecticides, *Pharmacol. Ther.*, **111**, 174–193.

Semenza, J.C., et al. (1997). Reproductive toxins and alligator abnormalities at Lake Apopka, Florida, *Environ. Health Persp.*, **105**, 1030–1032.

Solecki, R., et al. (2005). Guidance on setting of acute reference dose (ARfD) for pesticides, *Food Chem. Toxicol.*, **43**, 1569–1593.

Suntres, Z.E. (2002). Role of antioxidants in paraquat toxicity, *Toxicology*, **80**, 65–77.

Thorgeirsson, U.P., et al. (1994). Tumor incidence in a chemical carcinogenesis study of nonhuman primates, *Regul. Toxicol. Pharmacol.*, **19**, 130–151.

Watson, D.H., Ed. (2004). *Pesticide, Veterinary and Other Residues in Food*, C.H.I.P.S., Weimar, Texas, U.S.A.

Yeh, S.T., et al. (2006). Protective effects of *N*-acetylcysteine treatment post acute paraquat intoxication in rats and in human lung epithelial cells, *Toxicology*, **223**, 181–190.

# chapter fifteen

# Veterinary drugs and feed additives

Nowadays a number of veterinary drugs are used in large volumes and not as much for medical treatment of domestic animals as, *for prophylaxis of diseases and as growth stimulators.* The residues of active components and their metabolites in food of animal origin (edible tissues, milk, eggs, honey, etc.) are eaten by humans usually in very small amounts, but continuously. Hence, *these residues may cause harm to human health* in the long perspective. Over the years, no attention was paid to this issue, but lately, many of the earlier broadly used drugs have either been entirely banned in the European Union or have been allowed to be used only as medicine and not for the stimulation of a muscular mass gain. Legislatively, respective withdrawal periods during which pre-slaughtering administration of a drug is prohibited and systems of control and screening are built up. *Maximum residue limits* (MRL in mg/kg or μg/kg of raw foodstuff) have been established by the EU and *Codex Alimentarius* Commission for compounds allowed to be used as drugs. *To control the adherence to rules,* easy, rapid, and sensitive tests for the detection and characterization of chemical and veterinary drug residues are needed (Toldrá and Reig, 2006). A brief discussion of the groups of veterinary drugs and feed additives, most important for food toxicology, is subsequently presented.

## 15.1 Antibiotics

*Antibiotics or antibacterial drugs* are used in cattle breeding both as medicines and as growth stimulators. They improve the assimilability of nutrients and accelerate the growth of domestic animals (calves, sheep, swine, poultry, and fish). Active antibiotic residues can be found in meat, milk, and eggs. It is not possible to remove or to render them harmless in the raw food of animal origin since most antibiotics stand pasteurization well and also other types of heat treatment. Since some antibiotics are fat-soluble and some water-soluble, *they can be found both in high- and low-fat foodstuffs.*

It is well known that in addition to the curative effect, *antibiotics also have side effects.* Long-term use of antibiotics even in low doses may cause the appearance of microorganisms that are resistant to antibiotics and allergic reactions in sensitive humans. Antibiotics may also alter the enzymatic activity of the intestinal flora. It has been tried to separate human and animal antibiotics from each other. Sulfonamide antibiotics have exerted carcinogenicity. The most widely used antibacterial drug is *benzylpenicillin* (penicillin G).

In the past, *nitrofurane antibiotics like furazolidone, nitrofurantoine, furaltadone,* and *nitrofurazone* were widely used to treat salmonelloses and colibacterioses. Since nitrofuranes metabolize quickly in the animal organism, they are screened as metabolites like 3-amino-2-oxalodinone (AOZ) in the case of furazolidone. By now, it has been established that *nitrofuranes are carcinogenic* and their use has been banned in the EU. EU drug residue monitoring programs prescribe estimation of nitrofuranes as well as sulfonamides, penicillins, tetracyclins, fluoroquinolons, macrolides, and other antibiotics in the raw food material of animal origin. First, random samples of meat, milk, eggs, and so forth are analyzed by the microbiological agar-diffusion method, thereafter the samples that prove to be positive are subjected to a chromatographic analysis to confirm and quantify the particular antibiotic. Parts of the samples are immediately analyzed by chromatography to establish the content of a particular type of antibiotics. In Estonia, for example, mainly the presence of penicillin and gentamycin residues have been established in food raw material; drugs of honeybee streptomycin and tetracyclin have been found in honey.

*A special toxicological problem is chloramphenicol* (Figure 15.1) to which most of the aerobic and practically all the anaerobic bacteria are sensitive. Despite that, chloramphenicol residues are still found in the samples of the screening programs. *Chloramphenicol is a highly toxic substance* causing aplastic anemia, bone marrow suppression, and damage of the liver. Being mutagenic, chloramphenicol increases the risk of childhood lymphoblastic leukemia (Shu et al., 1987). The EU has, since 1994, prohibited the use of chlorampfenicol in the case of productive animals. Manufacture of oral chloramphenicol was stopped in the United States in 1991.

## 15.2 Hormones

*Hormones as anabolic chemical regulators* of physiological processes are used analogically to antibiotics as growth stimulators of agricultural animals.

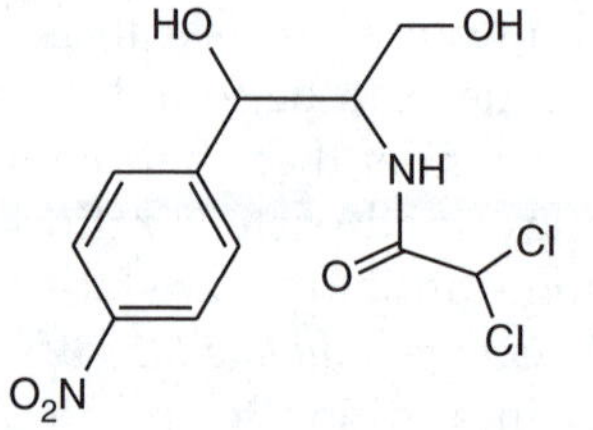

HO OH NH Cl O Cl $O_2N$

***Figure 15.1*** Structure of chloramphenicol.

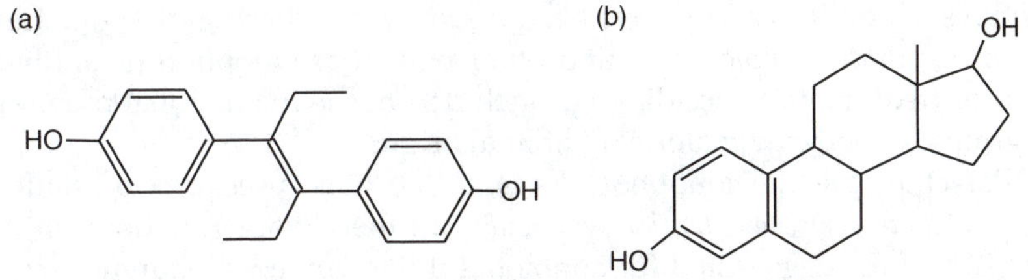

(a) (b)

HO OH HO OH

***Figure 15.2*** Structure of (a) diethylstilbestrol (DES) and (b) estradiol.

According to their chemical structure, hormones are divided into three main groups—*proteins, peptides, and steroids.* Since proteins and peptides are decomposed during digestion, the first two groups of hormones cannot have an essential toxicological value in humans. *Steroidal hormones* are stable in the digestive tract and they are well absorbed, and maintain their effect until metabolized into inactive compounds. They include natural hormones like *testosterone* and *estradiol,* and also synthetic substances with an analogical activity such as *esters of estradiol* and *testosterone, trenbolone, diethylstilbestrol* (DES) (Figure 15.2), *and zeranol* (group of resorcylic acid lactones). In the EU, the use of all steroidal hormones in cattle breeding is prohibited. Since estradiol and testosterone occur in animal organisms also naturally, MRL values are established for the estimation of their illegal use. While natural hormones are not expected, due to their short half-life, to have adverse effects in food, synthetic hormones are much more stable in the organism. Natural hormones like progesterone are used to regulate the gestation, while synthetic hormones, including *DES* belonging to stilbenes, are used as growth enhancers.

*DES* was used during the 1950s and later to accelerate the weight gain of bovines and sheep. But when this hormone was given to pregnant women to avoid miscarriages, both the teratogenic and carcinogenic side effect of DES appeared. Namely, in the case of some of the daughters of these women (the so-called DES-daughters), vaginal cancer appeared after they reached adult age, accompanied by structural changes of the reproductive tract, pregnancy complications, and infertility. In addition, it was established that the women who had used DES during pregnancy had an elevated risk of breast cancer.

*Zeranol* (7$\alpha$-zearalanol), an estrogenic hormone obtained from mycotoxin zearalenone (see Section 11.5) is highly hepatotoxic and produces hematological alterations at high doses. The use of DES and zeranol as growth stimulators is prohibited in the EU. In the EU countries, screening of meat for the content of DES, dienestrol, and hexestrol as well as for trenbolone, testosterone, and nor-testosteron is carried out.

## 15.3 Other veterinary drugs

### 15.3.1 Coccidiostatics

Coccidiostatics (anticoccidials) are added to feed to combat *coccidioses* caused by enteric protozoan parasites (*Coccidia*). At risk are poultry and rabbits.

Residues of coccidiostatics have been discovered in hen eggs. Estimation of *monensin, narasin, salinomycin,* and other polyether ionophorous antibiotics used for treatment of coccidioses is included in the EU national monitoring programs of food raw material of animal origin.

Polyether calcium ionophore *lasalocid,* the most used coccidiostatic has lately attracted *special attention,* especially in the United Kingdom. In 2003, over 12% of the eggs tested for antibiotics in the United Kingdom were contaminated with lasalocid—a toxic antibiotic. Some of the eggs were found to have much higher levels of lasalocid than ever recorded before in the United Kingdom. Over the past 6 years, there have been an increasing number of incidences of lasalocid residues found in eggs. A broader survey of spread of coccidiostatic residues in eggs in eight different European countries revealed that about 36% of the samples analyzed contained one or more of the nine anticoccidials in concentrations ranging from 0.1 to 63 $\mu$g/kg. *Salinomycin* and *lasalocid* accounted for more than 60% of all positive samples (Mortier et al., 2005).

*Lasalocid is permitted* for use in young birds destined to become layers, broiler chickens, quail, turkey, and pheasant feed, but it is illegal for use in laying-hen feed. Cross-contamination at the feed mill is considered to be the biggest source of the problem. However, the use of lasalocid in young birds near the laying age, the wrong feed delivered and fed on the farm, as well as failure to clean trucks between loads are all implicated.

Individuals at the greatest risk from lasalocid residues in food are

- Those on diets recommending higher egg consumption, such as the Atkins diet
- Unborn children, because of their poor ability to metabolize toxic chemicals
- Babies, ingesting lasalocid residues through breast milk or eating a certain infant formula
- Any young child with a heart condition
- The elderly, who suffer from cardiacarrhythmias, such as atrial fibrillation and other tachycardias
- Those who suffer from high blood pressure and some heart conditions

It is believed that there may be a potential link between *Sudden Adult Death Syndrome* and the consumption of lasalocid residues in food.

### 15.3.2 *Anthelmintics*

Anthelmintics have either a retarding or destroying effect on helmints. In the EU countries, food raw material of animal origin is analyzed for anthelmintics like *ivermectin, levamisole, fenbendazole, febantel,* and others according to a respective national monitoring program.

*Special attention is paid to ivermectin,* the most toxic anthelmintic (see also Section 15.1). Ivermectin, a macrocyclic lactone of the avermectine group has been approved for use in the United States in dogs, cats, horses, cattle, sheep, and swine. It is available in a variety of formulations. One formulation for use in cattle is an intraruminal sustained release device (IVOMEC SR Bolus). This is implanted orally in the rumen of cattle that are at least 12 weeks old (i.e., certain to be ruminating) at the beginning of the grazing season.

*The avermectins are neurotoxins,* which have been successfully used in the treatment of helmintic parasitic infections in a number of terrestrial farm animals and also in the treatment of river blindness in humans. Ivermectin has been administered as an oral treatment with feed and has been found to be effective for the treatment of sea lice on farmed Atlantic salmon. Ivermectin is poorly absorbed by fish, with a high percentage of the administered dose being excreted in the feces. The highest concentrations of absorbed ivermectin were found in the lipid-rich organs. The ivermectin remained in the tissues of the treated fish for a prolonged period of time and was excreted mainly in the unchanged form. Ivermectin can reach the marine environment via excretion with bile, unabsorbed via the fish feces, and by uneaten food pellets, and it has a strong affinity to lipid, soil, and organic matter. Risk assessments have shown that ivermectin is likely to accumulate in sediments and that the species therein would be more at risk than the species in the pelagic environment. Ivermectin has been shown to be toxic to some benthic infaunal species in single species tests, but there is no evidence that treatment of fish with ivermectin has affected multispecies benthic communities in the field situation (Davies and Rodger, 2000).

### 15.3.3 *β-agonists*

The use of β-agonists as growth stimulators is prohibited in the EU countries. *Clenbuterol* is allowed to be used to treat diseases of respiratory organs as well as birth complications. Acute clenbuterol toxicity resembles that of other β2-adrenergic agonists. Most reported cases of clenbuterol intoxication describe patients who have eaten livestock illicitly treated with clenbuterol. The symptoms of clenbuterol poisoning can be sustained sinus tachycardia, hypokalemia, hypophosphatemia, and hypomagnesemia after ingesting a reportedly small quantity of clenbuterol (Hoffman et al., 2001). Clenbuterol, as well as salbutamol and mabuterol are analyzed according to the EU national monitoring programs.

### 15.3.4 *Glucocorticoids*

In stress situations, hormones of the adrenal gland, like *cortisone,* are administered to domestic animals. Since these drugs have actually a broad area of action, they cannot be allowed to enter the food chain uncontrolled.

### 15.3.5 *Thyreostatics*

*Thiourea derivatives* that are used for this purpose are carcinogenic. The use of thyreostatics is either completely prohibited or allowed only for some medicinal purposes in the EU. Estimation of thiouracil, tapazole, methylthiouracil, propylthiouracil, and phenylthiouracil is included in the respective national monitoring programs of the EU countries.

A good review of safety, toxicology, and detection of veterinary drug residues in food, containing data about griseofulvin, $\alpha$-lactame antibiotics (penicillin and cephalosporins), macrolides, aminoglucosides, fluoroquinolones, sulfadimidine, carbadox, and olaquindox, furazolidine, chloramphenicol, ivermectin, tranquillizers, and carazolol can be found in the handbook edited by D.H. Watson (2004).

## References

Danaher, M., et al. (2006). Review of methodology for the determination of macrocyclic lactone residues in biological matrices, *J. Chromatogr. B: Analyt. Technol. Biomed. Life Sci.*, **844**, 175–203.

Davies, I.M. and Rodger, G.K. (2000). A review of the use of ivermectin as a treatment for sea lice [*Lepeophtheirus salmonis* (Krøyer) and *Caligus elongatus* Nordmann] infestation in farmed Atlantic salmon (*Salmo salar* L.), *Aquac. Res.*, **31**, 869–883.

Hoffman, R.J., et al. (2001). Clenbuterol ingestion causing prolonged tachycardia, hypokalemia, and hypophosphatemia with confirmation by quantitative levels, *J. Toxicol. Clin. Toxicol.*, **39**, 339–344.

Mortier, L., et al. (2005). Incidence of residues of nine anticoccidials in eggs, *Food Addit. Contam.*, **22**, 1120–1125.

Shen, H.-Y. and Jiang, H.-L. (2005). Screening, determination and confirmation of chloramphenicol in seafood, meat and honey using ELISA, HPLC-UVD, GC-ECD, GC-MS-EI-SIM and GCMS-NCI-SIM methods, *Anal. Chim. Acta*, **535**, 33–41.

Shu, X.O., et al. (1987). Chloramphenicol use and childhood leukemia in Shanghai, *Lancet*, **2**, 934–937.

Toldrá, F. and Reig, M. (2006). Methods for rapid detection of chemical and veterinary drug residues in animal foods, *Trends Food Sci. Technol.*, **17**, 482–489.

Watson, D.H., Ed. (2004). *Pesticide, veterinary and other residues in food*, C.H.I.P.S., Weimar, Texas, U.S.A.

# chapter sixteen

# Toxicants unintentionally entering food during processing, storage, and digestion

## 16.1 General

*The aims of food processing are increasing safety, quality, and palatability* as well as simplification of the end processing of food. Heat and smoke of fire and the low temperatures of cellars and caves were the oldest preservers of food. Over the centuries, various methods of food processing like drying, salting, fermentation, and pickling have been well known. Nowadays, the growing demand of consumers for healthy, nutritive, and diverse food urges manufacturers to improve the old well-tried food production technologies as well as to invent new ones, enabling better preservation of nutritive and sensoric properties. Novel technologies are, for example, radiation, use of pulsatory electric field, high-pressure pasteurization, and so forth. New food additives and packaging materials are being used.

During processing, practically an endless number of different chemical and physical processes occur that result in

- Partial or complete degradation or removal of nutritives that may reduce food utilizability and digestability
- Formation of new potentially toxic substances

The first problem can be solved by after-processing addition of nutritives; the second problem needs the help of food toxicology in the use of principally new technologies. Since the number of newly forming substances is, in principle, infinite, it is practically impossible to avoid a situation that some of these substances will be toxic. In addition, potentially toxic substances may be formed during storage of the processed food by heat, humidity, light, oxygen and catalysts, by contact with packaging materials, and so forth.

It is not possible to estimate the summary toxic effect of processed food as a simple sum of effects of various toxic, mutagenic, carcinogenic, and other components. During processing, a very high number of various chemical compounds are formed that often strengthen or weaken (antagonize) the effects of each other, most of which are unfamiliar for us. Therefore, the summary effect can be evaluated only in the real complex environment.

Examples of such processes are

1. *Nonenzymatic browning* (*Maillard reaction*). During food processing (drying, freezing, roasting, baking) chemical reactions occur between amino acids as well as other compounds containing an amino group and reducing saccharides (aldoses and ketoses), leading to the *formation of mutagenic* furanes, aminocarbonyls, pyrazines, and other promelanoid secondary amines, products of Amadori and Heyns that presumably inhibit the growth and hinder multiplication, damage the liver, cause allergies, have an undesirable role in aging, and so forth. Maillard reaction consists of the following main steps:
   a. *Reaction of the carbonylic group of a saccharide with free amino groups* of a protein or an amino acid; complex formed will be dehydrated (loss of a water molecule) and unstable *N*-substituted glucosamine (Schiff base) is formed.
   b. *Amadori rearrangement of N-glucosamine* into 1-amino-1-deoxy-2-ketoses through formation and isomerization of the immonium ion.
   c. Ketosamines formed may react further by one of the three possible ways:
      i. Simple *dehydration* (loss of two water molecules) with a formation of reductones and dehydroreductones, which are caramel and strong antioxidants in the reduced form. That is why dark beer has about five times higher reducing potential than pale beer.
      ii. *Formation of short-chain hydrolytic fission products* like diacetyl, acetol, pyruvic aldehyde, and so forth. The last may provide aldoles and high-molecular nitrogen-free polymers.
      iii. *Formation of Schiff base/furfural* by loss of three water molecules with a following reaction with amino acids and water and formation of *melanoids*, brown nitrogen-containing polymers. The compounds formed may be either sweet- or ill-smelling. The highest reactivity in the Maillard reactions have pentoses (e.g., ribose); hexoses react slower and disaccharides such as lactose are the slowest reactants. Among amino acids, lysine is the best in forming the dark color; that is why the milk proteins contain much lysine, and therefore, brown most intensively. Formation of melanoids can be hindered by a removal of one reactant—either saccharide or amino-compound from the mixture. Maillard reaction proceeds most intensively at water activities between 0.5 and 0.8.

      A good updated overview of the Maillard reaction and food connections has been published by Gerrard (Gerrard, 2006)
2. *Cooking meat at high temperatures and for a long time produces heterocyclic amines*, such as 2-amino-3,8-dimethylimidazol[4,5,-f]quinoxaline (MIQx), 2-amino-1-methyl-6-phenylimidazo[4,5,-b]pyridine (PhIP),

2-amino-3,4,8-trimethylimidazo[4,5,-f]quinoxaline (DiMeQx). Epidemiological studies have shown that these compounds may be associated with an increased risk of distal colon adenoma, independently of overall meat intake (Wu et al., 2006). A positive correlation between very well-done meat and prostate cancer risk has been also established (Cross et al., 2005). 2-amino-3-methylimidazo-[4,5,-f]quinoline (IQ) has proven to be a potent hepatocarcinogen in a 7-year animal study with nonhuman primates (Thorgeirsson et al., 1994). But mutagens other than heterocyclic amines, such as benzo[$\alpha$]pyrene (B[$\alpha$]P) (see Section 16.2), may also contribute to the meat-derived mutagenicity.

3. *Polyunsaturated fatty acids* (PUFAs) are, during storage and/or thermal processing, subjected to three different alterations: *autooxidation, thermal oxidation, and thermal polymerization*. The first process proceeds at temperatures below 100°C in either the presence of lipoxygenases or light and results in rancidification either by a free radical or single oxygen mechanism. Fatty acid hydroperoxides and epoxides formed can decompose into alkanes, aldehydes, and ketones. Reaction of peroxidation ends either by absorption of free radicals or by their polymerization into nonreactive products. Subtoxic concentrations of lipid peroxides can, like second messengers, stimulate signal transmission for $Ca^{2+}$-mediated protein phosphorylation, leading to cell proliferation, chemotaxis, and apoptosis. High concentrations of rancid fats (5% or higher) may cause a reduction of food use, diarrhea, weight loss, lycopenia, and hair loss. Hydroperoxides and/or their reaction products (epoxides, hydroxynonenal, malonic aldehyde) may break the transmission of a nervous impulse and give DNA-adducts revealing mutagenicity and carcinogenicity and enhance the probability of the development of cancer and atherosclerosis.
4. *Components of heated oils*, particularly monomeric cyclic fatty acids are toxic, as also fatty acid polymers that do not form adducts with urea, that is, they are evidently also cyclic or branched. To the previously described toxic effects of PUFA derivatives *hepatomegaly* (increase in the relative liver weight) induced by a peroxisome proliferation-like mechanism can be added. Those heat-borne fatty acid derivatives cause an elevation of some characteristic enzymes such as peroxisomal acyl-CoA oxidase or microsomal CYP4A1 and CYP2E1 activities (Martin et al., 2000).
5. On frying foods with high nitrite content, *carcinogenic nitrosamines* are formed (see Sections 3.2.8 and 16.5).
6. *Foods and beverages* like bread, yogurt, cider, malt beverages, wine, and sake, *prepared by yeast fermentation* contain, besides psycho- and vasoactive amines, mutagenic and carcinogenic *ethyl carbamate* (urethane) that is formed from arginine, aspartate, cyanogenic glucosides, and ethanol by the action of heat and light. The content of ethyl carbamate, recommended by FAO/WHO is below 10 ppb in nonalcoholic and 30–400 ppb in alcoholic beverages.

## 16.2 Polycyclic aromatic hydrocarbons

During incomplete combustion of organic materials like wood, coal, or oil, as a result of decomposition (pyrolysis) reactions, a large amount of various polycyclic aromatic hydrocarbons (PAH) are formed. *PAHs* are chemical compounds that consist of condensed aromatic rings and do not contain heteroatoms or carry any substituents. Depending on the particular molecular structure, the PAHs have a different and in many cases remarkable carcinogenicity to the test animals. PAHs are divided, on the ground of the number of fused benzene cycles (three or more) in the molecule into light and heavy PAHs. The second group contains four or more condensed benzene cycles. The most important PAHs are B[$\alpha$]P (*the most carcinogenic PAH*), *1,2-bensanthracene, chrysene, fluoranthrene, pyrene,* and others, altogether over 80 compounds, approximately 60% of which are carcinogenic to the mammals. As the reference substance, according to which the general content of PAHs in the environment as well as in food is expressed, mostly B[$\alpha$]P is used. Opinion has been expressed that since the total carcinogenicity of PAHs in the mixture can exceed this property of B[$\alpha$]P more than tenfold, it is not adequate to use the concentration of B[$\alpha$]P for assessment of food riskiness. Dependence of carcinogenicity on the fine structure of molecules is illustrated by the fact that benzo[*e*]pyrene, a stereoisomer of B[$\alpha$]P, has no detectable carcinogenicity.

*B[$\alpha$]P, like other carcinogenic PAHs, are actually procarcinogens*. It means that they turn into active carcinogens in the course of enzymatic metabolism to the *ultimate mutagen* 7,8-dihydroxy-9,10-epoxy-7,8,9,10-tetrahydro-B[$\alpha$]P (Figure 16.1, Section 2.3.2.1), sequentially catalyzed by cytochrome P450 monooxygenase complex (CYP), epoxide hydrolase, and again CYP (Shimada, 2006). This molecule intercalates in DNA and binds covalently to the $N^2$ position of the nucleophilic guanine bases. This binding induces mutations by perturbing the double-helical DNA structure. These perturbations can disrupt the normal process of copying DNA and lead to the formation of cancer. The mechanism of action of PAHs is similar to that of aflatoxin $B_1$, which binds to the $N^7$ position of guanine (Figure 11.1). The above-described ultimate mutagen is not the only product of the metabolism of the B[$\alpha$]P (Shimada, 2006).

*PAH molecules have the capability of self-amplification (induction) of their enzymatic activation* by receptor mechanism over Ah-receptor (AhR) that is

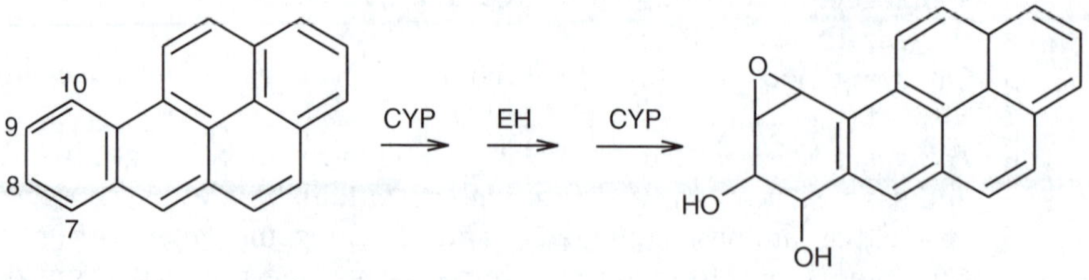

10
9
8
7
CYP
EH
CYP
O
HO
OH

*Figure 16.1* Three-step biotransformation of benzo[$\alpha$]pyrene, to its carcinogenic metabolite 7,8-dihydroxy-9,10-epoxy-7,8,9,10-tetrahydro-benzo[$\alpha$]pyrene. EH—epoxide hydrolase.

present in almost every organism (see Section 3.3.2). Very often, amplification proceeds in cooperation with dioxins (see Section 10.4). B[$\alpha$]P induces high concentrations of CYP1A1 by binding to the AhR in the cytosol. Upon binding, the transformed receptor translocates to the nucleus where it dimerizes with arylhydrocarbon receptor nuclear translocator (ARNT) and then binds xenobiotic response elements (XREs) in DNA, located upstream of certain genes. This binding, which increases the transcription of certain genes, notably of *CYP1A1*, is followed by an increased synthesis of CYP1A1 protein. Similarly CYP1A1 is induced by certain polychlorinated biphenyls (PCBs) and dioxins (Behnisch et al., 2001; Sagredo et al., 2006).

Food can be contaminated with PAHs:

- By atmospheric sediments (fruits and leaves in the gardens of industrial areas)
- During drying of cereal grains directly with combustion gases
- At smoking or barbecuing of meat or fish food over open horizontal coals
- At roasting of coffee beans, and so forth

In case of barbecuing on vertically laid coals or in an electrical or gas device in which the hot surface is located above the material to be processed, the formation of PAHs is minimal. This difference is caused by the circumstance that PAHs are formed by contact of dripping fat with the hot surface of coals, whence PAHs afterward arise with the smoke. Contemporary smoke generators enable to preclean the smoke, for example, by filtration before reaching the smoking chamber.

*PAHs are formed also during conservation of protein-rich foods* and during preparation of sugar caramel. They are found in fats and oils, in seafood and beverages, they are formed also at pickling and fermentation. PAHs have been detected, for example, in soy sauces.

For minimization of the content of PAHs in the food, it is recommended to remember the following:

- Possibly fat-free food should be processed at a lower temperature during a longer time and contact of food with an open fire, especially with a fire located below the object to be heated should be avoided.
- The intensity of the taste obtained through heating is not necessarily connected with the intensity of the brownish color of the processed food.
- At the same time, the processing method must guarantee the inactivation of all possible bacterial or endogenous toxins.

The allowable limit of B[$\alpha$]P content is 1 $\mu$g/kg of the final product in many countries. Actual concentrations of B[$\alpha$]P up to 6 $\mu$g/kg have been found in smoked sausages and up to 86 $\mu$g/kg in sausages heated on open horizontal coals. The allowable summary content of "heavy" and "light"

PAHs is somewhat higher in some countries. PAHs are, due to their high carcinogenicity, under continuous toxicological attention, and estimation of their content in various foodstuff is done during screening programs for the foodstuffs' safety.

Data about the content of PAHs in various foodstuffs can be found in the database of EPIC (European Prospective Investigation into Cancer and Nutrition—Jakszyn et al., 2004).

## *16.3 Alcohols*

Alcoholic beverages that have been consumed since prehistoric times for hygienic, medicinal, religious, and entertainment purposes contain *ethylic alcohol* or *ethanol*, the main product of fermentation that in case of ingestion in sufficiently high doses may cause lethal intoxication. Ethanol is quickly absorbed from the gastrointestinal tract into body fluids. To the extent of 90%, ethanol is metabolized first into acetic aldehyde by *alcohol dehydrogenase* (ADH) with a rate of 10–20 mL/h, further into acetic acid by *aldehyde dehydrogenase* (ALDH or AcDH), and finally into carbon dioxide and water (Figure 16.2). Metabolic energy, released at these oxidation reactions, is used for the synthesis of two macroergic molecules of nicotinamide adenine dinucleotide (NADH).

The *toxicity of ethanol manifests itself in many organs*, the symptoms depending on the dose. If consumed seldom and in small doses, ethanol may have even useful effects. *Most important are ethanol toxicities to the central nervous system* (*CNS*) (addiction and depression after acute doses), *to the developing embryo* (fetal alcohol syndrome) and *to the liver* (hepatomegaly, followed by cirrhosis). Medium doses of ethanol cause disorders of muscular coordination, slowed reaction, and at high doses unconsciousness and death (Figure 16.2). *The lethal dose is 300–500 mL of ethanol* (about 1 L of vodka) consumed by an adult within a period of less than 1 h. The scale of ethanol toxicity on the basis of its concentration in blood plasma extends from 500 to 1500 mg/L (0.5–1.0 promills—weak intoxication) to 5000 mg/L and more (5 and more promills—severe intoxication with coma and death). At the last-mentioned level, respiratory depression, hypotension, hypothermia, and hypoglycemia appear. The last adverse effect is caused mainly by inhibition of the glycogenesis process in the liver by ethanol.

*In case of chronic exposure to ethanol*, the main target of the toxic effect is the liver, but serious damage to the brain also occurs. During the development

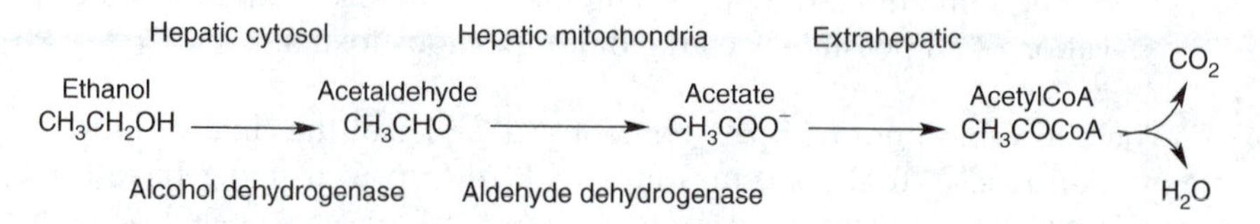


***Figure 16.2*** Metabolic oxidation of ethanol.

of *liver cirrhosis,* normal tissue is gradually replaced by collagen, causing a decrease in the efficiency of liver function. Large doses of ethanol may, due to accumulation of triglycerides, cause the formation of the so-called *fatty liver.* The last process is in turn caused by an increased synthesis of triglycerides in the liver due to the NADH/NAD ratio elevated by ethanol.

Lately, it has been shown that both acute and chronic ethanol consumption increase the production of *reactive oxygen species* (ROS), decrease cellular antioxidant levels, and hence, enhance oxidative stress in many tissues, especially in the liver. It is believed that the ethanol-induced oxidative stress plays a major role in the mechanisms by which ethanol produces a liver injury. One central pathway appears to be the induction, by ethanol, of the CYP2E1. This CYP isoform metabolizes and activates various potentially toxic substrates, including ethanol, to more reactive products. CYP2E1 is an effective generator of ROS (Dey and Cederbaum, 2006; Jimenez-Lopez, 2005).

The toxic effects of ethanol are caused by the alcohol itself and by its metabolite, acetic aldehyde. Other effects include malnutrition evoked by ethanol, release of endotoxins by enteric bacteria that in turn stimulates release of reactive compounds by Kupffer cells, rise of potency of other hepatotoxic agents, or a combination of various toxic effects. The International Agency for Research on Cancer (IARC) has considered ethanol to be a human carcinogen causing cancers of the oral cavity, larynx, esophagus, and liver. Mechanism of ethanol carcinogenicity is still unclear. Ethanol is also a teratogen.

*Alcoholism or alcohol addiction is a serious form of chronic toxicity of ethanol.* Alcoholism is a complex disease or disorder consisting of an enforced excessive usage of ethanol regardless of well-known adverse consequences. This addiction is usually accompanied by tolerance to the toxic effects of ethanol and to specific symptoms (tremors and mental disorders) appearing after cessation of the consumption. Men fall ill with alcoholism more often than women do; significant differences exist between various ethnic and race groups. The heritabilities of alcoholism of men and women are rather equal, respectively, 40% and 60%, indicating that the intersex differences in the rate of alcoholism are largely of social origin.

*Attempts to indicate genetic loci favoring alcoholism* have elucidated several candidate genes, but their role in the formation of alcoholism, excluding genes of ADH and ALDH is still unclear. The data collected for genes of ADH and ALDH are vice versa rather single-valued—some functional polymorphisms, characteristic of Asian populations, and likely existing also in the case of Europeans, protect against alcoholism. Three genes—*ADH2*2*, *ADH3*3*, and *ALDH2*2* guarantee a defense against alcoholism either by acceleration of formation or retardation of further metabolism of acetaldehyde. High aldehyde levels formed in these cases cause extremely unpleasant sensations that will keep one from alcohol consumption. East Asian people, homozygous to the allele *ALDH2*2* of ALDH are, for example, practically safe against alcoholism (Iwahashi and Suwaki, 1998).

Studies of the ADH and ALDH allele rates of the Inuits of Alaska have revealed an existence of five national groups, differing from each other in the frequency of the alleles. A completely surprising and so far unexplainable discovery was that allele ALDH2*2 defending East Asian people against alcoholism works neither in the case of Alaskan indigens nor in the case of American Indians. Both these populations of mongoloid origin easily fall ill with alcoholism (Segal, 1999).

Ethanol is essentially more toxic to women than to men (Baraona et al., 2001). The reasons are as follows:

1 Since the volume of body liquid is smaller in the case of women, the concentration of blood ethanol after drinking the same quantity of ethanol is higher in their organs.
2 In the case of women, the activity of gastric ADH is substantially lower than in the case of men. Women who have fallen ill with alcoholism have the activity of gastric ADH very close to zero. Hence, in the case of women, more ethanol is absorbed in blood after ingestion of the same amount of alcohol. For example, ethanol hepatotoxicity is much higher in the case of women and they have, more often, liver cirrhosis and hepatitis.

The other aliphatic alcohols are more toxic than ethanol. The reasons are a slower metabolism and formation of more toxic substances at their biotransformation.

*Methanol or methyl alcohol* is, like ethanol, oxidated by hepatic ADH into especially reactive and toxic *formaldehyde* and further to *formic acid* (formates), that, in sufficiently high doses, can cause depression of the nervous system, metabolic acidosis, blindness, and death. Methanol may have a toxic effect on the CNS even by percutaneous absorption. Inhalation of high doses of methanol may cause headache, dizziness, anesthesia, and irritation of the respiratory organs. Methanol is generally not a mutagen or sensibilizer; its teratogenic effect manifests itself in the case of high doses. *Ethanol is an efficient antidote of methanol* blocking its metabolism and enabling excretion of methanol in the unchanged form. Methanol can be found in small quantities in alcoholic beverages, where the basic components are water and ethanol, but methanol intoxications are caused mainly by its erroneous consumption instead of ethanol.

Methanol is significantly less toxic to most experimental animals than to humans, since most animal species metabolize methanol differently. Nonprimate species do not ordinarily show symptoms of metabolic acidosis or the visual effects that have been observed in primates and humans due to accumulation of formates. In the case of rats, further oxidation of formates is dependent on the hepatic tetrahydrofolate levels and on the activity of the enzyme 10-formyl-tetrahydrofolate dehydrogenase (EC 1.5.1.6). The activity of this enzyme is much lower in human than in rat liver due to a reduced amount of enzyme protein in human liver. In the case of mice, even three different routes of the formate oxidation exist (Cook et al., 2001).

Intoxications caused by the other aliphatic alcohols (propanols, butanols, etc.) are very rare.

## 16.4 Bacterial toxins

It must be stressed that *60–90% of the food-borne intoxications are caused by pathogenic bacteria.* In case of eating bacterially contaminated food, there might develop in the organism

1 Intoxication or poisoning (*Clostridium botulinum, Staphylococcus aureus*)
2 Illness, connected with high levels of food contamination by spores of the facultative pathogens (*Clostridium perfringens, Bacillus cereus*)
3 Infection of *Salmonella* spp., *Shigella* spp., *Campylobacter jejuni*, or *Listeria monocytogenes*
4 Illness with unclear etiology (*Proteus* spp., *Escherichia coli, Pseudomonas* spp.)

The toxic effects connected with enteral bacteria are caused by their *enterotoxins*, which can be divided into two groups:

- *Exotoxins*, excreted by bacteria into the environment
- *Endotoxins* that a bacterium releases on disintegration of its cell

The following discussion contains a short characterization of the bacteria producing exotoxins or endotoxins with the main stress on the physiological properties of the respective toxins.

### 16.4.1 Exotoxins

*Exotoxins*, produced mainly by Gram-positive bacteria are mostly *very toxic antigenous proteins*, becoming activated after a certain latency period. Here belong toxins excreted by *Clostridium botulinum, Cl. perfringens, S. aureus*, and *B. cereus*.

#### 16.4.1.1 Staphylococcus

Worldwide, the most common intoxications with food of animal origin are caused by enterotoxins of *Staphylococcus aureus*. *S. aureus* is omnipresent in the environment as a motionless spore not forming coccus, which excretes at least five enterotoxins (A-SEA, SEB, SEC, SED, and SEE) with different antigenic properties. Superantigens secreted by *S. aureus* are extremely potent toxins that overstimulate the host immune system by binding to the major histocompatibility complex (MHC) class II and T-cell receptors and activating a large population of T-cells. These superantigens are the causative agents in acute diseases, food poisoning, and toxic shock syndrome, and also in more chronic conditions such as inflammatory skin diseases (Hong-Geller and Gupta, 2003). A-SEA is the most efficient causative agent of food poisonings.

A-SEB stimulates the immune system, causing fever and other symptoms of common cold. *S. aureus* also produces a number of other proteinaceous substances such as DNA-ase, lipases, hemolysins, fibrinolysin, and hyaluronidase, which can be toxic to some animal species. Although all the strains of *S. aureus* are potentially pathogenic, production of the enterotoxin depends on the presence of DNA-ase and coagulase.

*Enterotoxins of S. aureus are relatively simple thermostable proteins* with $M_w$ in the interval of 25–35 kDa. *S. aureus* grows at temperatures 7–48°C, but the formation of toxins increases by approaching the more optimal growth conditions, that is, pH = 4.5; $t$ = 20–45°C. Food containing $5 \times 10^5$ bacterial cells per gram is hazardous. And even food where the content of bacterial cells is still lower may contain toxin molecules, because thermal processing sufficient for destruction of cells may not be strong enough to denaturate the already formed toxin molecules.

*Staphylococcal intoxications can be caused* by various readymade foods; most typical examples are poultry and animal meat, potato salad, shellfish, confections, sandwiches, and readymade vegetable salads that have been kept for a long time at room temperature and somehow contaminated with staphylococci. The other microorganisms causing food putrefaction would have inhibited the multiplication of staphylococci if the raw material had not been thermally processed.

*First symptoms of the intoxication* (nausea, vomiting, abdominal cramps, and diarrhea) will appear during 1–6 h after eating food contaminated with staphylococci. Sweating, dehydration, weakness, salivation, anorexia, and shock may also occur. Recovery happens usually during 1–3 days if treatment including correction of water balance is applied. Deaths are rare. *To avoid intoxications,* it is necessary to minimize, by use of respective methods of food hygiene, the probability of contamination of thermally processed food with staphylococci.

#### *16.4.1.2 Clostridia*

*16.4.1.2.1 Clostridium botulinum. Botulism,* caused by thermolabile proteinaceous neurotoxin of *Cl. botulinum,* a thermostable spore-forming Gram-positive anaerobic bacillus, *is by far the most dangerous bacterial intoxication.* In several cases, botulism, also known as "sausage poisoning," has a fatal end, and the therapy of the survivors may last for months. The name of the species as well as that of the intoxication stem from the Latin word *botuli,* meaning sausage.

*Botulinum toxin is one of the most acutely toxic substances known,* its peroral $LD_{50}$ is 0.01 $\mu$g/kg bw in the case of rats but as low as 0.0003 $\mu$g/kg bw in the case of intravenous administration to mice. Less than 1 $\mu$g of botulin toxin is needed to kill a man by peroral administration.

The spores of the bacterium are very persistent, but the toxin is destroyed at temperature 79°C within 20 min or at 85°C within 5 min. Botulinum toxin is stable in the acidic environment of the stomach, where the toxic component is defended against the action of gastric juice and pepsin by the nontoxic component of the toxin. In the duodenum, the toxin is activated by trypsin, followed by absorption into the lymph.

*Botulinum toxin is a mixture of seven neurotoxic proteins* (A, B, C, D, E, F, and G) with molecular weights of about 150 kDa. Toxin molecules are two-chain polypeptides with a 100-kDa heavy chain joined by a disulfide bond to the 50-kDa light chain. This light chain is a protease that, like a Zn-endopeptidase attacks one of the fusion proteins at a neuromuscular junction, preventing vesicles from anchoring to the membrane to release acetylcholine. By inhibiting acetylcholine release, the toxin interferes with nerve impulses and causes paralysis of muscles. Further, the muscle will act as if the nerve has been removed from it—the process results in the paralysis of the victim. The heavy chain of the toxin is particularly important for targeting the toxin to specific types of axon terminals. *Fortunately, there is an antitoxin to botulism* (toxoid or inactivated toxin) that is also used to treat humans participating as "guinea pigs" in the scientific research of botulism (Kessler and Benecke, 1997).

*Symptoms of the intoxication appear typically 12–36 h after eating contaminated food*; the latency period may extend from 4 h in the case of very large doses to 4 days in the case of very small doses. The first symptoms are nausea, vomiting, and diarrhea, which are followed mainly by neurological disorders like headache, dizziness, blurred vision, dysphonia, coordination disorders, weakness of face muscles, dysphagia, dysarthria, loss of light reflex, respiratory difficulties, and pharyngeal paralysis. *Paralysis starts in the head* and moves downward all over the body. Despite the severity of the symptoms, the victim is conscious and with a normal body temperature. Death follows usually within 3–5 days but sometimes after 24 h or even after 2–3 weeks through respiratory failure. Survivors may be partially paralyzed for 6–8 months; convalescence requires formation and growth of new fibrils outside of the old ones.

Especially harmful are the spores of *C. botulinum* to infants younger than 12 months, whose intestinal microflora is not sufficiently developed yet to destroy the spores. Feeding infants with honey where quite often germinative spores of *C. botulinum* can be found is hazardous. Connection between fatal infant botulism and honey has been proved (Tanzi and Gabay, 2002).

*C. botulinum is widely distributed in the environment,* especially in raw fruits, and vegetables, and in the digestive tract of animals. The main causative agents of human botulism are neurotoxins A and B. Type E can be isolated mainly from soil, but also from seawater, marine, and lacustrine sediments as well as from fish viscera. Studies have revealed that *C. botulinum* is capable of producing the toxin, for example, in mushrooms that are in a polyvinyl film or vacuum package. It is important to watch that the permeability characteristics of the packaging material would have led to a minimum formation in the package of anaerobic conditions where the spores of *C. botulinum* grow well. Besides that, storage of the product at temperatures below 3°C is obligatory.

Cases of food-borne botulism are rare, but they still occur at times. So in 2006, a poisoning event with four cases caused by an infected commercial carrot juice happened in the States of Florida and Georgia and in Toronto, Canada (Anonymous, 2006). In most cases (even 72% according to the data of United States), botulism still occurs due to consumption of insufficiently processed homemade meat preserves that contain toxins of B and E types. Of 263 cases, out of 160 food-borne botulism events in the United States, 102 (39%) cases and 58 events occurred in Alaska. All the Alaska cases were attributable to traditional Alaskan native foods (Sobel et al., 2004).

Sufficiently high humidity, pH exceeding 4.6, and anaerobic conditions for some time—these are the conditions optimal for accumulation of a necessary amount of botulinum toxin in the food, earlier contaminated with *C. botulinum*. In case of eating such a food without reheating, botulism may develop. *Botulism is excluded if* the spores of *C. botulinum* are destroyed by heating before their germination into vegetative culture starts. The minimal necessary heating time at canning of products of low acidity is 2.4 min at 121°C that inactivates 98.1% of the spores at their concentration of $10^{12}$. *The other appropriate inhibitor of botulinum toxin formation in the meat-containing foods is the nitrite ion*—inversely proportional relationship exists between the concentration of the nitrite ion added to food and the probability of formation of the botulinum toxin. Besides meat preserves, the problem of botulism (toxins A and B) has arisen in the case of canned vegetables (beans, corn, leaf-rich plants, particularly pepper). Fewer cases of botulism have occurred with canned fruits. From fish products, toxin E, and from the liver paste, toxin F have been isolated. Toxins C and D that cause avian botulism are harmless for humans.

*Food-borne botulism can be avoided* by proper conservation of foods prepared for long-term storage both in the industry and at home, by boiling of garden vegetables for at least 3 min before serving and by throwing away all swollen and injured preserves after boiling them. When smoking fish, the central part of the fish should be heated to 80°C minimum for 30 min, followed by an immediate freezing of the product. To avoid the growth of *C. botulinum*, the pH value should be set low and chemical preservatives such as sodium chloride, nisin, polyphosphates, lactic acid, sodium- or potassium nitrates, or potassium nitrite should be added. *It is not recommendable to overuse nitrates and nitrites* since particularly nitrites may prove to be the source of product toxicity, first of all carcinogenicity (see Section 16.6). The ideas of Paracelsis are also valid in the case of the botulinum toxin—serotypes A and B are available for clinical use, and have been shown to be safe and effective for the treatment of dystonia, spasticity, and other disorders connected with muscle overactivity (Comella and Pullman, 2004).

*16.4.1.2.2 Clostridium perfringens* is widely distributed in the digestive tract of both humans and animals. Food-borne intoxications caused by *C. perfringens* are connected with sporulation of the bacterium in the large intestine. *Enterotoxin* (*lecithinase or α-toxin*), released during the sporulation may cause accumulation of liquid in the intestines. Lecithinase possesses both *a lethal-necrotic* and *hemolytic* activity. Proteinaceous enterotoxin of *C. perfringens* with a monomer mass of about 30 kDa forms after pentamerization channels in the cellular membrane, facilitating the entrance of redundant $Ca^{2+}$ ions and water into the cell and causing cell death. Entrance of the toxin into the bloodstream is followed by release of $K^+$ ions from hepatocytes and death caused by the *hyperkalemic heart attack*. $LD_{50}$ of the toxin is 0.1 μg/kg bw in the case of rats.

Out of five antigenically and toxicologically different types (A, B, C, D, and E) of *C. perfringens*, type A is the one most connected with human gastroenteritis. Type C, which has caused only two outbreaks of enteritis with some fatal cases in Europe, produces two different toxins—lethal necrotic and hemolytic. Since for the growth of *C. perfringens*, all amino acids and growth factors are needed, *only meat and fish products can cause this poisoning.* Most often, these are roast beef, stewed beef, meat sauce, and pies for type A and pork as well as meat of other animals and fish for type B. The foods contaminated with this bacterium have mostly been heat-processed within less than 1 h at a temperature lower than 100°C, and thereafter kept either warm or cooled down slowly. The spores surviving the heat shock will multiply faster in the food and are able to produce more enterotoxin in the intestines than the spores not subjected to thermal treatment. An overview of the toxicity mechanism of *C. perfringens* has been published by Smedley et al. (2004).

Owing to a wide distribution of *C. perfringens* in nature, it is very difficult to completely avoid contamination of food, but multiplication of the organism as well as toxin production must be minimized by heating the product at a sufficiently high temperature, fast cooling, and avoiding long-term reheatings before eating.

*C. perfringens* is capable of causing in a non-food-borne way diseases like gangrene, appendicitis, childbed fever, and enteritis.

#### *16.4.1.3* Bacillus cereus

Bacillus cereus is a spore-forming bacillus, which is widespread in nature. It can be found in soil, cereal grains, milk, herbs, spices, and other dried foods. Cases of illness after consumption of contaminated mustard or cress shoots have also been reported. Some strains of *B. cereus* are capable of growing even at negative Celsius temperatures.

Since *B. cereus* synthesizes two types of enterotoxins (*emetic and diarrheal*), altogether at least seven various toxins, it also causes two types of intoxications. The *diarrheal toxin*, produced by actively growing bacterial cells, is a protein that can be activated by proteolytic enzymes like trypsin or pronase as well as by storage for 30 min at 56°C or a higher temperature. Diarrheal intoxication can result from eating various meat and vegetable foods, fish, milk, and ice cream contaminated with *B. cereus*. The diarrheal toxin increases the permeability of the cellular membranes for various substances. The *emetic toxin*

(Cereulide) with an unknown mechanism of action is stable at heating even at 121°C during 90 min, to alterations of pH, and to the action of trypsin or pepsin. The main sources of the emetic toxin are various cereal grains and fried rice served in Chinese restaurants.

Most of the outbreaks of food-borne intoxications caused by *B. cereus* have happened in Northern and Eastern Europe. A review of *B. cereus* has been published by Schoeni and Wong (2005).

#### *16.4.1.4 Campylobacters*

The bacteria of species *Campylobacter jejuni* and *C. coli* are the leading causative organisms of human bacterial enteritis called *campylobacteriosis*. The thermophilic campylobacters are microaerophils. The most important species, *C. jejuni,* has an optimal growth temperature of 42°C, but is sensitive to drying and freezing. The main source of campylobacters is meat of slaughtered animals, contaminated with feces. As a source of contamination, humans as well as drinking water or soil can be pointed out. The direct vectors of the infection are most often poultry, raw or rare (underdone) meat, or unpasteurized milk products. Cross-contamination during production as well as sales of meat and milk products is also widespread.

Campylobacters produce the so-called *cytolethal distending toxin* (CDT), causing cell death through a direct injury of DNA. CDT is a protein, and its molecule consists of three subunits (CdtA, CdtB, and CdtC), with the subunit CdtB having nuclease properties. There is little information about the other subunits (Ceelen et al., 2006).

*C. jejuni* causes, most often, acute self-limiting enteritis with a duration of 1–10 days. The infected site is mostly the small intestine. The symptoms are abdominal pain, diarrhea, fever, and in some cases, also convulsions and vomiting. The symptoms may appear repeatedly during several weeks. Complications such as peritonitis, septicemia, and arthritis can follow. CDT poisoning may cause various gastric and duodenal ulcers.

#### *16.4.1.5 Listeria*

*Listeria monocytogenes* causes an extremely serious bacterial infection *listeriosis* mainly in the case of humans with the immunodeficiency syndrome, newborns, elderly people, and pregnant women. In the latter case, listeria may cause abortions and early births and in the case of newborns, meningitis, respiratory complaints, and skin nodules.

*L. monocytogenes* is a small Gram-positive saprophytic and facultative anaerobic bacillus widely spread in nature. It has been isolated from birds and mammals including humans, from soil, water, plants, and so forth. It is capable of growing at temperatures 1°C–45°C. The virulence of listeria is based on the parasitism that follows the entrance of the bacteria into the host phagocyte. Listeria is able to move from the phagosome to the host cell cytosol. This process is substantially mediated by listeriolysin O, the essential determinant of pathogenicity of listeria. Listeriolysin O is a protein with a molecular weight of about 56 kDa, belonging to the same family as pore-forming cytolysins such as streptolysin O, perfringolysin, and pneumolysin produced, respectively, by *S. pyogenes*, *C. perfringens*, and *S. pneumoniae*.

*It is possible to fall ill with listeriosis* by consuming foods contaminated with *L. monocytogenes* such as milk, soft cheeses, and other milk products, poultry and animal meat, vegetable salads, and so forth. Listeriosis can cause meningitis, bacteremia, septicemia, endocarditis, conjuctivitis, chronic tympanitis, and a number of other diseases; gastrointestinal symptoms appear only in the case of one-third of the intoxications. Listeriosis is a relatively rare disease but with a relatively high mortality rate (about 30%). Listeriosis is fortunately curable with antibiotics (Jemmy and Stephan, 2006).

### 16.4.2 Endotoxins

Endotoxins are produced mainly by Gram-negative bacteria (about 300 cognate organisms). *Bacterial endotoxins are complex formations* strongly attached to the bacterial cell wall, consisting of proteinaceous, polysaccharidic, and lipidic components. They are relatively thermostable and have no latency period. The bacteria producing endotoxins live as parasites in the intestines of the vertebrates (including humans) and cause typhoid and paratyphoid fevers (typhus), salmonellosis, bacterial dysentery, and other infective diseases. In the case of food-borne infections, the viable bacteria occur in the intestines in the content of food. Most important sources of the infection are egg products, frozen poultry, minced beef, confections, and cocoa. Despite further multiplication of such bacteria either in the gastrointestinal tract or in some other organs, the bacterial dose necessary to evoke the intoxication depends on the particular microorganism.

Salmonellas (over a thousand species) are the most well-known endotoxin-producing bacteria. Salmonellas are able to live both in the indigenous microflora of the host organism and freely in nature. They can grow both in aerobic and anaerobic conditions and live on various substrates. Salmonellas grow well in various foods and also in water, contaminated by feeds or feces. Their growth becomes completely inhibited when the environmental pH is below 3.8, temperature below 7°C, and activity of water below 0.94. Salmonellas are destroyed by thermal processing (boiling or frying) of food and pasteurization of milk (72°C; 15 s) or fruit juices (70–74°C; ≤20 s).

*Salmonellas have a wide range of potencies*—they are able to produce both a relatively light gastroenteritis and potentially lethal typhus. *They have caused several global pandemics during the last decades* and it is believed that, salmonellas together with campylobacteria cause most of the food-borne diseases, for example, in the United States. The most severe salmonelloses are caused by *Salmonella enteritidis* and *S. typhimurium*.

Gastroenteritis caused by salmonella starts 24–48 h after the entry of the bacteria into the organism with fever, nausea, and vomiting, followed by abdominal cramps and diarrhea. The latter is the dominant symptom during 1–4 days and usually resolves spontaneously in the course of 7 days. The fever persists during half of the disease period; the summary spectrum of symptoms can extend from relatively light to a severe dysentery-like syndrome. Dehydration, endangering the life of very young as well as very old patients may also occur. In most cases, other therapy besides general support is not needed.

The so-called *cross talk between host and bacterial cells* has been proposed as the toxicity mechanism of salmonellas. Bacterial cell respond to the vicinity of the host cell with an activation of a special type III protein secretion system. The latter is a needle-shaped organelle stretching through both (inner and outer) cellular membranes. The whole 100 nm long complex is designed for transportation of bacterial effector proteins into a nonfagocytary host cell, with a task to prepare, by interference of the cellular signalization system, a favorable ground for endocytotic invasion of the whole bacterial cell. Here belong substantial alterations in the membrane and cytoskeleton structures, preinflammatory activation of the synthesis of cytokines (IL-8), and so forth. The latter process actually evokes profuse diarrhea. Unlike such intracellular pathogens as *Listeria* or *Shigella,* which rapidly move further into cytosol, Salmonella stays in the endocytotic vesicle during its entire intracellular life cycle. Therefore, Salmonella is not easily discoverable by the immune system of the host organism. Cytotoxic effects leading to apoptosis are induced by Salmonella in the macrophages (Galan, 2001).

*Bacterial dysentery or shigellosis is elicited by Shigella dysenteriae*; the respective bacterial toxin (*shigatoxin* or *verotoxin*) is a multisubunit protein with a summary molecular weight of 32 kDa. One of its subunits (A with $M_w$ = 32 kDa) is responsible for the toxic effect, the other five B-subunits with molecular weight of 7.7 kDa are involved in binding the toxin to the target cells. Shigatoxin attacks the endothelial cells of blood vessels, B-subunits bind to the specific cellular component Gb3, and the whole toxic complex enters the cell. In the cell, the subunit A inactivates the ribosome, the protein synthesis is stopped, and the cell dies. Like the A-subunit of the plant toxin ricin (see Section 8.1), the A-subunit of shigatoxin is also *N*-glucosidase. Blood vessel endothelium must continuously regenerate itself or else bleeding starts. Shigatoxin threatens only small vessels found in the digestive tract, kidneys, and lungs, but not the arteries or veins. Special targets for the toxin are the endothelial cells of the glomerular blood vessels, the destruction of which leads to a renal insufficiency, and often to the *lethal hemolytic uremia*. Since *S. dysenteriae* enters the organism mainly through contaminated food or water, the first symptoms of the shigella intoxication or dysentery appearing after some hours are bloody diarrhea, abdominal pain, vomiting, and bloody urine in the case of the hemolytic uremia. Food-borne intoxications with shigatoxin also damage the lungs and nervous system. An immunomethod can be used for diagnosing. A contemporary review of the main aspects of shigella as a food-borne pathogen has been written by Warren et al. (2006).

$LD_{50}$ of shigatoxin is below 20 $\mu$g/kg bw in the case of both intraperitoneal and intravenous administration to mice and 200 $\mu$g/kg bw if intraperitoneally administered to rabbits. Some enterohemorrhagic strains of *Escherichia coli* are analogically toxic.

## 16.5 *Biogenic vasoactive amines*

Here belongs, first of all, *histamine* [2-4′-aminoethyl)imidazole], but also *cadaverine* (1,5-diaminopenthane), *putrescine* (1,4-diaminobutane), *spermine, spermidine,* and other low-molecular vasoactive amines that are formed during the processing and storage of food.

Histamine is factually included into the normal physiology of mammals; it can be found, for example, in the stem cells or basofils. Histamine exerts its effect in the respiratory, cardiovascular, gastrointestinal, and hematoimmunologic systems and skin by binding to the cell membrane receptors $H_1$, $H_2$, and $H_3$.

Histamine poisoning occurs if

- The regulation mechanism of histamine formation in the body becomes defective.
- Too much of histamine metabolism inhibitors (a number of drugs, ethanol, etc.) have been administered.
- Too much histamine has been ingested with food.

The most common symptoms of histamine intoxication are connected with the cardiovascular system. An excess of histamine causes a pathological broadening of peripheral blood vessels resulting in nettle rash, hypotonia, flush, and pulsatory headache. Dizziness, weakness, burning sensation in the mouth and throat, and dysphagia may also occur. Contraction of the intestinal smooth muscle induced by histamine may cause vomiting, diarrhea, and abdominal cramps. Pain and itching connected with urticaria may be caused by the stimulation of sensoric and motor neurons by histamine. To evoke the symptoms of intoxication, at least 70–1000 mg of histamine is necessary to be ingested during one meal.

*Histamine is formed from amino acid histidine* by the action of bacterial enzyme decarboxylase (Figure 16.3). The other vasoactive amines such as

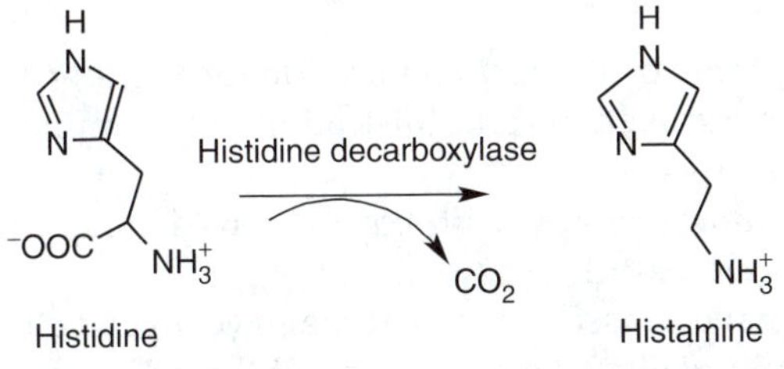

***Figure 16.3*** Enzymatic synthesis of histamine from amino acid histidine.

cadaverine from lysine, putrescine from ornithine, spermidine and spermine from arginine, tyramine from tyrosine, and tryptamine from tryptophan are analogically formed from the respective amino acids. A number of bacteria like *Proteus* spp., *Klebsiella* spp., and *Morganella morganii* are capable of synthesizing vasoactive amines.

There are many foods that either contain histamine or cause the organism to release it. Poisonings connected with biogenic amines may occur after consumption of fish, cheese (most of all, Swiss cheese), wine, and other fermented foodstuffs. Fruits and vegetables also contain histamine and other endogenous biogenic vasoactive amines. If in the case of fish especially high concentrations of the vasoactive amines are formed due to putrefaction, then in the case of wine, cheese, sauerkraut, or canned cucumber or pepper, we have a normal result of the respective fermentation processes. Histamine-rich foods such as cheese, sausages, sauerkraut, tuna, tomatoes, and alcoholic beverages may contain histamine up to 500 mg/kg (Wohrl et al., 2004).

### 16.5.1 *Scombroid poisoning*

Histamine is connected also with the so-called *scombroid fish poisoning* that emerges after eating either rotten or bacterially contaminated fish (Lehane and Olley, 2000). The rotten fish may have retained its initial look and smell. Intoxication develops rapidly—either practically immediately or within half an hour and lasts usually for 3 h, but may last for several days too.

*Symptoms of scombroid fish poisoning* resemble the symptoms of food allergy mediated by immunoglobulin E. Poisoning begins with a flush in the face, neck, and upper part of the arms, with a tingle or burning sensation in the mouth, with eruption on the upper body and hypotonia, often followed by headache and itching of the skin. Finally, dizziness, nausea, vomiting, diarrhea, and abdominal pain may appear. In severe cases, tachycardia, heart palpitation, respiratory distress, and shock may also occur. Commonly, only a part of the enumerated symptoms appear in case of every scombroid poisoning. The poisoning can be diagnosed on the basis of these symptoms, by their quick onset, and by the success of the antihistaminic therapy. The diagnosis is confirmed by the elevated histamine content in the suspicious food measured during some hours. Histamine as the main toxic factor in the poisoning will be indicated by the following facts:

- The symptoms of intoxication are identical either to the organism's reaction to intravenous administration of histamine or to the respective allergic reaction.
- Antihistaminic therapy with corticosteroids and antihistamines H1 and H2 is successful.
- Fish that have caused the syndrome have an elevated content of histamine (minimally 0.2 mg in 100 g of fillet or 200 ppm).

*Probably other vasoactive amines* as well as other substances are participating in the development of scombroid fish poisoning. Although scombroid poisoning is generally connected with high histamine levels in fish, it is nothing similar to any ordinary histamine intoxication. The symptoms of scombroid poisoning are often mixed up with those of Salmonella infection or food allergy. It is fully understandable, because histamine is also a potent primary allergen.

Since scombroid poisoning is often light and self-recovering, usually no special antihistaminic therapy is needed. The nature of a particular intoxication depends on the individual and other factors, considerably more severe and requiring a special therapy in the course of the illness in the case of allergic or elderly people and patients who are using drugs, histamine inhibitors, detoxifying enzymes like isoniazide, or inhibitors of monoamine oxidases.

*The most harmful fish* belong to the families *Scombridae* (tuna, mackerel, etc.) and *Scomberesocidae* (Atlantic saury, etc.), but also nonscombroid fish like mahi-mahi (*Coryphaena hippurus*), sardines, anchovy (*Engraulis encrasicholus*), herring, marlins, bluefish (*Pomatomus saltatrix*), and others. In the muscles of all these fish, there is a relatively high content of free histidine reaching 1 mg/kg in herring and even 15 mg/kg in tuna. The other fresh fish contain less than 1 mg/kg of histidine.

In histidine-rich fishes, the histamine synthesis from histidine, catalyzed by bacterial decarboxylases, starts immediately after death. It begins even earlier than *postmortem* proteolysis release of additional histidine from the proteins. In the case of some fishing techniques, a fish will die before it has been pulled out of water. Therefore, the concentration of histamine and other biogenic amines may greatly increase without an appearance of any of the organoleptic indicators of the putrefaction. Therefore, it is not possible to recognize a hazardous fish by smell or taste.

*The main bacteria causing scombroid poisoning* by decarboxylaton of histidine and other mentioned amino acids belong to the family *Enterobacteriaceae*; the most important of them are *Morganella morganii* and *Klebsiella pneumoniae*. *Clostridium perfringens, Hafnia alvei, Lactobacillus bucneri, L. delbrueckii*, and *Vibrio* sp. may also participate in development of scombroid poisoning. Putrefaction and bacterial production of ammonia as well as biogenic amines occur especially fast at elevated temperatures. When a sufficient bacterial population has formed, the enzymatic process of histamine formation will continue although much slower even at temperatures below 0°C.

It is interesting that eating of histamine-containing fish may cause stronger toxic effects than peroral administration of the same amount of pure histamine. The latter is largely metabolized during absorption through the intestinal wall or in the liver, and even at high doses will cause relatively weak effects. There is a hypothesis that scombroid poisoning is caused either by exogenous histamine potentiated by the other biogenic vasoactive amines and maybe by other toxicants or by endogenous histamine, released from the stem cells by the influence of some still unknown substances formed during the fish putrefaction process. Actually, scombroid poisoning may take place upon eating any food containing amino acids necessary for the synthesis of biogenic amines and which is contaminated by suitable bacteria.

The synthesis of histamine is not the only way of histidine metabolism. In the fish putrefaction process, an even more important role is assigned to *catabolic formation of glutamate* catalyzed by most of the bacteria. The first step of this metabolism is the separation of ammonia from histidine by histidine ammonia lyase (HAL) or histidinase with *the formation of urocanic acid*. The latter has been found in sufficiently higher concentrations than histamine (4.74 and 0.19 mg per 100 mL of mackerel respectively) after storage of fish for 8 days at 0°C. Urocanic acid, side by side with histamine, may be one of the causative agents of scombroid poisoning. Biogenic amines like putrescine or cadaverine are candidates for these agents as well.

None of the convenient food processing methods including freezing, canning, or smoking is capable of decomposing the substance or substances causing scombroid intoxication. It is much easier to avoid the formation of these agents in fish. Fish should be cooled as fast as possible, recommendable below 10°C, a storage lasting for more than 4 h must be carried out at 0°C or lower temperatures.

*Analysis*. High-performance liquid chromatography (HPLC) is the standard method for estimation of histamine and other biogenic vasoactive amines; methods of immunoanalysis are also used. The generally acceptable toxic threshold concentration of histamine is 1 mg/g of fish fillet, although the exact lowest observable adverse effect level (LOAEL) of histamine evoking the scombroid poisoning is unknown. The U.S. Food and Drug Administration (US FDA) considers 200 and 500 mg/kg of fish fillet as the indicator of putrefaction of tuna and as the indicator of hazard, respectively.

## 16.6 Nitrates, nitrites, and nitrosamines

*Nitrate ions* ($NO_3^-$) *occur in our food from two main sources*—as physiological constituents of vegetables and as food additives. Of the more consumed garden vegetables, radish, red beet, lettuce, and corn salad *contain the highest amounts of nitrates* (on average 1400–2000 mg/kg). Cabbage, carrots, spinach, and rhubarb contain 200–1000 mg/kg of nitrate ion, and other vegetables as well as fruits and berries contain less (Belitz and Grosch, 1999, p. 467). The nitrate-ion content in the food of plant origin depends largely on fertilization and other conditions. Vegetables make up 99% of the total daily consumption of nitrates (10–150 mg/person). Nitrate, added to meat to give it the characteristic taste and pink color and to avoid becoming rancid as well as to avoid development of *Clostridium botulinum* spores, provides less than 0.1 mg daily. The acceptable daily intake (ADI) value of nitrates is 3.64 mg/kg.

*Nitrates by themselves are not hazardous to humans* in the actually consumed amounts. The problem is that *nitrates are reduced into nitrites* by mammalian microbial systems. *Nitrite ions are acutely toxic,* their $LD_{50}$ for humans is 22 mg/kg bw, the allowable highest concentration in meat products is 200 ppm, and the ADI value is 0.135 mg/kg. The reduction of nitrates starts in the mouth, where 8% of food nitrates are reduced by oral bacteria. This process is facilitated by nitrate ions returning to the mouth from the intestines by blood and saliva. Nitrites are, in very small concentrations, also used as meat additives. Nitrite ions have two known adverse effects on the organism:

1. *Nitrites oxidize hemoglobin to methemoglobin* that is not capable of binding oxygen. When the methemoglobin concentration in the blood increases to a very high level, anoxia may appear in the tissues. In case of use of water containing over 30 mg/L of $NO_3^-$, originating from soil, fertilizers, and so forth and vegetables and meat with a high nitrate content in infant formulas and other foods, life-threatening methemoglobinemia may occur, especially in infants. In 1980s, widespread intoxication of infants with over ten fatalities were recorded in Minnesota, South Dakota, and Iowa with nitrates originating from spring water. These waters contained about 150 mg/L nitrate ions, which were converted into nitrites in the organisms of the infants. An essentially higher sensitivity of small children to the effect of nitrates is likely caused by a lower level of enzyme cytochrome $b_5$ reductase in their erythrocytes.
2. *Nitrites,* both those added to meat before canning and those formed in the mouth from nitrates, produce *N-nitrosamines* and *N-nitrosamides* in favorable conditions like that of an acidic environment of the stomach or cooking of proteinaceous foods by reaction with secondary amines or amides, (Figure 16.4) respectively. *These compounds are efficient mutagens and rodent carcinogens.* Problems arise if either too much nitrate or nitrite is erroneously added to a meat product or the ions are not evenly dispersed in the product. Formation of nitrosamines proceeds over diazonium salts.

*Nitrosamines are found in foods such as fish, meat, and cheese* that have been supplemented with nitrites to guarantee better preservation, and also in drugs, cosmetics, tobacco, tea, beer, and so forth. They are formed also during food frying or smoking (see Section 16.1). Levels of nitrosamines have

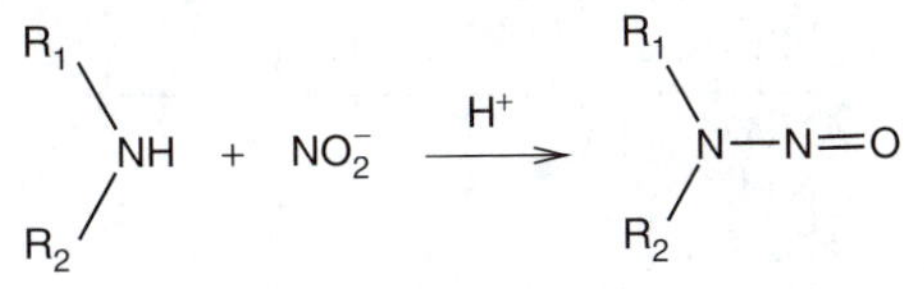

$$R_1R_2NH + NO_2^- \xrightarrow{H^+} R_1R_2N{-}N{=}O$$

***Figure 16.4*** Formation of nitrosamines. $R_1$ and $R_2$ are a methyl-, ethyl- or some other alkyl group or both together a heterocyclic group like pyrrolidyl or piperidyl.

decreased in food products in recent years owing to changes in food processing such as continued reduction of allowable nitrite levels during preservation, suspension of the use of nitrate for certain food groups, or an increased use of nitrosation inhibitors, such as ascorbate or erythorbate.

The rule of thumb is that the less that food has been processed and preservatives added, the lower the content of nitrosamines. *Most widespread is N,N-dimethylnitrosoamine* (*NDMA*; $R_1 = R_2$ = methyl) that is also the most potent carcinogen. NDMA may cause tumor formation in the liver, kidneys, bladder, esophagus, stomach, small intestine, pancreas, and the respiratory organs. Nitrosamine is most widespread in cheese, beer, and nitrate-nitrite-processed meat. The other ubiquitous nitrosamines are *N,N*-diethylnitrosoamine (NDEA), *N,N*-di-*n*-butylnitrosoamine (NDBA), *N*-nitrosopiperidine (NPIP), and *N*-nitrosopyrrolidone (NPYR), the concentration of the latter increases during the cooking of meat.

*Nitrosamines are the so-called secondary carcinogens*; they are not active in their original molecular form, but are converted *in vivo* into instable hydroxyalkylic compounds that form an alkylating alkylcarbonium ion very active toward DNA. Since so far, no really efficient alternative has been invented for inhibiting growth of botulism bacterium in food, it is not possible to prohibit completely the use of nitrates and nitrites as food additives (Table 16.1). But, on the other hand, it is possible to inhibit the formation of nitrosamines from nitrites by adding reducers such as ascorbic acid, cysteine, gallic acid, tannins, sodium sulfite, and so forth to food.

In Estonia, the upper limits for the summary contents of the two most toxic nitrosamines—NDMA and NDEA have been established—2 $\mu$g/kg in meat and meat products, 3 $\mu$g/kg in beer, fish and fish products, and 4 $\mu$g/kg in smoked meat products.

Unlike nitrosamines, nitrosamides are primary carcinogens, that is, they react directly. Carcinogenesis may also be caused by the nitrite ion itself.

Information concerning the content of nitrates, nitrites, and nitrosamines in various foodstuffs can be found in the database of project EPIC (Jakszyn et al., 2004).

***Table 16.1*** Acute Toxicities of the Main Food Nitrosamines (Rat, Oral Administration)

| Nitrosamine | $LD_{50}$ ($\mu$g/kg) |
|---|---|
| NDMA | 40 |
| NPIP | 200 |
| NDEA | 280 |
| NPYR | 900 |
| NDBA | 1200 |

## 16.7 Acrylamide

Acrylamide or propenic acid amide (AcA) has also been called a cooking carcinogen. AcA has been widely used for decades in various spheres of life, mostly (99.9%) for manufacturing of polyacrylamide (PAA). PAA, which by itself is of low toxicity to mammals, is used in purification of wastewaters, in manufacturing of paper and cellulose, and in chemical laboratories for preparation of analysis gels. PAA is also used in the cosmetic industry for preparation of shampoos and perfumes as well as a soil conditioner in agriculture.

The geno- and neurotoxicity of AcA have been well known for a long time; its no observed adverse effect level (NOAEL), established on the basis of genotoxicity and carcinogenicity to rats, is 0.1 mg/kg bw. NOAEL of peripheral neuropathy is 0.5 mg/kg bw and that of reproductive toxicity 2 mg/kg bw, again for rats. AcA is not mutagenic in bacterial systems, but demonstrates its genotoxicity in various tests. AcA is rapidly absorbed by all possible routes, and, due to its high solubility in water distributed everywhere in the organism including milk and placenta. Metabolism of AcA occurs along two main routes—oxidation to the main metabolite glycidamide and conjugation with glutathione. Glycidamide is mutagenic to bacteria and causes an unplanned DNA synthesis in human placental cells. AcA causes the thyroid gland tumor of both female and male rats and testicular and mammary gland tumors of rats. Both AcA and glycidamide are secreted rapidly, with $\tau_{1/2}$ in the organism about 2 h.

Until recently, no data was available concerning the occurrence of AcA in food. In April 2002, a sensational message concerning the discovery of AcA in various foodstuffs was published by the Swedish Food Agency. Thereafter, thorough investigations of the occurrence of AcA in various foods were carried out in Sweden, Norway, France, New Zealand, the Netherlands, the United States, and other countries. These studies entailed also a respective human risk assessment mainly from the viewpoints of neurotoxicity and carcinogenicity of AcA (FDA/CFSAN, 2002).

The results of a Dutch study of 344 "suspicious" food products revealed that the highest mean AcA contents (in $\mu$g/kg) were in deep-fried mashed potatoes (1270), potato crisps (1249), cocktail snacks (1060), and gingerbread (890). Much lower levels of AcA were found, for example, in deep-fried chips (351), children's biscuits (283), toast bread (183), and corn flakes (121). Infant formulas, rye bread, muesli, salted peanuts, and other popular foods contained much lower levels of AcA. The mean AcA exposure of the National Food Consumption Survey (NFCS) participants was estimated to be 0.48 $\mu$g/kg bw per day. The authors concluded that the *risk of AcA neurotoxicity is negligible*, but the *additional cancer risk may be higher* (Konings et al., 2003).

Shaw and Thomson, (2003) have tried compare the NOAEL value of AcA with a typical daily consumption of AcA in industrially developed countries (0.3–0.5 $\mu$g/kg bw) obtained mainly by eating mostly starch-rich fried or roasted foods, outstanding in their AcA content, like potato crisps and hot chips. This comparison reveals that the amount of AcA consumed is about three times lower than the hazardous amount. But it should also be kept in mind that, for example, children consume, due to their special dietary habits, possibly 2–3 times more AcA. Obviously, *this problem calls for additional investigation*.

The mechanism of AcA formation during preparation of specific foods is not fully understood yet, but most likely, it is a process resembling the Maillard reaction (Section 16.1). It has been estimated that carbohydrates and amino acids (particularly asparagine) are the necessary initial compounds and heating at temperatures 130–180°C or even higher is obligatory. The longer that food is processed, the higher the AcA content in the served food.

Readers who are interested in more thorough data about the toxicity of food-borne AcA are directed to review the respective report of SCF—*Final opinion of the Scientific Committee on Food on new findings regarding the presence of AcA in food* (at http://europa.eu.int/comm/food/fs/sc/scf/out131_en.pdf).

## 16.8 Chlorinated propanols

Chlorinated propanols or chloropropanols are found primarily in the delicacies containing acidically hydrolyzed plant proteins (Tritcher, 2004). *Typical examples are protein-rich soybean extracts or soy sauces* as well as winged bean sauces that have been treated (hydrolyzed) at high temperatures with concentrated hydrochloric acid. Chloropropanols have also been found in many other foods and food additives like roasted or baked cereal grain products. Chloropropanols may be formed in food also by interaction of chloride ions with lipids. Although theoretically, formation of seven various chloropropanols is possible, 3-chloro-1,2-propandiol (3-MCPD) and 1,3-dichloro-2-propanol (DCP) are the main representatives of the food-borne chloropropanols. Analyses of the 100 various soy sauces performed in the United Kingdom showed that the content of 3-MCPD was between the limit of detection (LOD) and limit of quantitation (LOQ—0.02 mg/kg) in the case of seven samples and above LOQ in the case of 25 samples. Sixteen out of these twenty-five samples contained more than 1 mg/kg of 3-MCPD, the highest content of 3-MCPD being 93 mg/kg. Of the other product groups, the highest concentration of 3-MCPF was found in breads and biscuits (up to 0.134 mg/kg) and in thermally processed fish and meat (up to 0.08 mg/kg). Studies of homemade products showed that both grilled and roasted bread and cheese contain 3-MCPD also, which may enter food due to its contact with the packaging material.

*Oral $LD_{50}$ of 3-MCPD is 150 mg/kg bw in the case of rats*. Animal tests have shown that chloropropanols evoke formation of tumors. As a result of repeated short-term administration of 3-MCPD doses higher than 1 mg/kg bw, alterations were observed in the sperm morphology of rats as well as other mammals, and the male individuals suffered from decreased fertility. Oral administration to rats and mice of 3-MCPD in doses over 25 mg/kg bw caused injuries of the CNS. Dose-dependent progressing nephropathy, tubular hyperplasia (the most sensitive adverse reaction), and adenomes were formed during a chronic toxicity test of 3-MCPD administered orally to rats. Hyperplastic and neoplastic injuries appeared in the Leydig cells of the testes, uterus, and pancreas of the rats.

*In vitro* mutagenicity tests with bacterial and mammal cells also gave, in most cases, positive results, but only at high doses (0.1–9 mg/L). Since several *in vivo* tests were negative, 3-MCPD is considered to be a nongenotoxic carcinogen. So far, no results of epidemiological or clinical studies are available.

As seen clearly from the experimental results, *no toxic effects should occur in case of reasonable nutrition*. Taking into account local dietary habits and customs, and the popularity of the soy sauces and contents of chloropropanols discovered mainly in these sauces, the medium daily dose of 3-MCPD was estimated to be 140–290, 230–630, and 540–1100 $\mu$g/person, respectively, in the United States, Australia, and Japan. On the other hand, the tolerable daily intake (TDI) of 2 $\mu$g/kg bw was obtained for 3-MCPD, considering the results of animal chronic toxicity tests (first of all, formation of tubular hyperplasia), nongenotoxic mechanism of action, enabling the application of the principle of biological threshold concentration, and the safety or uncertainty factor 500 for transferring the results of animal tests to humans (see Section 6.2.4). Comparing the TDI value with the estimates of real consumption, it becomes obvious that *the enthusiastic consumers of soy sauce must compare their real consumption against the safety margin*.

## 16.9 Phthalates

Phthalates or diesters of orthophthalic (*o*-benzenedicarboxylic) acid are synthetic compounds that have never been used alone but are included in other products. Almost 90% of phthalates produced are used as plasticizers in such flexible polyvinylchloride (PVC) products as plastic bags, packaging materials, and storage containers for food, floor covering materials, soft toys, intravenous tubules, and so forth. Phthalates are used also in some pesticide compositions, glues, inks, detergents, lubricating oils, plastic clothes, and furniture, together with polyvinylacetate in skin care products. All over the world, 5 million tn of phthalates were produced in 1998. Most of them, especially those of the esters of higher alcohols, are fat soluble and hence, capable of concentrating in foodstuffs like butter, margarine, and cheese and bioaccumulate in the adipose tissue of humans and water organisms. Owing to their rapid metabolism, the metabolites are excreted with urine and feces. Lower phthalates do not concentrate in the food chain.

*Di-(2-ethylhexyl) phthalate* (DEHP) and *diisononylphthalate* (DINP) are the two main phthalates of higher alcohols, which are used as plasticizers in flexible vinyl products such as PVC packing plates and soft toys. Naturally, the aforementioned compounds can get into an infant's mouth and into contact with food. Use of these compounds in direct contact with food is prohibited in a number of European countries. In the United States, the manufacturers of toys voluntarily promised to reduce the use of phthalates in their products up to 3% and they have done it already. US FDA considers DEHP as an indirect food additive. Of the phthalates, dimethylphthalate (DMP), butylbenzylphthalate (BBP), dibutylphthalate (DBP), and dioctylphthalate (DOP) have found practical use.

Residues of DEHP and DINP have been discovered in fish, water (including the drinking water of cities), and in sediments, but they can also bioaccumulate. Chronic exposure of fish to phthalates may have adverse effects on the health of humans eating the fish. Human exposure to the residues of phthalates may occur through drinking water, the dust of floor covering materials, foodstuffs packaged into PVC-materials, and blood transfusion when phthalate-containing PVC-bags are used for donor blood collection. Links have been discovered between PVC floor covering materials and infant bronchoconstriction.

*Absorption, distribution, metabolism, and excretion*. In case of oral administration of a dose of 30 mg to humans, 11–15% of phthalates is excreted with urine, the summary absorption is presumably 25–25%; in the case of rodents, most of the dose is excreted with bile. At low doses, most of DEHP is hydrolyzed already in the intestines by lipases into mono-(2-ethylhexyl)phthalate (MEHP) and 2-ethylhexanol. Respective lipases have also been found in the pancreas, liver, lungs, skin, and blood plasma. After absorption, DEHP reaches the blood, liver, kidneys, and reproductive organs, where it is metabolized into MEHP. The latter process is actually a set of reactions, resulting in the formation of about 30 various metabolites. The further most important route is the oxidation of MEHP and conjugation of the products with glucuronic acid for excretion.

*Toxic effects*. Most of the respective data originate from animal studies with rats and mice. Toxic effects of various phthalates on the reproductive organs and liver (including the formation of hepatic cancer), their teratogenicity and antiandrogenicity, depend on both the test animal and particular phthalate. *In vivo* experiments have shown that small doses of DBP damage the reproductive organs of male rats like an endocrine disrupter. *DEHP is toxic to the developing embryo*. BBP has elicited estrogenicity in the human breast cancer cell culture; both BBP and DBP had an estrogenic effect on several cell lines of the breast cancer and were bound to the estrogen receptor of rainbow trout. Both are also agonists of the estrogen receptor and BBP can act also as an antiandrogen.

*DEHP influences carbohydrate metabolism in the liver* as well as the oxidase system of rodents but not of monkeys. The mixed-function oxygenase (MFO) system consists of CYP and a number of hydratases and hydroxylases. Peroxisome proliferators including phthalates induce isoenzymes of the CYP4A subfamily. DEHP also has an effect on the membrane proteins and lipids.

Oral administration of DEHP has provided the following data:

- Single doses up of 10 g are not lethal for humans. DEHP is most probably not able to cause an acute death of humans.
- Large doses (5–10 g) may cause complaints in the gastrointestinal tract.
- Rats and mice (especially males) are most susceptible to the hepatotoxic effects of DEHP; the primary physiological response is hepatic hyperplasia. Consequently, DEHP is an epigenetic causative agent of hepatic tumor of rodents by the peroxysome proliferation mechanism.
- Long-term experiments performed with rodents with daily doses of 100 mg/kg bw showed a toxic effect of DEHP to the reproductive organs of males. MEHP is likely the toxic metabolite.
- Primates are less susceptible to the adverse effects of DEHP than rodents.

A profound overview of DEHP hepatotoxicity has been published by Rusyn et al. (2006).

*The most important human exposure to phthalates still occurs obviously by food* that has been contaminated either during production and processing or through contact with packaging material containing phthalates. Respective investigations performed in the United Kingdom produced the following results (MAFF, 1996a,b):

1. *Infant milk compositions*. All 15 compositions studied contained phthalates; the highest summary concentration was 10.2 mg/kg. Following the manufacturers' instructions, an infant would have obtained a daily dose of 0.13 mg/kg bw of phthalates.
2. *Other foods*. The highest content of phthalates in the milk products (114 mg/kg) was found in one soft cheese sample. Phthalates were found also in cakes, fats, and candies. The highest contents of DBP were measured in broth granules—62 mg/kg, in vegetable fat—8.4 mg/kg, in cakes with a chocolate glaze—5.8 mg/kg and the highest contents of DEHP were estimated in cookies—25 mg/kg and in vegetable fat—11 mg/kg.

In addition to the possible human exposure through inhaled air (pulmonary route), there are still only very few people who are really endangered by phthalates (still about thousand-fold reserve of safety). A considerably higher, although short-term, exposure to phthalates takes place at transfusion of both donor blood and platelets (daily dose 2.1–27.5 mg/kg bw) and especially at extracorporeal oxygenation of newborns (daily dose 42–140 mg/kg bw).

## 16.10 Bisphenols

Epoxide resins, polymers synthesized by condensation of epichlorhydrine with bisphenol A (BPA), are widely used as a protective coating of plastic packaging material of foods including cans. The yield of a polymerization reaction is never 100% and the resin obtained always contains monomer residues—in this case, molecules of *unpolymerized diglycidyl ether of bisphenol A* (BADGE). Various studies have shown that BADGE may migrate from the packaging material into canned meat, fish canned with vegetable oil, vegetable material, milk products, and canned beverages and fruits up to the concentrations on ppb level. The highest estimated summary concentration of BADGE and products of its hydrolysis (including those containing chlorine atoms) in canned food so far is 12.6 mg/kg. According to the Directive of the EU Commission 2002/72/EEC, the highest allowable specific migration limits (SML) of BPA and BADGE in food are 3 and 0.02 mg/kg, respectively. Later, the last limit was raised to the level of 1 mg/kg for BADGE and products of its hydrolysis.

The aforementioned compounds have exerted genotoxicity and estrogenic activity (endocrine disruption) in *in vitro* tests at concentrations of $10^{-4}$ M, whereby the endocrine disruption takes place without participation of the classical estrogen receptor. At typical levels of food contamination, BADGE and products of its decomposition exerted also *in vivo* toxicity to the reproductive system.

Obviously, additional investigation of the toxicity of BADGE is needed (Nakazawa et al., 2002).

## References

Anonymous. (2006). Outbreak news. Botulism, Canada and United States, *Wkly Epidemiol. Rec.*, **81**, 386.

Baraona, E., et al. (2001). Gender differences in pharmacokinetics of alcohol, *Alcohol. Clin. Exp. Res.*, **25**, 502–507.

Belitz, H.-D. and Grosch, W. (1999). *Food Chemistry*, 2nd edn., Springer, linn, lehekülg või pt.

Behnisch, P.A., Hosoe, K. and Hasaka, S. (2001). Bioanalytical screening methods for dioxins and dioxin-like compounds—a review of bioassay/biomarker technology, *Environ. Int.*, **27**, 413–439.

Brynestad, S. and Granum, P.E. (2002). *Clostridium perfringens* and foodborne infections, *Int. J. Food Microbiol.*, **74(3)**, 195–202.

Ceelen, L.M., et al. (2006). Cytolethal distending toxin generates cell death by inducing a bottleneck in the cell cycle, *Microbiol. Res.*, **161**, 109–120.

Cha, W., Fox, P. and Nalinakumari, B. (2006). High-performance liquid chromatography with fluorescence detection for aqueous analysis of nanogram-level *N*-nitrosodimethylamine, *Anal. Chim. Acta*, **566**, 109–116.

Cross, A.J., et al. (2005). A prospective study of meat and meat mutagens and prostate cancer risk, *Cancer Res.*, **65**, 11779–11784.

Comella, C.L. and Pullman, S.L. (2004), Botulinum toxins in neurological disease, *Muscle Nerve.*, **29**, 628–644.

Cook, R.J., Champion, K.M. and Giometti, C.S. (2001). Methanol toxicity and formate oxidation in NEUT2 mice, *Arch. Biochem. Biophys.*, **393**, 192–198.

Dey, A. and Cederbaum, A.I. (2006). Alcohol and oxidative liver injury. *Hepatology*, **43**, Suppl 1, S63–S74.

Fayle, S.E. and Gerrard, J.A. (2002). *The Maillard Reaction*, Royal Society of Chemistry, Cambridge, UK.

FDA/CFSAN. (2002). FDA draft action plan for acrylamide in food, Available from: http://www.cfsan.fda.gov/~dms/acryplan.html

Galan, J.E. (2001). Salmonella interactions with host cells: type III secretion at work, .*Annu. Rev. Cell. Dev. Biol.*, **17**, 53–86.

Gerrard, J.A. (2006). The Maillard reaction in food: Progress made, challenges ahead—Conference Report from the Eighth International Symposium on the Maillard Reaction, *Trends Food Sci. Technol.*, **17**, 324–330.

Hong-Geller, E. and Gupta, G. (2003). Therapeutic approaches to superantigen-based diseases: a review, *J. Mol. Recognit.*, **16**, 91–101.

Iwahashi, K. and Suwaki, A. (1998). Ethanol metabolism, toxicity, and genetic polymorphism. *Addict. Biol.*, **3**, 249–259.

Jakszyn, P., et al. (2004). Development of a food database of nitrosamines, heterocyclic amines, and polycyclic aromatic hydrocarbons, *J. Nutr.*, **134**, 2011–2014. Available from: http://epic-spain.com/libro.html

Jemmi, T. and Stephan, R. (2006). *Listeria monocytogenes*: food-borne pathogen and hygiene indicator, *Rev. Sci. Tech.*, **25**, 571–580.

Jimenez-Lopez, J.M. and Cederbaum, A.I. (2005). CYP2E1-dependent oxidative stress and toxicity: role in ethanol-induced liver injury. *Expert Opin. Drug Metab. Toxicol.*, **1**, 671–685.

Kessler, K.R. and Benecke, R. (1997). Botulinum toxin: from poison to remedy, *Neurotoxicology*, **18**, 761–770.

Konings, E.J., et al. (2003). Acrylamide exposure from foods of the Dutch population and an assessment of the consequent risks, *Food Chem. Toxicol.*, **41**, 1569–1579.

Lehane, L. and Olley, J. (2000). Histamine fish poisoning revisited, *Int. J. Food Microbiol.*, **58**, 1–37.

MAFF. (1996a). Food surveillance information sheet number 82: Phthalates in food, http://archive.food.gov.uk/maff/archive/food/infsheet/1998/no168/168phtha.htm

MAFF. (1996b). Food sureveillance information sheet number 83: Phthalates in infant formulae, http://archive.food.gov.uk/maff/archive/food/infsheet/1998/no168/168phtha.htm

Martin, J.-C., et al. (2000). Cuclic fatty acid monomers from heated oil modify the activities of lipid synthesizing and oxidizing enzymes in rat liver, *J. Nutr.*, **130**, 1524–1530.

Nakazawa, H., et al. (2002). In vitro assay of hydrolysis and chlorohydroxy derivatives of bisphenol A diglycidyl ether for estrogenic activity, *Food Chem. Toxicol.*, **40**, 1827–1832.

Pegg, R.B. and Shahidi, F. (2000). *Nitrite curing of meat. The N-nitrosamine problem and nitrite alternatives*, Food & Nutrition Press, Inc., Trumbull, Connecticut, USA.

Rusyn, I., Peters, J.M. and Cunningham, M.L. (2006). Modes of action and species-species effects of di-(2-ethyl)phthalate in the liver, *Crit. Rev. Toxicol.*, **36**, 459–479.

Sagredo, C., et al. (2006). Quantitative analysis of benzo[$\alpha$]pyrene biotransformation and adduct formation in Ahr knockout mice, *Toxicol. Lett.*, **167**, 173–182.

Segal, B. (1999). ADH and ALDH polymorphism among Alaskan Natives entering treatment for alcoholism, *Alaska Med.*, **41(2–12)**, 23.

Schoeni, J.L. and Wong, A.G. (2005). *Bacillus cereus* food poisoning and its toxins, *J. Food. Prot.*, **68**, 636–648.

Shaw, I. and Thomson, B. (2003). Acrylamide food risk, *Lancet*, **361**, 434.

Shimada, T. (2006). Xenobiotic-metabolizing enzymes involved in activation and detoxification of carcinogenic polycyclic hydrocarbons, *Drug Metab. Pharmacokinet.*, **21**, 257–276.

Smedley, J.G. III, et al. (2004). The enteric toxins of *Clostridium perfringens, Rev. Physiol. Biochem. Pharmacol.*, **152**, 183–204.

Sobel, J., et al. (2004). Foodborne botulism in the United States, 1990–2000, *Emerg. Infect. Dis.*, **10**, 1606–1611.

Tanzi, M.G. and Gabay, M.P. (2002). Association between honey consumption and infant botulism, *Pharmacotherapy*, **22**, 1479–1483.

Thorgeirsson, U.P., et al. (1994). Tumor incidence in a chemical carcinogenesis study of nonhuman primates, *Regul. Toxicol. Pharmacol.*, **19**, 130–151.

Tritcher, A.M. (2004). Human health risk assesment of processing-related compounds in food, *Toxicol. Lett.*, **149**, 177–186.

Warren, B.R., Parish, M.E. and Schneider, K.R. (2006). Shigella as a foodborne pathogen and current methods for detection in food, *Crit. Rev. Food Sci. Nutr.*, **46**, 551–567.

Wohrl, S., et al. (2004). Histamine intolerance-like symptoms in healthy volunteers after oral provocation with liquid histamine, *Allergy Asthma Proc.*, **25**, 305–311.

Wu, K., et al. (2006). Meat mutagens and risk of distal colon adenoma in a cohort of U.S. men, *Cancer Epidemiol. Biomarkers Prev.*, **15**, 1120–1125.

# *chapter seventeen*

# *Food additives*

## 17.1 Traditional food additives

### 17.1.1 General principles

Special toxicological problems may arise with food additives, that is, with compounds intentionally added to food for improvement of its various characteristics. Some of the food additives such as sodium chloride, acetic acid, or various spices have been used already for centuries. Despite that, to date, the physiological effects of sodium chloride are being investigated, leading often to discoveries of new adverse effects on human health. By now, a purposeful addition of various compounds to food has become really extensive, that is, about 2500 food additives are in use. In Europe, to keep track of them and for the sake of brevity, every food additive has a so-called E-number. In the unified scheme, many food additives not officially confirmed in Europe but allowed, for example, in the United States or Australia or anywhere else also have, with the help of Codex Alimentarius Committee, the numbers. In addition to substances that have stood the test of time and are labeled as *generally regarded as safe* (GRAS), new food additives are continuously appearing, the physiological effects of which are not always very clearly established.

*Food additives can be divided*, according to their sources, into three main groups:

1. Substances, isolated from edible plants or from other living material, for example, alginates (E 401), agar (E 406), and carragenan (E 407) isolated from seaweeds, lecithin (E 322) from soybeans, pectin (E 440) from fruits, and so forth
2. Substances contained in foodstuffs, but the production of which by chemical synthesis is cheaper, such as antioxidant ascorbic acid or vitamin C (E 300), yellow dye $\beta$-carotene (E 160a), and so forth
3. Substances not found in nature and which are obtained only synthetically, such as sweetener saccharin (E 954) or antioxidant *t*-butylhydroxyanisole (E 320)

In the system of E-numbers, the additives have been divided into classes according to the main purpose of their use (see Table 17.1).

*Table 17.1* Classification of Food Additives According to Their Purpose

| Purpose of the additive and interval of E-numbers | Examples |
|---|---|
| Colorants E 100–E 199 | Tartrazine (E 102), nitrites(E 249, 250), anthocyanins (E 163), carotenes (E 160a), lycopene (E 160d) |
| Antioxidants E 300–E 399 | Buthylated hydroxytoluene (BHT) (E 321), ascorbic acid (E 300), sulfites (E 221–E 228) |
| Stabilizers E 400–E 499 | Plant gums (E 410–E 415) |
| Flavors starting with E 500 | Cinnamic aldehyde |
| Flavor enhancers starting with E 500 | Glutamates (E 620–E 625) |
| Preservatives E 200–E 299 | Nitrites (E 249–E 250), nitrates (E 251–E 252), benzoic acid (E 210) |
| Emulsifiers E 400–E 499 | Polyoxyethylene-sorbitan fatty acid esters (E 432–E 436) |
| Thickeners E 400–E 499 | Agar (E 406), carrageenan (E 407) |
| Acids/alkalis starting with E 300 | Citric acid (E 330), acetic acid |
| Bleaches starting with E 500 | Benzoylperoxide (E 928) |
| Sweeteners starting with E 500 | Saccharine(E 954), neohesperidine DC (E 959) |

In addition to the classes listed in Table 17.1, a number of other groups of additives such as anticaking, antifoaming and foaming agents, fillers, color stabilizers, glazing agents, propellants, and so forth exist.

*The function of food preservatives and the need for them are evident*—reduction of the probability of bacterial or fungal infections. Preservatives reduce both biological and chemical degradation and prolong the storage time of food. A special group of preservatives are the antioxidants that inhibit the oxidation of food components and also the cellular components of the consuming organism (see also Section 3.3.6). Addition of colorants is substantiated, first of all, from the manufacturer's viewpoint—making the product more attractive. Besides these groups, *the list of food additives contains substances of which the actual function is not very obvious.*

Many producers are advertising their goods as additive-free or as containing only natural additives. The last statement, even if true, is by far not a sufficient argument. As we have seen already, *natural compounds may be at least of the same toxicity as synthetic ones.* In principle, natural food additives should have also been tested from the toxicological point of view before starting their utilization. The respective tests ordinarily use a lifelong exposure of the test animals to various doses of the substance studied, the highest dose exceeding by several times the doses presumably ingested by humans. These tests are not always sufficient—it is well known that test animals may not have a response equal to that of humans to the effects of the substance studied. Organisms may vary in absorption, distribution, as well as metabolism of substances. Use of extremely high doses may, due to saturation of the metabolic and secretion routes, lead to accumulation of the substance in the organism of a test animal. This problem arose, for example, at testing of *saccharin*, complicating the interpretation of the test results. Although the doses of food additives are usually small, in the case of many people, the exposure can be lifelong and in the case of others, sporadic. All these variations are very difficult to simulate with laboratory animals.

In real life, cross-reactivities can occur between various food additives as well as between food additives and intrinsic components of the food. A good review of that issue has been published by Scotter and Castle (Scotter and Castle, 2004). For example, nitrites are one of the oldest preservatives to be added to meat and meat products to give a nice pink color via the reaction of the nitrite ion with myoglobin. Later it was discovered that nitrites inhibit bacterial growth. Now we also know that nitrites may, by reaction with food-intrinsic amines, produce carcinogenic nitrosamines (see Section 16.5). Despite that negative effect, the use of nitrites as food additives is still allowed in most countries, since their antibacterial effect overweighs the carcinogenicity of the nitrosamines formed from nitrites. The latter chemical process has been attempted to be suppressed by addition of vitamins C and E to the meat products.

As with any other substances, the toxicity of which is necessary to be assessed, three types of toxicological tests are to be carried out with the food additives: *in vitro* tests with cell cultures, *in vivo* tests with lower organisms, and *in vivo* tests with higher animals (see Chapter 5).

JECFA (*Joint FAO/WHO Expert Committee on Food Additives*) has established *allowable daily intakes (ADI) for food additives* that were later accepted by many other committees and countries. ADI is the quantity of a chemical compound that an individual can be daily exposed to during all his or her lifetime without any harm to his or her health. ADI values are usually determined on the basis of results of long-term, less often acute *in vivo* animal tests, taking into account also human observations. The no observable adverse effect level (NOAEL) [(or no-observed-effect-level (NOEL)] values obtained from these tests performed with the most susceptible animal species are multiplied with the safety or uncertainty factor, *usually* $10 \times 10 = 100$ (see also Section 6.2.4). The respective internationally recognized *in vivo* tests have two main objectives:

1. Estimation of the most essential toxic effects of a substance by identification and study of target tissues
2. Estimation of respective NOAEL (see Section 6.2.4) necessary for calculation of ADI

*For estimation of ADIs for some food additives* (e.g., of erythrosine, cantaxanthine, and tin chloride), results of human tests, *combined with safety factor 10*, were used. Although this approach provides most accurate results, by ethical considerations, it can be applied only after very thorough animal tests.

Sometimes, it is quite complicated to evaluate whether a particular physiological effect (e.g., loss of weight) is actually adverse or not. In these cases, a conservative approach has been used by JECFA. This approach provides that an effect should be considered to be adverse and it should be taken into account at estimation of ADIs in the absence of reliable opposing data.

In addition to the food additives, ADIs have been estimated for the pesticide and veterinary drug residues in the food (see Chapters 14 and 15).

*Analogical to ADI parameter is the tolerable daily intake* (*TDI*). TDI marks the quantity of *a contaminant* in food or water that can be consumed during a lifetime without any risk to human health. Contaminants such as heavy metals differ from residues by the circumstance that their presence in food or water has never had any useful purpose. ADIs mark the substances that have been added to food either as a result of purposeful human action (like food additives) or through pesticides and veterinary drug residues (Nassredine and Parent-Massin, 2002).

Next, we will have a look at the toxicity of food additives by groups.

### 17.1.2 Colorants

As a result of recent toxicity tests, several earlier permitted food colorants have been withdrawn from use. An example is *butter yellow* or *methyl yellow* (*N,N*-dimethyl-4-aminoazobenzene), for which hepatic carcinogenicity was proved (Mori et al., 1986). Problems have arisen also with *tartrazine* (E 102—Yellow Dye No. 5), another widely used synthetic food colorant (Figure 17.1). Tartrazine has caused more allergic and intolerance reactions in the case of asthmatics and individuals having aspirin intolerance than all the other azo-dyes. Tartrazine may cause migraine, blurred vision, itching, and common cold. In combination with benzoic acid, another food additive (E 210), tartrazine evokes superactivity and neurobehavioral disorders (excitability, restlessness, somnipathy, concentration difficulties) of children. Several metabolites of tartrazine are formed in the stomach and mutagenicity of the urine of the test animals having ingested tartrazine has been demonstrated (Henschler and Wild, 1985). Use of tartrazine as a food colorant is prohibited, for example, in Norway and Austria.

*Allergic hypersensitivity reactions can be caused also by other azo-dyes* like quinoline yellow (E 104), sunset yellow (E 110), carmine (E 120), azorubine (E 122), amaranth (E 123), ponceau 4R or bright red (E 124), red G (E 128), and allura red AC (E 129). The red colorant erythrosine (E 127) may disturb

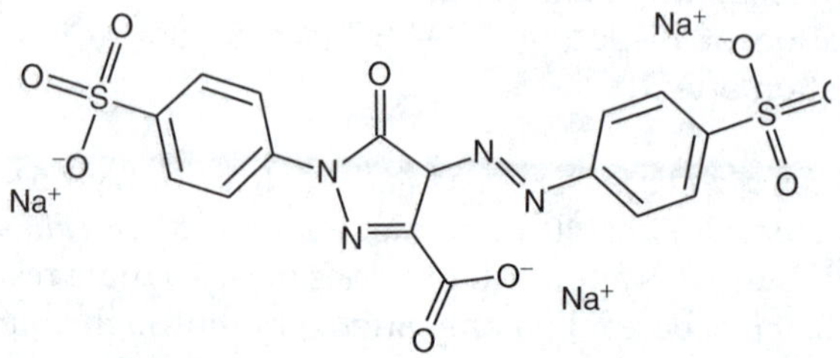

***Figure 17.1*** Structure of tartrazine.

the functioning of the thyroid gland, has caused cancer of this gland in the animal experiments, and may cause supersusceptibility to UV-rays and worsen allergies. Allergics may have additional trouble with consuming blue colorants like patent blue (E 131), indigocarmine (E 132), or brilliant blue FCF (E 133) as well as green S (E 142), brilliant black (E 151), brown FK4 (E 154), or litholrubine (E 180).

Using the Comet-test, it has been shown that out of 39 chemicals continuously used as food additives, colorants are the most genotoxic. Seven colorants (amaranth, allura red, new coccine, tartrazine, erythrosine, phloxine, and rose bengal) induced a damage of gastrointestinal cells already at concentrations as low as 10 and 100 mg/kg. Four of them (amaranth, allura red, new coccine, and tartrazine) induced a damage of rectal DNA already at doses close to the respective ADIs. *Consequently, the food colorants should be considered seriously from the viewpoint of their toxicity* (Sasaki et al., 2002).

### *17.1.3 Sweeteners*

ADI values have been estimated for four permitted artificial sweeteners, the highest ADI (40 mg/kg) has aspartame, the lowest (5 mg/kg) saccharin and its salts. Natural compounds like sorbitol or xylitol have no ADI values. Nevertheless, it should be borne in mind that synthetic sweeteners as all other purely synthetic food additives are xenobiotic substances, the real effects of which can become clear over decades and lifetimes of several generations. Therefore, they are recommended to be used only when no alternatives exist (in case of diabetes, substantiated saccharose-free diets).

From a toxicological viewpoint, the histories of *synthetic sweeteners cyclamate* (E 952—cyclamic acid and its sodium and potassium salts) and *saccharin* (E 954) are very interesting and instructive. The acute toxicity of saccharin and its sodium, potassium, as well as calcium salts is low ($LD_{50}$ in the interval 5–17.7 g/kg bw). They are not metabolized and the volunteers consuming saccharin over months did not feel any side effects. The first two long-term studies confirmed the safety of saccharin; the following two investigations, later announced as nonadequate, showed a weak carcinogenicity of saccharin. Since saccharin is often used together with cyclamate (mostly in the ratio 1:10), in one of the following studies, mixtures of these sweeteners were used up to doses of 2.5 g/kg bw. Because it causes bladder cancer, the use of cyclamate, which is 30–50 times sweeter than saccharose (sucrose), calorie-free and heat-stable, and works synergistically with other sweeteners, was banned. Since the repeated study, carried out in Canada, showed that saccharin is capable of causing bladder cancer and therefore, it was also banned both in Canada and in the United States, followed by a violent negative reaction of consumers for whom saccharin was practically the only food sweetener. Further epidemiological studies gave contradictory results. *At the moment, the use of saccharin as a food sweetener is principally permitted both in the United States and in Europe.* But in Europe, the highest allowable content of saccharin is fixed at 100–320 mg/kg, depending on the foodstuff. In the United States, saccharin products should be labeled, indicating possible adverse effects. Saccharin is used in more than 90 countries.

*Cyclamate has been also partly rehabilitated*; it is permitted to be used in many countries like Canada and more than 50 countries in Europe, Asia, South America, and Africa. Nevertheless, the finding that cyclohexylamine, a metabolite of cyclamate caused murine testicular atrophy (Roberts et al., 1989), was disquieting. Cyclamate may also facilitate the development of photodermatitis. A special study showed that a long-term feeding of nonhuman primates with high doses of cyclamate did not affect the general health of most of these animals. No clear evidence of compound-related testicular changes and tumors was detected in the cyclamate monkeys after more than 20 years of dosing (Takayama et al., 2000). Since 1970, however, the use of cyclamate has been banned in the United States. Although 75 subsequent studies have failed to show that cyclamate is carcinogenic, the sweetener has yet to be reapproved for use in the United States. Nevertheless, usage of both saccharin and cyclamate (Figure 17.2) are considered to be unadvisable during pregnancy as well as nursing.

The synthetic peptidous sweetener and flavor enhancer *aspartame* (methyl ester of L-aspartyl-L-phenylalanine—E 951) provides, by metabolism, noxious compounds such as diketopiperazine as well as methanol and formaldehyde. An organism is capable of neutralizing small doses of aspartame, that is 160 times sweeter than saccharose, but problems may arise at higher doses. Still, the last series of tests has shown that aspartame is harmless to human health, even in case of several times higher doses than normally consumed. *An exception is the individuals suffering from phenylketonuria*, a rare heritable disease diagnosed already at birth. These patients must limit the administration of amino acid phenylalanine, included into the peptide chain of aspartame (Figure 17.3) (Butchko et al., 2002). Persons who are sick with phenylketonuria will hopefully be soon relieved by *alitame*, a completely new sweetener, which is ten times sweeter than aspartame and has no aftertaste. In some countries, alitame is already available in the market under the tradename Aclame. Alitame is a dipeptide like aspartame, consisting of aspartic acid and modified alanine, and not containing problematic phenylalanine.

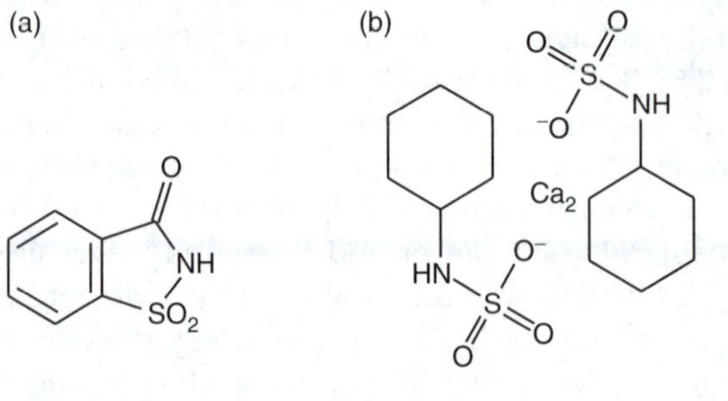

(a) O NH $SO_2$

(b) O O S NH $^-$O $Ca_2$ HN O$^-$ S O O

*Figure 17.2* Structure of (a) – saccharin (H linked to N is the main generator of sweetness), (b) Ca-cyclamate.

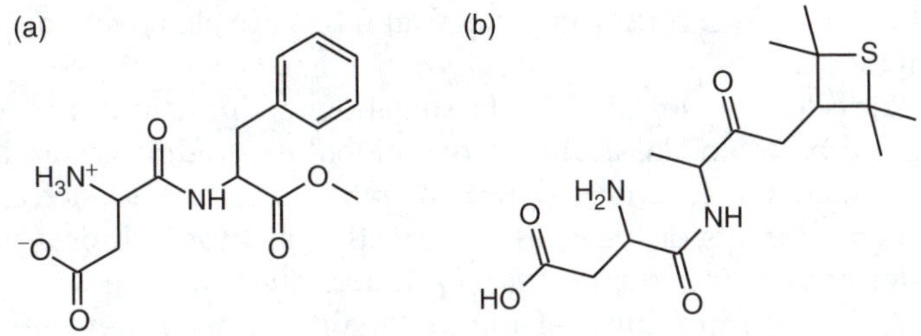

*Figure 17.3* Structure of (a) aspartame, and (b) alitame.

### 17.1.4 Preservatives

*Sorbic acid* and its sodium, potassium, or calcium salts (E 200, 201, 202, and 203) may cause skin irritations, burning sensation in the mouth, and flush around the mouth, and may exacerbate asthma or allergic diseases. Symptoms of supersensitivity may appear in humans who are allergic to aspirin. *Benzoic acid* and its sodium, potassium, or calcium salts (E 210, 211, 212, and 213) may also cause an occurrence of more violent symptoms of allergy and asthma, as well as headache, migraine, and stomach irritation. *The antifungal preservative biphenyl* (E 230) used for treating the peel surface of oranges, lemons, and other citrus fruits may diffuse through the peel into sarcocarp, causing supersensitivity, nausea, irritation of eye and nose, and vomiting upon eating the fruit. *The preservative hexamethylenetetraamine* (HMTA—E 239), the use of which is prohibited, for example, in Finland, is decomposed in the organism into ammonia and formaldehyde. The latter may elicit allergy and cancer.

### 17.1.5 Antioxidants

*Gallic acid esters* like *propyl-, octyl-, or dodecylgallate* (E 310, 311, 312), very efficient antioxidants, may cause respiratory distress in the case of asthmatics and people with aspirin intolerance. E 311 is also the causative agent of the so-called syndrome of burning mouth. *Butylhydroxyanisole* (BHA—E 320) and *butylhydroxytoluene* (BHT—E 321), often used together, cause asthma, allergic common cold, and eczemas. Both have caused tumors in animal tests and other issues.

## 17.2 Functional additives

### 17.2.1 General principles

During the last decade, the food market has been supplemented by products containing substances (mixtures, components) that have been added to classical food to increase its health-promoting ability. *These novel foods are called functional foods.* Since no clear regulative system exists for these new types of food, the laws concerning classical foods are also being applied for functional foods.

Regardless of the legislative classification, in practice, it is possible to distinguish between classical or conventional food additives and health-promoting functional additives. If the conventional additives, such as acidity regulators or colorants or bleaches, are mostly used for technical purposes or for better advertising, the *functional additives must have a positive effect on the health of the consumer*, thus raising the quality of his or her life. In principle, it is possible to live without these additives. In most cases, people do not directly perceive their positive effects. Consumption of functional foods begins to reveal itself after years when, being of old age, one begins to notice that he or she is still in a good shape and has developed neither cancer nor coronary arterial diseases (CHD). Humans live fuller, longer lives after consuming these bioactive food additives, which correct the physiological processes of the organism. These additives help reduce the risk of falling ill with various diseases. *Functional food additives are not drugs*, they do not cure illnesses directly, but help delay the genesis of a disease or slow down its development (Kruger and Mann, 2003).

*Various natural substances* can be regarded as health-promoting functional food additives or nutraceuticals. A short list of natural sources of the functional additives is presented in Table 17.2.

It is possible to isolate these active compounds from the aforementioned and other sources, mostly of plant origin, and add them to foods, where they are suitable. This activity has been one of the main forces pushing further the development of plant chemistry as well as methods of instrumental chemical analysis during the last decade. More and more substances have

***Table 17.2*** Examples of Natural Functional Foods Exerting Health-Promoting Effects

| Functional food | Known health-promoting substance(s) | Potential health benefit |
|---|---|---|
| Green and black tea | Catechins | Reduction of tumor risk, antioxidants |
| Broccoli | Sulforaphane and other isothiocyanates; indoles | Reduction of tumor risk |
| Garlic and other plants of genus *Allium* | Allylsulfides | Reduction of tumor and CHD risk, cholesterol level and blood pressure, antibiotics |
| Various berries and fruits | Flavonoids like catechins and anthocyanins | Strong antioxidants |
| Red grape wine | Polyphenols, first of all *trans*-resveratrol and quercetin | Reduction of tumor and CHD risk |
| Soya foods | Soya protein | Reduction of cholesterol level |
| Tomatoes and tomato products | Lycopene and others | Reduction of tumor risk, antioxidant |

been found in plants that can be regarded as functional food additives. The sources can be both long-known dietary plants and others. In many cases, ideas for seeking new functional additives originate from (especially oriental) ethnopharmacological knowledge. However, this knowledge based mainly on severe effects of various substances, including natural, on human health (death, acute severe intoxications) may be insufficient to prognosticate acutely weaker effects of low doses of the substances evoking a physiological response during a longer period. *The health-promoting substance must not turn into a health-damaging substance in the new dietary environment.* To demonstrate its safety, the following principles should be taken into account:

- *Functional food additives are* already by their nature and objectives of use *biologically active substances* and they may, depending on the quantity, elicit physiological effects of various intensity from suboptimal over therapeutic (health promoting) to obvious toxicity. For estimation of the doses corresponding to various responses, mechanisms of both pharmaceutical and toxicological action of the substance(s) must be known.
- *Functional food additives are a very heterogeneous group of substances;* they can be
  a. A single conventional or synthetic substance or substance isolated from a definite material
  b. A complicated plant extract, containing the necessary bioactive compound
  c. A product, obtained from a novel source using also a novel method like biotechnology

In any case, for safety assessment of a product, versatile knowledge of its chemical composition is necessary. Every product type has its specific safety problems that must be solved.

- *Possible exposure to a functional health-promoting food additive must be compared with the safety margin at its administration*: depending on the particular compound, historical knowledge concerning this compound, and/or results of scientific research (animal toxicology, absorption, distribution, metabolism, and excretion—altogether ADME, clinical tests) must be used for the estimation of the safe level of the substance; in some cases, the safety area between planned level of administration and level of potential toxicity can be very narrow.
- As in the case of drugs, *chemical reactions may occur between an added bioactive substance and the other food components*—the likelihood of such possible complication must be also assessed. It must be taken into account that unlike drugs, contact with a functional food additive can be much longer and thereby uncontrolled.

As already mentioned, in all cases, it is very important to know, as exactly as possible, the chemical composition of the additive. In the case of a single compound, when identification and quantification only of this compound is necessary, use of a single analytical method may be sufficient. In the case of more complex mixtures such as plant extracts, usually the application of a whole complex of methods is necessary to identify and quantify satisfactorily a large number of substances belonging to various classes. These can be

1. Either firmly established or supposedly functionally active compounds
2. Assisting substances that help strengthen the effect and that are consequently also necessary, congener substances
3. Unavoidable additives and contaminants

For example, a "profile" of the main components of the product can be sketched out of thin layer chromatography (TLC) and high performance liquid chromatographic (HPLC) data; additional methods are needed to identify and quantify the active and other components.

For a chemical compound obtained by novel processes like biotechnology, the key question can be the estimation of "substantial equivalency" with an earlier known compound that has been a food component, sometimes for centuries. Methods must be elaborated enabling "profile" or "fingerprints" of the new compound. The criteria of the substantial equivalency are

- Nutritional equivalency
- Characterization of the chemical equivalency of the active additives
- Characterization of the unavoidable additional substances (contaminants)

In the case of herb extracts often used as remedies already for centuries, it is necessary to point to historical documents describing use of the extract (e.g., Ayurveda, handbooks of Chinese, Japanese, and other oriental medical practices) as well as to refer to contemporary scientific literature in respect to an earlier use of this plant and additives in traditional foods and remedies.

As with any other new substances, three types of toxicological tests should be carried out with functional food additives: *in vitro* tests with cell cultures, *in vivo* tests with lower animals, and *in vivo* tests with higher animals. At least part of these tests should be carried out both separately with the additive and with the final complex food (Glei et al., 2003).

*Functional foods* must contain traditional nutritive components as well as compounds like vitamins, polyphenols, and dietary fibers and on the other hand, they *must not contain anything harmful to the eater*. This sets very high dual demands to the elaborators of functional foods and to the safety assessors in respect of properties of the additives. First, their positive effect should be clearly shown and, second, absence of a toxic effect both in case of normal and undue consumption should be solidly estimated (or the possible risk groups defined). The situation is still somewhat complicated since contradictory data is available about the so-called phytoprotectants. For example, it has been shown that polyphenols as well-known antioxidants are capable of producing free radicals and exerting pro-oxidative properties (Glei et al., 2003).

### 17.2.2 Functional additive–drug interactions

Nowadays, in case of simultaneous consumption of a real drug and functional food additives, special toxicological issues may occur. So it is known that about 20% of Americans take the prescribed drugs simultaneously with various herbal products and elevated doses of some vitamins. About 30% of modern drugs like morphine, atropine, digoxin, varfarine, theofylline, and many others originate from botanical sources.

The most well-known problematic sources of the functional food additives in the United States are gingko, Saint John's wort, ginseng, kava, garlic, and coneflowers (Kruger and Mann, 2003). Here must be added ephedra, the use of which is not very wide, but which can cause a number of serious health disorders.

1. *Gingko tree* (*Gingko biloba*). Interaction between this food additive and anticoagulants, aspirin, and drugs affecting platelets is possible. Since gingkolides are potent inhibitors of the platelet aggregation factor (PAF), they can, in cooperation with aspirin, nonsteroidal anti-inflammatory drugs (NSAID) or varfarine, elicit serious bleeding disorders. Data concerning subarachnoid and intracerebral bleeding in case of anticoagulant therapy are available.
2. *Saint John's wort* (*Hypericum perforatum*) has an increasing twofold effect on cytochrome P450 monooxygenase complex (CYP) activity in the liver that *may reduce the plasma level of such drugs as* digoxin, theophyllin, cyclosporine, varfarin, and ethinylestradiol/desogestrel. Data exists about bleeding caused by Saint John's wort together with oral contraceptives. Decrease of the plasma level of cyclosporine by the influence of Saint John's wort has caused the rejection of an acute heart transplant as well as a kidney transplant by the organism. Formation of the so-called "serotonin syndrome" has been observed in case of simultaneous administration of Saint John's wort and synaptic serotonin reuptake inhibitors.
3. *Ginseng* (*Panax ginseng*) elicits hypoglycemic effects; one must be careful in case of simultaneous administration with insulin or hypoglycemic drugs.
4. *Garlic* (*Allium sativum*). Since garlic has effects on platelet aggregation as well as fibrinogen, it may produce blood coagulation

disorders in case of usage with anticoagulants like aspirin, NSAIDs, and dipyridamole.

5. *Coneflowers* (*Echinace purpurea, E. pallida, E. angustifolia*) are hepatotoxic if used together with such hepatotoxic drugs as anabolic steroids, methotrexate, or ketoconazole. The immunosystem-stimulating effects of coneflowers may combine with the effect of antirejection drug cyclosporine used in case of organ transplantation.
6. *Ephedra* (*Ephedra*). Ephedra, because it contains adrenalin agonists like ephedrin and cognate alkaloids, has positive ionotropic and cronotropic cardiac effects. Use of ephedra has caused several deaths. Ephedrin may, in cooperation with drugs blocking the ability of monoamine oxidase (MAO) to degrade adrenalin, stimulate adrenergic neurotransmitter system by an increase of adrenalin concentration. Ephedrin may also, by interaction with volatile anesthetics such as halotane and cardiac glycosides, cause cardiac arrhythmia. Ephedrin toxicity may increase in case of a simultaneous exposure to the central nervous system stimulants like caffeine or decongestants. In most countries, the use of ephedrin as a food additive is prohibited. About ephedrin and its toxicity, see also Section 8.3.4.

A more profound consideration of food–drug interactions can be found in the respective handbook (McGabe et al., 2003).

## References

Butchko, H.H., et al. (2002). Aspartame: review of safety, *Regul. Toxicol. Pharmacol.*, **35**(2 Pt 2), S1–S93.

Glei, M., et al. (2003). Initial *in vitro* toxicity testing of functional foods rich in catechins and anthocyanins in human cells, *Toxicol. In Vitro*, **17**, 723–729.

Henschler, D. and Wild, D. (1984). Mutagenic activity in rat urine after feeding with the azo-dye tartrazine, *Arch. Toxicol.*, **57**, 214–215

Kruger, C.L. and Mann, S.W. (2003). Safety evaluation of functional ingredients, *Food Chem. Toxicol.*, **41**, 793–805.

Liu-Stratton, Y., Roy, S. and Sen, C.K. (2004). DNA microarray technology in nutraceutical and food safety, *Toxicol. Lett.*, **150**, 29–42

McGabe, B.J., Frankel, E.H. and Wolfe, J.J., Eds. (2003). *Handbook of Food–Drug Interactions*, CRC Press, Boca Raton, FL.

Mori, H., et al. (1986). Genotoxicity of a variety of azobenzene and aminoazobenzene compounds in the hepatocyte/DNA repair test and the Salmonella/mutagenicity test, *Cancer Res.*, **6**(4 Pt 1), 1654–1658.

Nasreddine, L. and Parent-Massin, D. (2002). Food contamination by metals and pesticides in the European Union. Should we worry? *Toxicol. Lett.*, **127**, 29–41.

Roberts, A., et al. (1989). The metabolism and testicular toxicity of cyclohexylamine in rats and mice during chronic dietary administration, *Toxicol. Appl. Pharmacol.*, **98**, 216–229.

Sasaki, Y.F., et al. (2002). The comet assay with 8 mouse organs: results with 39 currently used food additives, *Mutat. Res.*, **519**, 103–119.

Scotter, M.J. and Castle, L., (2004). Chemical interactions between additives in foodstuffs: a review, *Food Addit. Contam.*, **21**, 93–124.

Takayama, S., et al. (2000). Long-term toxicity and carcinogenicity study of cyclamate in nonhuman primates, *Toxicol. Sci.*, **53**, 33–39.

Walker, R. and Lupien, J.R. (2000). The safety evaluation of monosodium glutamate, *J. Nutr.*, **130**, Suppl. 4S, 1049S–1052S.

# chapter eighteen

# Vitamins

## 18.1 General

*Vitamins* or *amines of life* (from Latin *vita* = life) are low-molecular bioactive exogenous substances mainly used by an organism as structural–functional domains (cofactors) of complex enzymes. Consequently, vitamins are indispensable for enzyme catalysis and for the normal functioning of an organism. Vitamins do not provide energy and are not building blocks. *Animals (including humans) are in relation to most vitamins auxotrophs;* that is, their organisms are not able to synthesize vitamins but must import them. Some exceptions are the participation of human enteral microflora in furnishing the organism with vitamins such as pantothenic acid, niacin, vitamin K, and so forth, and the synthesis of vitamin $D_3$ in skin cells with the help of UV-radiation. Provided that food contains a sufficient amount of carotene, a human organism is able to synthesize vitamin A from provitamin. *It is more exact* to say that humans receive necessary vitamins through their gastrointestinal tract.

*Vitamins are minor components of food* that an organism needs in very small amounts (daily micrograms or milligrams) for growth, reproduction, and normal functioning; the need for any vitamin is more or less exactly known. Nowadays over 20 vitamins are known, they are divided into *fat-soluble* (vitamins A, D, E, and K) and *water-soluble* (vitamin $B_6$, vitamin $B_{12}$ or cyanocobalamine, vitamin C or ascorbic acid, biotin, folic acid, niacin, pantothenic acid, vitamin $B_2$ or riboflavin, and vitamin $B_1$ or thiamine) vitamins. Fat-soluble vitamins, marked with one and the same letter, are usually complexes of the so-called vitamers, characterized by similar structures. Depending on the particular vitamin, their store in the human organism is sufficient for 4–40 days. Consequently, the supply of vitamins through the digestive tract must be almost unbroken.

*In case of a partial deficiency of vitamins,* pathogenic situations called *hypovitaminoses* are formed in the organism with symptoms, depending on the vitamin. General symptoms of hypovitaminoses are tiredness, pain in muscles and joints, heart palpation, and so forth. Complete absence of a vitamin causes *avitaminosis* with much more specific symptoms.

A situation may occur that an organism contains sufficient amounts of a vitamin, but still the symptoms of hypovitaminosis appear. This situation is caused by *vitamin agonists*, substances that somehow block, inhibit, or attenuate the action of a vitamin. An example can be protein *avidin*, contained in uncooked hen eggs (see Section 12.4) and antipyridoxine factor *linatine* from linseeds (Klosterman et al., 1967). The other vitamins also have specific antagonists.

Since vitamins are biologically active substances, a question may arise—what will happen if an organism gets too much of them, mainly with foods enriched with vitamins? Do the vitamin overdoses lead to toxic responses?

*Hypervitaminoses*, caused by vitamin overdoses are rare, but their incidence is continuously rising in the developed countries. More and more foods with an enhanced content of one or another vitamin are prepared. More and more frequently, people start to use, either in case of relatively insignificant health disorders or just in case, special vitamin preparations enabling administration of enormous doses of vitamins. *Vitamins A and D may very easily cause hypervitaminoses*, which very rarely happens with vitamin C, riboflavin, pantothenic acid, or biotin. Lately, an investigation has begun to study the health effect of the so-called megadoses of vitamins C and E. Indirect data indicates that huge doses of these vitamins reduce the risk of coronary arterial diseases (CHD) and delay aging. At the same time, there is also indirect data pointing, vice versa, to a carcinogenicity of vitamin C megadoses (Lee et al., 2001).

Further, we will consider the hypervitaminosis caused by particular vitamins.

## 18.2 *Vitamin A: Phenomenon of smokers*

A normal daily need for *vitamin A* or *retinol* is, depending on the sex and age, 0.4–1.3 mg. The human organism assimilates retinol (Figure 18.1a) in the form of its provitamins—as *retinyl esters* from eggs and milk and as *carotenoids* from plants such as carrot. The organism converts *carotene* (Figure 18.1b), the dimerized form of retinol, into various vitamers ($A_1$, $A_2$) of vitamin A in the liver only if the organism really needs them. Therefore, the overdoses of plant carotene are generally relatively nontoxic in comparison with the animal retinyl esters, notwithstanding the skin of the enthusiastic drinkers of carrot juice that turn yellow due to accumulated carotene.

*Deficiency in vitamin A* causes nyctalopia as well as a pale and dry skin. Agonists of vitamin A are, for example, citral contained in orange and lemon peels, enzyme lipoxidase of raw soybeans and vitamin E, that, inhibiting the effect of vitamin A elicits, like vitamin A deficiency, alterations in the eyes.

*After a long-term continuous consumption of high doses of vitamin A* symptoms may appear such as tiredness, loss of appetite, dermatitis, and muscular pain, all of which will disappear after stopping administration of vitamin A overdoses. Since the vitamers of vitamin A have an action mechanism similar to the steroid hormones, a continuous administration of daily doses exceeding 10,000 IU (20 mg) may also evoke symptoms of serious intoxication, such as hydrocephalus, severe headaches, and ostealgia, loss of

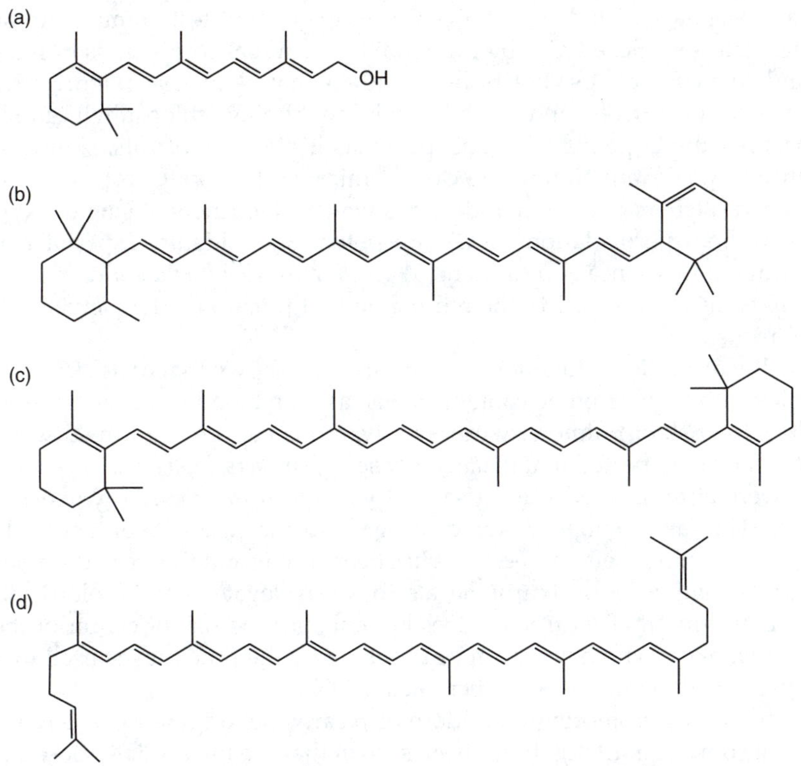

*Figure 18.1* Structures of carotenoids: (a) retinol, (b) $\alpha$-carotene, (c) $\beta$-carotene, and (d) lycopene.

hair, skin pigmentation, vomiting, and constipation. Vitamin A megadoses can be obtained by eating large amounts of specific foods, for example, 10 mg of retinol is present in 30 g of bovine liver, 500 g of hen eggs, or 2500 g of mackerel. *Especially high* (about 2,000,000 units per 100 g—45 times more than in the bovine liver) *is the vitamin A content in the liver of the polar bear* and other arctic animals.

*Both an incomplete and excessive use of vitamin A* and nicotinic acid *may cause teratogenesis*. Pregnant women, who take daily over 10 mg of vitamin A, risk having a child with defects of the face, heart, thymus, and nervous system. Studies carried out in the United States revealed that one out of 57 birth defects may be caused by improper use of vitamin A, consisting in administration of the vitamin without equivalent doses of synergistic foodstuffs that support the use of vitamin A by the organism. No problems of teratogenicity have occurred at the daily doses of the vitamin below 10,000 units. A recent review of vitamin A toxic effects has been published by Penniston and Tanumihardjo (2006).

*Vitamin A is an antipromoter of cancer*, that is, by inhibiting the pathological division of cells, vitamin A prevents formation of skin, lung, bladder, breast, esophageal, and stomach cancer in the case of test animals. An analogical effect is elicited also by carotenoids, the initial compounds of retinol, found in many plants. It has been shown that *β-carotene* (Figure 18.1c), contained in carrots and in all foods containing chlorophyll, inhibits cancers of the large intestine and pancreas in the case of rats, skin cancer induced by UV-radiation in the case of mice, and so on. Carotenes, which are good catchers of the free radicals as well as quenchers of singlet oxygen causing lipid peroxidation (LPO), are potent antioxidants. Especially outstanding in this sense is β-carotene. *A potent antioxidant and anticarcinogen is also lycopene* (Figure 18.1d), the red pigment of tomato (*Lycopersicon esculentum*) fruits.

Already earlier epidemiological studies had shown that diets rich in vegetables with high carotene content as well as high blood levels of β-carotene reduce the risk of pulmonary cancer. Given this background, the discovery was totally unexpected that *in the case of hearty smokers, high doses of β-carotene* that were administered during several years *increased the risk of pulmonary cancer*. The same tendency was discovered in the case of workers dealing with highly carcinogenic *asbestos*. It has been also found that fervid smokers, who take high doses of β-carotene, also have an elevated risk of colon cancer. It is interesting that β-carotene does not enhance the risk of cancer of these earlier smokers who have given up smoking by the time they started to use the high provitamin doses (Rietjens et al., 2005).

The health promoting daily dose of β-carotene, originating solely from natural sources, is 10 mg. It has been shown that the lowest daily dose causing adverse effects on the smokers is 20 mg. Therefore, the fervid smokers must be very careful with elevated doses of β-carotene.

The reasons why β-carotene acts in this way are not fully understood yet. *According to one hypothesis*, β-carotene stimulates the metabolic phase I enzymes, particularly cytochrome P450 monooxygenase complex (CYP), which in turn activates the potentially carcinogenic components of cigarette smoke as well as other secondary carcinogens. CYP1A1 and CYP1A2 similarly catalyze activation of aromatic amines, polychlorinated biphenyls (PCBs), dioxins, and polycyclic aromatic hydrocarbons (PAHs), CYP2A of butadiene, hexamethylphosphoramide, and nitrosamines, CYP2B1 of olefins and halogenated hydrocarbons and CYP3A of aflatoxins, 1-nitropyrene, and PAHs into potent genotoxic compounds.

*A second hypothetical mechanism* is based on the alteration of normal retinoid signalization by the products of oxidative degradation of β-carotene in the case of smokers, favored by a high concentration of oxygen in the lungs. To crown it all, the reactive oxygen species (ROS), originating either from tobacco smoke or asbestos may catalyze the oxidation of β-carotene into toxic metabolites. *Combinations of the two mechanisms* have also been suggested (Rietjens et al., 2005).

The case of smokers and carotene shows expressively how complicated and often with unexpected results may be the simultaneous effect of several foreign substances upon the organism.

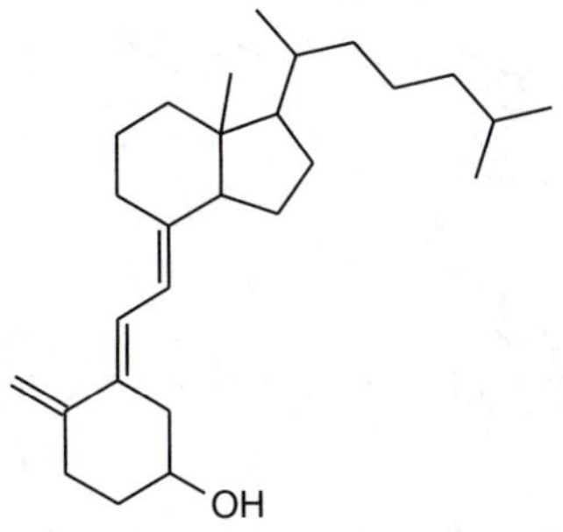

***Figure 18.2*** Structure of cholecalciferol or vitamin $D_3$.

## 18.3 *Vitamin D*

*Vitamin D or calciferols* are a group of fat-soluble *hormonal substances* with an antirachitic effect and similar structure that only historically belong among vitamins. Depending on age, the daily need of humans is 0.005–0.01 mg, the highest dose marking persons over 60. *Vitamin $D_3$ or cholecalciferol* (Figure 18.2) is formed in the skin from 7-dehydroxycholesterol (provitamin $D_3$, analog of cholesterol). Analogically, *vitamin $D_2$ or ergocalciferol* is formed from ergosterol. Both metabolites are responsible for absorption of calcium from the intestines and accumulation in the organic part of the bones by stimulation of the synthesis of the calcium-binding protein.

In most foods, excluding fish like herring and cod-liver as well as halibut-liver oil, the content of vitamin $D_3$ is low.

Since vitamin D vitamers are actually hormones, *people must be extremely careful* with vitamin D preparations. The lethal dose of vitamin D is 13 mg/kg for dogs. Doses exceeding 25 mg (infants) or 100 mg (adults) may cause damage of the kidneys and nervous tissue. Toxic doses of vitamin A are only somewhat higher than its recommended daily allowance (RDA).

*Symptoms of acute intoxication* with vitamin D, appearing at single doses of 15 and 50 mg in the case of infants and adults, respectively, are nausea, vomiting, loss of appetite, pain in the joints, alternating constipation, and diarrhea. *A long-term ingestion of doses of 1–2 mg* may cause chronic poisonings by accumulation of calcium crystals in various organs as carbonate and phosphate. The symptoms of this intoxication are growth retardation; demineralization of bones; irreversible accumulation of calcium in the soft tissues of the body that may damage the heart, lungs, and kidneys; hypercalcemia; hypercalcinuria; hypertension; and acute pancreatitis. D-vitamin exerts also a teratogenic effect resulting in mental underdevelopment and growth retardation.

## 18.4 *Vitamin E*

*Vitamin E* is the common name of eight natural fat-soluble vitamers or *tocols—$\alpha$-, $\beta$-, $\gamma$-,* and *$\delta$-tocoferols* and *$\alpha$-, $\beta$-, $\gamma$-* and *$\delta$-tocotrienols*. Every vitamer

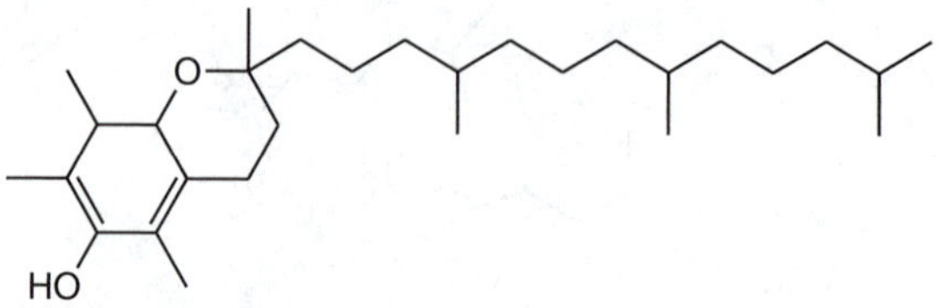

*Figure 18.3* Structure of $\alpha$-tocoferol.

has its own specific biological activity. In most cases, the main active substance is *$\alpha$-tocoferol* (Figure 18.3). New facts concerning the activity of the other tocols are continuously appearing. Unlike the other vitamins, the synthetic form of $\alpha$-tocoferol, prepared from isophytol that can be found in various food additives, is not identical to its natural relative but contains seven additional epimers of tocoferol. This is the reason why the synthetic vitamin is much less active than natural vitamin E. No data about possible side effects of the synthetic epimers exist.

*The richest natural sources of tocols* are vegetable (wheat germ, soybean, corn) oils, margarines, nuts, seeds, cereal grains, and green vegetables. *Vitamin E is the most important lipophilic antioxidant in the human organism.* Its main action site is the cellular membrane, the integrity of which is defended by vitamin E. As an antioxidant, vitamin E cooperates with its water-soluble associate vitamin C. Vitamin E also protects fats, including "poor" cholesterol, found in the low-density lipoprotein (LDL), against oxidation. In addition, vitamin E behaves as an enzyme cofactor, having several other functions in the organism. Like aspirin, vitamin E fluidifies blood, but without the adverse side effects characteristic of aspirin. Daily administration of 400 IU of vitamin E reduces the risk of CHD by 75%.

*Vitamin E is relatively nontoxic.* The organism needs daily 3–13 mg of tocols. Administration of large doses, exceeding 400 mg, may result in toxic effects like stomatitis, creatinuria, and urticaria, deterioration of the brain blood supply, headache, vision disorders, diarrhea, and so forth. *Vitamin E is an agonist of vitamin A.* Animal tests have shown that very large doses of vitamin E may disturb absorption of the other fat-soluble vitamins. As a result, a deficiency in vitamins A, D, and K may appear. A definite adaptation to large doses of vitamin E occurs. Recent epidemiological studies have revealed that long-term use of vitamin A preparations may somewhat increase the risk of CHD in the case of elderly people suffering from diabetes mellitus or from some disease of blood vessels (Lonn et al., 2005).

Consumption of vitamin E is really indicated in cases of

1. People whose metabolism is not capable of absorbing fats from the gastrointestinal tract. They include patients with cystic fibrosis or with Crohn's disease, or patients with partial removal of stomach.
2. Early newborns with a very low birth weight (below 1500 g).

3. People with inherited abetalipoproteinemia that causes disorders in nerve impulses, as well as muscular weakness and retinal degeneration.

## 18.5 Vitamin K

Vitamin K includes naphthoquinonic antihemorrhagic vitamins $K_1$, $K_2$, and $K_3$, the central functions of which are connected with blood coagulation.

*Vitamin $K_1$* or *phyloquinone* or *phytomenadione* (Figure 18.4) is necessary for synthesis and activation of a number of proteinaceous coagulation factors (prothrombin, proconvertin, Christmas-factor), and the daily need is 1–4 mg. Vitamin $K_1$ is found in green leafy vegetables such as spinach, cabbage, or cauliflower as well as in veal or porcine liver. Vitamin $K_1$ is relatively nontoxic and doses exceeding 500 times the RDA (0.5 mg/kg) have no adverse effects, but a continuous administration of very high doses may elicit hemolytic anemia or jaundice in newborns. In the case of small children, at high doses of phytomenadione, formation of pigments in the brain, nerve damage, and hemolysis may occur. Injection of 10 mg of vitamin $K_1$ to an infant on three consecutive days may have a lethal outcome. The latter tragic situation took place in the first years of the vitamin $K_1$ study, when phytomenadione was injected into newborns to prevent hemorrhage. At present, newborns are protected against that disease by a modification of vitamin $K_1$.

*Menadione, a vitamin K precursor* (sometimes called also vitamin $K_3$), has a finite toxicity resulting from its reaction with sulfhydryl groups; it can cause hemolytic anemia due to G6PD deficiency (see Section 8.10), hyperbilirubinemia and kernicterus in infants. Menadione supplements have been banned by the FDA because of their high toxicity. *Menadione should not be used to treat vitamin K deficiency* (Singh and Duerksen, 2003).

*Vitamin $K_2$ or menaquinone* is synthesized by many bacteria. Menaquinone is not considered to be toxic, but its ancestor *menadione* is toxic.

## 18.6 Vitamin $B_2$

*Vitamin $B_2$ or riboflavin* is produced by intestinal microflora and is absorbed easily. Riboflavin is continuously needed by the organism for usage of

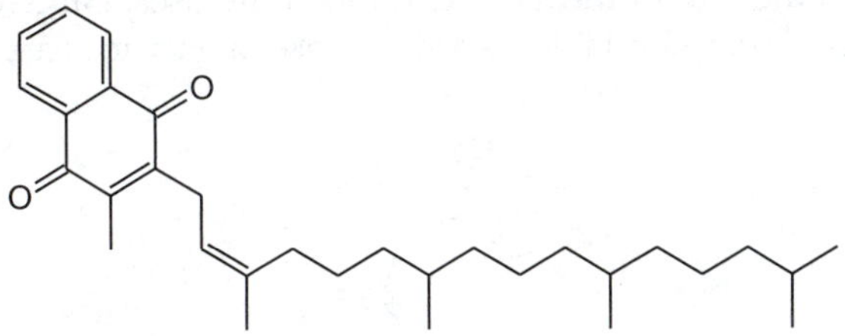

**Figure 18.4** Structure of phytomenadione or vitamin $K_1$.

oxygen as well as activation of vitamin $B_6$ (pyridoxine). Riboflavin helps produce niacin and is necessary for the functioning of the adrenal glands. Vitamin $B_2$ is used also in the formation of erythrocytes, in production of antibodies, and so forth. *Vitamin $B_2$ deficiency* may cause dizziness, somnipathy, dermatitis, loss of hair, growth retardation, eye diseases, inflammations of the mouth and tongue, skin injuries, and so forth. The daily need for this vitamin is 1.2–1.6 mg. *Riboflavin has a low toxicity;* even doses exceeding the RDA by several times have no essential toxicity.

## 18.7 Vitamin $B_6$

*Vitamin $B_6$ or pyridoxine* is clearly toxic at doses exceeding RDA (0.3–2.5 mg, depending on age) by about 1000 times. Excessive ingestion of pyridoxine causes severe sensory neuropathy. A daily dose of 2–5 g may cause walking difficulties and smarting pain in the feet. Long-term huge doses may cause difficulties in handling of small objects and dullness of hands. Two months after administration of the giant doses ends, a recovery starts, taking 2–3 years for complete recovery. The mechanism of toxic action of pyridoxine has not been fully elucidated. Pyridoxine-produced neuropathy is characterized by necrosis of dorsal root ganglion (DRG) sensory neurons and degeneration of peripheral and central sensory projections, neurons with a large diameter being particularly affected (Perry et al., 2004).

## 18.8 Vitamin C

*Vitamin C or ascorbic acid* (Figure 18.5) is a water-soluble antioxidant, and quencher of the free radicals in the water-containing parts of the cell. Ascorbic acid acts synergistically with fat-soluble antioxidant vitamin E regenerating the stable reduced form of tocols. The RDA of vitamin C is 40 mg and 60 mg, respectively for infants and adults. Large doses of vitamin C are toxic to people, who have either a heritable or acquired disposition to the formation of renal or bladder stones. These stones usually consist of calcium oxalate, found in larger amounts in cocoa, chocolate, rhubarb, and spinach. Part of the ingested ascorbic acid is metabolically decomposed with the formation of an oxalate. For example, a daily dose of 3 g of vitamin C doubles the amount of the oxalate passing through the kidneys as well as the bladder that may be accumulated in these organs in the case of the aforementioned diseases. Large doses of ascorbic acid are not recommendable also in the case of gastric ulcer.

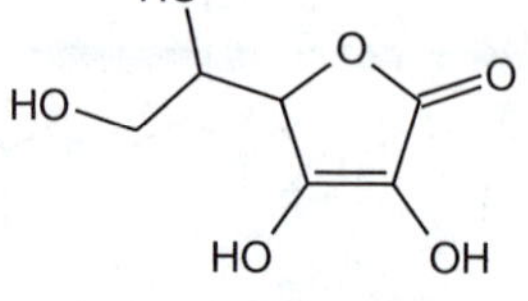

HO
O
O
HO
HO OH

*Figure 18.5* Structure of vitamin C or ascorbic acid.

A few years ago, a scientific discussion was started about the carcinogenicity of ascorbic acid megadoses. Lee with coauthors have shown, that ascorbic acid *in vitro* promotes, rather than inhibits, the formation of genotoxic LPO products from the lipid hydroperoxides, (Lee et al., 2001). Sowell and coworkers have been successful in rehabilitation of vitamin C, showing that ascorbic acid has dual properties, being both the promoter of formation and *in vivo* nucleophilic scavenging of electrophilic LPO products, thus rendering them harmless (Sowell et al., 2004). Obviously, the discussion, which is facilitated by a fast progress in the sensitivity of analytical techniques, is continuing. In any case, the *use of extremely high doses of vitamin C must be regarded with suspicion.*

## 18.9 Vitamin $B_3$

*Vitamin $B_3$ (also vitamin PP) or niacin* exists in two forms, as *nicotinic acid* (3-pyridine-carboxylic acid) and as *nicotinamide* (Figure 18.6), both satisfying the requirements of adult humans. RDA of niacin is 15–19 mg in the case of adults. Niacin is toxic at doses exceeding the RDA by about 100 times. It may cause flushing of the face, nausea, diarrhea, and hepatic injuries. Despite these complications, high daily doses (1.5–4 g) of nicotinic are used for reduction of blood cholesterol level. Nicotinic acid decreases the level of the so-called "bad" LDL-cholesterol, simultaneously increasing the level of the "good" high-density lipoprotein (HDL) cholesterol and reducing the plasmatic level of triglycerides (Tavintharan and Kashyap, 2001).

## 18.10 Diagnosing and therapy of vitamin intoxications

Diagnosis of vitamin intoxication is usually formulated by the food consumed and case history of the patient. Very important is the questioning of the patient. In the case of some, especially water-soluble, vitamins, the diagnosis can be confirmed by specific tests on the presence of these vitamins in plasma or urine. The unabsorbed megadoses of many vitamins evoke an accumulation of water in the intestines that will cause diarrhea.

The prime condition of the therapy of a vitamin poisoning is to stop the administration of excessive doses of the vitamin. Vitamin D intoxication requires additional measures for reduction of the calcium level in the blood; otherwise, it may cause hypercalcemia. Severe hypercalcemia can be cured

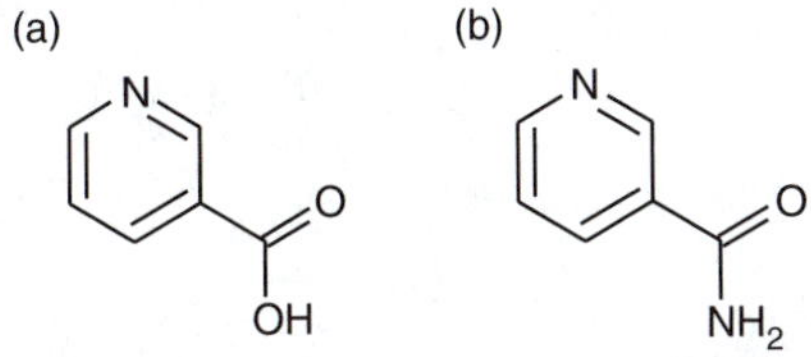

(a) N O OH

(b) N O $NH_2$

***Figure 18.6*** Forms of niacin: (a) nicotinic acid, and (b) nicotinamide.

by an intravenous administration of 2–3 L of 0.9% solution of sodium chloride over 1–2 days.

Mostly, the prognosis of vitamin intoxication therapy is good. Exceptions are severe poisonings with vitamins A, $B_6$, and D. It is impossible to cure the deposition of vitamin D in soft tissues and birth injuries caused by vitamin A. The injuries of the nervous system, caused by giant doses of vitamin $B_6$, are curable, but a complete recovery may take more than a year.

*Vitamin intoxications are easily avoidable.* One must just stop using vitamin preparations, either completely or use them under strict medical control. For example, if the doctor has prescribed vitamin D preparations, it will be obligatory to keep an eye on the plasma level of calcium.

## References

Klosterman, H.J., Lamoureux, G.L. and Parsons, J.L. (1967). Isolation, characterization, and synthesis of linatine a vitamin $B_6$ antagonist from flaxseed (*Linum usitatissimum*). *Biochemistry*, **6**, 170–177.

Lee, S.H., Oe, T. and Blair, I.A. (2001). Vitamin C-induced decomposition of lipid hydroperoxides to endogenous genotoxins, *Science*, **292**, 2083–2086.

Lonn, E., et al. (2005). Effects of long-term vitamin E supplementation on cardiovascular events and cancer: a randomized controlled trial, *JAMA*, **293**, 1338–1347.

Penniston, K.L. and Tanumihardjo, S.A. (2006). The acute and chronic toxic effects of vitamin A, *Am. J. Clin. Nutr.*, **83**, 191–201.

Perry, T.A., et al. (2004). Pyridoxine-induced toxicity in rats: a stereological quantification of the sensory neuropathy, *Exp. Neurol.*, **190**, 133–144.

Rietjens, I.M., et al. (2005). Review—Molecular mechanisms of toxicity of important food-borne phytotoxins, *Mol. Nutr. Food Res.*, **49**, 131–158.

Singh, H. and Duerksen, D.R. (2003). Vitamin K and nutrition support, *Nutr. Clin. Pract.*, **18**, 359–365.

Sowell, J., Frei, B. and Stevens, J.F. (2004). Vitamin C conjugates of genotoxic lipid peroxidation products: Structural characterization and detection in human plasma. *Proc. Natl. Acad. Sci. USA.*, **101**, 17964–17969.

Tavintharan, S. and Kashyap, M.L. (2001). The benefits of niacin in atherosclerosis, *Curr. Atherosc. Rep.*, **3**, 74–82.

# chapter nineteen

# Glossary

**abasia:** Inability to walk due to impaired muscular coordination. The term covers a spectrum of medical disorders such as choreic, paralytic, spastic, and trembling abasia.

**absorption:** Movement of substances from the site of administration to the general blood circulation. Absorption is a multistep process including diffusion through biomembranes. During absorption, the substance can be modified by decomposition or metabolism.

**acute:** (in toxicology) A short-term (single or multiple during 24 h) exposure to the effect of a toxic substance and usually a violent response of the organism.

**additive:** (in toxicology) Summary toxic effect of components of a mixture that equals the arithmetic sum of the effects of components.

**ADI—acceptable daily intake:** The highest daily amount of a chemical substance (food additive, pesticide, veterinary drug) that humans are expected to contact with during all of the lifespan without any adverse effect to his or her organism.

**aerobic/anaerobic:** Process, occurring respectively in conditions with or without access to air.

**agonist:** A molecule that selectively binds to a specific receptor triggering a response in the cell. An agonist mimics the action of an endogenous biochemical molecule (such as hormone or neurotransmitter) that binds to the same receptor.

**AhR—arylhydrocarbon receptor:** A protein, binding reversibly aromatic hydrocarbons including toxic substances such as dioxine, PCBs, PAHs. As a result of this binding, for example, enzymes metabolizing xenobiotics are induced.

**alkaloids:** Nitrogen-containing alkaline heterocyclic compounds that protect plants against the attack of microorganisms, pests, and herbivorous animals.

**allergy:** A reaction of an organism to the effect of a foreign substance consisting in appearance of a supersensitivity reaction mediated by the immune system. Allergy may be mediated either by antibodies or by cells.

**Alzheimer's disease:** A neurodegenerative disease characterized by progressive cognitive deterioration, the most common type of dementia.

**anaphylaxis:** Immediate immunologic reaction consisting of contraction of smooth muscles and broadening of capillaries, emerging at release of pharmacologically active endogenous substances (e.g., of histamine) after administration of a xenobiotic material. The severest type of anaphylaxis—**anaphylactic shock**—may lead to death in minutes if untreated.

**androgens:** Male sexual hormones (e.g., testosterone).

**anemia:** Lowering of the hemoglobin content of blood, usually accompanied by reduction in erythrocyte count and hematocrit.

**aneuploidy:** A condition in which the number of chromosomes is abnormal due to extra or missing chromosomes; in other words, it is a chromosomal state where the number of chromosomes is not a multiple of the haploid set.

**anoxia:** Absence of oxygen in the tissues.

**antagonism:** (in toxicology) Weakening of the effect of one substance (agonist) by another substance (antagonist). Antagonists are often used as antidotes. Biological mechanisms behind antagonism can be very various.

**antibody:** A specialized protein or immunoglobulin (Ig) that lymphoid tissue of an animal produces against recognized antigen. Mammal immunoglobulins are divided into classes such as IgA, IgD, IgE, IgG, and IgM. Cross-linking of antibodies on the surface of antigen leads to activation of the complement system.

**antidote:** A substance, which in case of timely usage, may substantially reduce the adverse effect of a toxicant.

**antigen:** A macromolecule that an animal organism recognizes as foreign, starting to synthesize antibodies specific to the antigen. An antigen is recognized by the antigen receptor of the T-cells or B-cells or by antibodies.

**apoptosis:** Programmed cell death that can be evoked also by definite foreign compounds.

**ARfD—acute reference dose:** An estimate of acute daily oral exposure of a toxic substance to the human population (including sensitive subgroups) that is likely to be without any appreciable risk of adverse effects during a lifetime. It can be derived from a NOAEL or LOAEL, with uncertainty factors applied to reflect limitations of the data used. Generally used in EPA's noncancer health risk assessments.

**ataxia:** Disorder of muscular coordination; it is divided into static and dynamic ataxia.

**ATP—adenosine triphosphate:** A multifunctional nucleotide, rich in energy, widely used by cells to transfer chemical energy into metabolic.

**AUC—area under the curve** in coordinates: Plasma (blood) concentration of a substance and time from start of the exposure to this compound.

**bioaccumulation:** Accumulation of a substance in an appropriate tissue of an organism, occurs especially in case of lipophilic substances.

**bioavailability:** A measure of the rate and extent of a foreign compound that reaches the systemic circulation and is available at the site of its action.

**biomagnification:** Process whereby the concentration of a foreign substance increases as one moves toward the top of the food chain. As a result, the highest is the concentration of the substance in the organism of the top predator.

**biomarker:** A specific substance formed in an organism as a response to the exposure to a foreign substance that is usable as a proof of this exposure or study of the mechanism of action of the xenobiotic.

**biotransformation:** Chemical alteration of a substance, especially of a drug, within the body by the action of enzymes, consists of phases I, II, and (III).

**blood–brain barrier** = **hematoencephalic barrier:** A complex of structural systems in the brain tissue wall preventing passage of many compounds from blood into the brain.

**botulism:** A fatal paralytic intoxication evoked in case of eating of food contaminated with the toxin of the bacterium *Clostridium botulinum.*

**carcinogen:** A substance that is able to cause a malignant tumor if administered to an organism.

**carcinogenicity:** A property of substances to cause abnormally rapid and uncontrolled multiplication of somatic cells leading to formation of a malignant tumor in the organism exposed to a carcinogen.

**CHD**—coronary heart diseases.

**chromosome:** Organelle composed of DNA and proteins, visible in the light microscope.

**chronic:** (in toxicology) A long-term (up to a lifetime) exposure to the effect of a toxic substance and response of the organism to this exposure.

**cirrhosis:** A disease of the liver, kidneys, lungs, heart, and other internal organs consisting of perishing of cells specific to the organ and replacement with a contracting cicatrix tissue. Cirrhosis may be caused by chronic inflammations, intoxications, or metabolic disorders.

**clastogenesis:** Breakage of chromosomes resulting in gain, rearrangement, or disappearance of pieces of chromosomes.

**CNS—central nervous system:** Consisting of the brain and bone marrow in the case of the vertebrates.

**cocarcinogen:** A substance or factor that will not promote cancer by itself but potentiates the cancer promoting effect of a carcinogen.

**Codex Alimentarius Commission:** A subsidiary body of the Food and Agriculture Organization (FAO) and the World Health Organization (WHO) of the United Nations. Codex Alimentarius is the major international mechanism for encouraging fair international trade in food while promoting the health and economic interest of consumers. More than 160 countries are members of the Codex Alimentarius Commission. The Codex Alimentarius Commission accomplishes its work through Committees hosted by the member governments.

**conjugation:** Joining of a special endogenous moiety like glucuronyl or sulfate group to the molecule of a xenobiotic during phase II of the biotransformation. In general, conjugation enhances the excretability of a foreign compound and reduces its toxicity.

**Crohn's disease:** A chronic inflammatory disease of the intestines, causing primarily ulceration of the small and large intestines, also called granulomatous enteritis or colitis.

**CYP:** The cytochrome P 450 monooxygenase complex (EC 1.14), the most important enzyme family of the biotransformation phase I.

**cytochrome $a_3$:** A hem-containing enzyme, part of the cytochrome[c]oxidase complex, the terminal enzyme in the mitochondrial electron transport chain.

**DAD—diode array detector:** A type of UV-Vis detector in chromatographic systems that simultaneously measures absorbance across a broad spectrum of light.

**dermatitis:** Inflammation of the skin.

**dinoflagellates:** Single-celled sea algae possessing two flagella (organelles of movement).

**distribution:** (in toxicology) Reversible uptake of a substance from general circulation to tissues.

**disulfide bridge** or **bond:** Covalent S–S bond that has an important role in stabilization of the tertiary structure of many proteins.

**dose:** Total amount (in milligrams, grams, moles, etc.) of a biologically active foreign compound that is in direct contact with an organism; in the case of a toxic compound, dose is a very important toxicity factor. A dose may be acute, subacute, subchronic, or chronic.

**$ED_{50}$:** Effective dose-50. Dose of a substance, pharmacologically effective for 50% of the population exposed to the substance or that elicits 50% of the response expected for a population.

**electrophile:** A chemical particle that has a preponderance of electrons together with a partial positive charge and a partially or totally empty electron orbital caused by different electronegativities of atoms. Electrophiles are liable to interact with **nucleophiles.**

**elimination:** Irreversible removal of a substance from general circulation into the organs, followed by removal from body either by excretion or metabolism.

**ELISA—Enzyme-linked immunosorbent assay:** A sensitive immunoassay that uses an enzyme linked to an antibody or antigen as a marker for the detection of a specific protein, especially an antigen or antibody. It is often used as a diagnostic test to determine the exposure to a particular infectious agent.

**encephalopathy:** A degenerative disease of the brain.

**endocrine disrupter:** A foreign substance that causes disorders in the endocrine function, leading to adverse effects in an animal or its offspring.

**endocytosis:** A complex way of transportation of substances or particles or droplets into the cell.

**endogenous:** Substances that originate from within an organism, tissue, or cell.

**enterohepatic recirculation:** The cycling of substances from the blood into the liver, then into the bile and gastrointestinal tract. It can be followed by reuptake of the compound into the blood possibly after definite chemical or enzymatic degradation.

**enzyme:** Proteinaceous biocatalyst or enhancer and regulator of biochemical reactions in organisms.

**epidemiology:** A branch of medical science studying formation and distribution of diseases in populations.

**epigenetic:** (In case of description of carcinogenicity or mechanisms of carcinogenesis) A carcinogen or mechanism, not connected with mutagenicity or genotoxicity.

**estrogen:** Female sexual hormone (e.g., estradiol).

**EU SCF**—European Commission Scientific Committee on Food.

**exposure:** (in toxicology) Direct contact of a toxicant with the surface (including inner surfaces) of an organism.

**FAO:** Food and Agricultural Organization of the United Nations.

**favism:** Hemolytic disease forming after eating of broad beans (*Vicia faba*), caused by glucose-6-phosphate dehydrogenase (G6DH) deficiency.

**FDA:** See US FDA.

**fibrosis:** Accumulation of collagen in a tissue.

**Fick's law:** At a constant temperature, the rate of diffusion of a compound across a membrane is proportional to the concentration gradient and the surface area.

**flavonoids:** Plant polyphenolic compounds, good antioxidants, are divided into flavones, flavanones, flavonoles, isoflavones, anthocyanines, chalcones, aurones, and other compounds.

**food chain:** An imaginary chain in nature, every higher link of which eats the one below and itself being eaten by the one above. At the bottom of the food chain, there are bacteria and plants, at the top are carnivores, including humans.

**free radical:** An atom or molecule that has an unpaired electron(s). Free radicals are usually very reactive.

**GC-MS:** Gas chromatography-mass spectrometry. Method for separation and qualitative as well as quantitative estimation of chemical compounds.

**gene:** The smallest functional unit of genome, a section of DNA. Genes are divided into structural (code the structure of polypeptide chain of a protein synthesized on the basis of the particular gene) and regulatory genes.

**genome:** A set of genes containing the genetic material of an organism.

**genotoxic:** A substance that adversely interacts with the genetic material of a cell or an organism.

**glomerulus:** A functional unit of the kidney consisting of small bunches of capillaries, projecting into Bowman's capsule, serving to collect the filtrate from the blood of those capillaries and direct it into the kidney tubule.

**glutathione:** The tripeptide glutamyl-cysteinyl-glycine with free reduced SH-group. Contained mainly in the liver and also in most of the other tissues; has an important role in cellular defense against oxidants as well as in detoxification of foreign compounds.

**GLP—good laboratory practice:** System of standard operating procedures recommended to be followed to avoid production of unreliable and erroneous laboratory data.

**GRAS—generally recognized as safe:** A US FDA designation that a chemical or substance (including certain pesticides) added to food is considered as safe under the conditions of its intended use by experts, and so exempted from the usual Federal Food, Drug, and Cosmetic Act (FFDCA) food additive tolerance requirements.

**half-life:** The time during which the concentration of a compound has been decreased by half.

**hazard:** Intrinsic property of a single substance or mixture to cause adverse effects in an organism exposed to it.

**hematoencephalic barrier = blood–brain barrier:** The barrier system separating the blood from the parenchyma of the central nervous system.

**hemodialysis:** A process, consisting of removal of a toxic compound from the blood of a patient by diffusion across a semipermeable membrane.

**hemoglobinuria:** Presence of hemoglobin in the urine.

**hemolytic anemia:** A pathological condition in which the erythrocytes undergo uncontrolled decomposition.

**hemoperfusion:** A process, consisting of the removal of a toxic compound from the blood of a patient by binding to activated charcoal or some other sorbent.

**hemorrhage:** Uncontrolled escape of blood from blood vessels.

**homeostasis:** The ability of biological systems to maintain the balance of biochemical reactions and energy and avoid harmful deviations of the basic system parameters from the equilibrium.

**hormone:** A chemical messenger from one cell (or group of cells) to another. All multicellular organisms including plants possess hormones.

**HPLC—high performance liquid chromatography:** Method for separation and qualitative as well as quantitative estimation of chemical compounds, based on differences in distribution of various compounds between mobile liquid and solid or pseudoliquid stationary phases.

**hydrophobicity = lipophilicity:** A property of molecules to repel water molecules. Hydrophobic nonpolar molecules are poorly soluble or practically insoluble in water, but well soluble in hydrophobic solvents as hydrocarbons and their halogenides and lipids, including cell membranes.

**hyperplasia:** Abnormal increase of the number of cells of an organ or tissue that may be caused either by increased cellular division or reduction of cell death.

**hypertrophy:** Abnormal increase of cellular dimensions. For example, fat vacuoles may increase.

**hypoglycemia:** The physiological condition caused by lower than normal concentration of glucose in blood.

**hypoxia:** The physiological condition caused by lower than normal concentration of oxygen in blood.

**IARC**—International Agency for Research on Cancer.

**idiosyncratic:** (in toxicology) An adverse reaction to a definite compound characteristic only for a small group of individuals, caused by abnormality of the organism(s) of this (these) individuals.

**immunostimulation:** Potentiation of immune reaction above normal level. Immunostimulation may be useful (e.g., to overcome immunosupression) or adverse (may cause allergy/supersensibility or autoimmunity) for an organism.

**immunosuppression:** Retardation of immune reaction by chemical compounds or physical factors below the normal level.

**immunotoxicity:** Situation in which a chemical compound or physical factor changes the structure or functioning of an immune system.

**induction:** (Considering metabolism of xenobiotics) Increasing biotransformation activity of enzymes by exposure to a foreign substance. Generally follows acceleration of the metabolism of the causative agent, without additional synthesis of the enzyme.

**inflammation:** Nonspecific multistep defense reaction of an organism, consisting of infiltration of leucocytes into peripheral tissue with a following release of a number of molecules evoking physiological defense mechanisms.

**initiation:** The first step of the multistep process of carcinogenesis consisting of chemical reaction between carcinogen and DNA molecules.

**intraperitoneal (i.p.):** A route of administration of a compound to an animal by direct injection into the intraperitoneal (abdominal) cavity.

**ischemia:** Physiological condition consisting of reduction or complete blocking of the blood flow into a tissue caused by narrowing of the arteries and resulting in ischemic damage of the tissue.

**isoenzyme:** One of several forms of an enzyme, preferably catalyzing similar but still distinct reactions.

**JECFA (FAO/WHO)**—Joint FAO/WHO Expert Committee on Food Additives.

**LC-MS—liquid chromatography-mass spectroscopy:** Method for separation and identification of chemical compounds, based on their separation by liquid chromatography followed by mass spectroscopic detection and identification.

**$LD_{50}$:** Acute dose of a compound causing the death of 50% of the test animals.

**local toxicity:** Toxicity, affecting only the site of application or exposure.

LOD—limit of detection: The lowest quantity of a substance that can be distinguished from the absence of that substance within a stated confidence limit (generally 1%).

**LOQ—limit of quantification:** The lowest concentration at which quantitative results can be reported with a stated confidence limit.

**macromolecule:** A large polymeric molecule such as protein, nucleic acid, or polysaccharide that has been composed of low-molecular units or monomers such as amino acids, nucleotides, and monosaccharides.

**macrophage:** A large phagocytic cell, a component of the reticuloendothelial system.

**metabolic activation:** Conversion of relatively stable and inactive molecules of a substance into highly reactive, mostly electrophilic particles, capable of causing serious damage of cellular macromolecules. The term is used also to denote the conversion of therapeutically nonactive prodrugs into active drugs.

**methemoglobin/methemoglobinemia:** The oxidized hemoglobin molecule that has lost its capability to transport oxygen/disease caused by high percentage of oxidized hemoglobin in blood.

**microsomes:** Fraction of cellular organelles containing fragments of the smooth endoplasmatic reticulum.

**mitochondrion:** Organelle responsible for cellular respiration and several other essential metabolic processes.

**monooxygenase:** Enzyme system participating in oxidation of substances (e.g., cytochrome P450).

**MTD—maximally tolerated dose:** Highest dose of a substance in food that can be fed during 90 days to an animal without appearance of any toxicity symptoms than other symptoms of developing cancer.

**mutagen:** A physical or chemical agent that causes in case of contact heritable alterations or mutations in the genetic material of an organism. As a result, cancer may develop in the organism.

**mycotoxins:** Toxins produced in the raw materials of food or feed by various molds.

**NADH/NAD$^+$:** Coenzyme reduced/oxidized nicotinamide adenine dinucleotide.

**NADPH/NADP$^+$:** Coenzyme reduced/oxidized nicotinamide adenine dinucleotide phosphate.

**nematodes (or roundworms):** One of the most common phyla of animals, with over 20,000 different described species (over 15,000 are parasitic), that are ubiquitous in freshwater, marine, and terrestrial environments.

**neoplasia:** Pathological process causing growth of abnormal tissue of neoplasm. Neoplasms form usually either benign or malignant tumor tissue.

**nephritis:** Inflammation of kidneys.

**nephron:** A functional unit of the kidneys, producing urine. Consists of long tubule divided into sections where reabsorption of substances into the bloodstream takes place.

**neutrophil:** The most abundant type of white blood cells that form an integral part of the immune system.

**NOAEL/NOEL—no observed (adverse) effect level:** The highest dose or exposure level at which no (adverse) effect has been observed in the case of any of the test animals. NOAEL values are used as the basis for calculation of **ADI**s.

**nucleophile:** A chemical particle, which has an excess of electrons together with a negative partial charge caused by different electronegativities of atoms. Nucleophiles tend to react with **electrophiles.**

**octanol–water partition coefficient** ($K_{ow}$)**:** Ratio of the solubilities of a substance in *n*-octanol and water. The higher the $K_{ow}$, the more hydrophobic or lipophilic the substance is.

**OECD**—Organization for Economic Cooperation and Development.

**osteoporosis:** A disease consisting of disappearance of the mineral and matrix phases of bones; the result is disposition to bone fractures.

**oxidation:** Removal of electron(s) from an atom of a molecule; process can be accompanied by joining of oxygen atom with the molecule oxidated.

**paresthesia:** Abnormal sensation such as tingling.

**Parkinson's disease:** A degenerative disorder of the CNS, which often impairs the sufferer's motor skills and speech.

**PCR—polymerase chain reaction:** A method of analysis in which definite sections of DNA are selectively and cyclically multiplied by the help of enzyme DNA-polymerase.

**peripheral neuropathy:** Damage to nerves of the peripheral nervous system.

**peroxidases:** Enzymes catalyzing oxidation reactions using hydrogen peroxide as an oxidizing agent. They are widely distributed in tissues, including neutrophils.

**peroxisomal proliferators:** Compounds that alter the number and properties of the peroxisomes. Peroxisome proliferators may cause tumor formation in rodents but not in humans.

**peroxisomes:** Membrane-bound organelles in the cells of animals, plants, fungi, and protozoa that contain enzymes synthesizing and decomposing hydrogen peroxides. In the peroxisomes, metabolism of lipids, sterols, and purine as well as peroxidative detoxification take place.

**phago/pinocytosis:** The uptake of solid (phago) or liquid (pino) particles into the cell; specific types of **endocytosis.**

**phases I, II and III:** Consequent phases of metabolism of xenobiotics in the organism.

**plasma:** Blood from which the cells have been removed.

**ppb:** Parts per billion: number of the same type of particles per $10^9$ summary particles in a solution (e.g., $\mu$g per kg)

**ppm:** Parts per million: number of the same type of particles per $10^6$ summary particles in a solution (e.g., $\mu$g per g)

**ppt:** Parts per trillion: number of the same type of particles per $10^{12}$ summary particles in a solution (e.g., ng per kg)

**promotion:** The third step of the multistep development of carcinogenesis, following **initiation** and **transformation.**

**pulmonary edema:** Accumulation of pulmonary liquid in the air spaces of the lungs.

**RDA—recommended daily allowance:** The levels of intake of essential nutrients that, on the basis of scientific knowledge, are judged by the Food and Nutrition Board of National Academy of Sciences of USA to be adequate to meet the known nutrient needs of practically all healthy persons.

**receptor:** A biochemically active and sensitive molecular complex, mostly located on the outer surface of the cellular membrane, binding of a suitable molecule (ligand) to which starts a chain of reactions altering the biochemical (physiological) state of the cell. Receptor may be located also in the cytosol.

**reproductive toxicology:** Branch of toxicology studying the toxic effects of xenobiotics on reproductive and neuroendocrine systems of the adult animals, embryos, germs, newborns, and prepubertal mammals.

**RfD—reference dose:** An estimate of a daily oral exposure to the human population (including sensitive subgroups) that is likely to be without an appreciable risk of deleterious effects during a lifetime. It can be derived from a NOAEL, LOAEL, with uncertainty factors generally applied to reflect limitations of the data used. Generally used in EPA's noncancer health assessments.

**ribosomes:** Cellular organelles, attached to the endoplasmatic reticulum, involved in protein synthesis.

**ROS—reactive oxygen species:** Oxygen-containing inorganic and organic ions, free radicals, and peroxides that are highly reactive due to the presence of unpaired valence electrons.

**singlet oxygen:** Excited highly reactive atom of oxygen.

**subacute:** Exposure to a toxicant of duration intermediate between acute and subchronic.

**subchronic:** Exposure to a toxicant of duration intermediate between subacute and chronic.

**superoxide** ($^{*}O_2^-$)**:** Oxygen molecule with an extra unpaired electron, anion radical.

**synergism:** (in toxicology) A situation, when summary toxic effect of a mixture of substances is higher than the sum of toxic effects of the components.

**systemic toxicity:** Toxic effect affecting a system in the organism usually remote from the site of application or initial exposure.

**$TD_{50}$—toxic dose-50:** The dose, that is toxic to half of the population exposed to this toxicant or elicits half of the expected maximal toxic response.

**teratogen:** A substance causing toxic effects to the germ or embryo after entrance into the maternal organism.

**thiol group:** The same as a SH or sulfhydryl group.

**thyroid hormones:** Tyrosine-based hormones like thyroxine ($T_4$) and triiodothyronine ($T_3$), synthesized in the thyroid gland.

**TI—therapeutic index:** Ratio $ED_{50}/TD_{50}$.

**TLC—thin layer chromatography:** Subtype of liquid chromatography, in which the solid or pseudoliquid stationary phase is a thin layer spread on a glass or plastic plate.

**tolerance:** A physiological phenomenon, consisting of reduction of a compound's toxicity after its repeated dosage.

**toxicokinetics:** Study of the rates of processes occurring with toxic substances in the organism.

**transformation:** The second step of the multistep development of carcinogenesis, following **initiation**.

**urticaria or hives:** Most benign form of anaphylaxis.

**US EPA**—United States Environmental Protection Agency.

**US FDA**—The Food and Drug Administration: An agency of the United States Department of Health and Human Services; it is responsible for regulating food (humans and animals), dietary supplements (human and animal), cosmetics, medical devices (human and animal) and radiation-emitting devices (including nonmedical devices), biologics, and blood products in the United States. It was formed in 1930.

**WHO:** World Health Organization.

**xenobiotic:** A substance foreign to a biological system (organism).

# Index